Kant and the Exact Sciences

MICHAEL FRIEDMAN

Kant and the Exact Sciences

HARVARD UNIVERSITY PRESS
Cambridge, Massachusetts
London, England

Printed in the United States of America
This book has been digitally reprinted. The content
remains identical to that of previous printings.
First Harvard University Press paperback edition, 1994

Library of Congress Cataloging-in-Publication Data
Friedman, Michael, 1947–
Kant and the exact sciences / Michael Friedman.
p. cm.
ISBN 0–674–50035–0 (cloth)
ISBN 0–674–50036–9 (pbk.)
1. Science—Philosophy. 2. Kant, Immanuel, 1724–1804—
Contributions in science. I. Title.
Q175.F893 1992
501–dc20 91–25022
CIP

For my philosophical parents

Norman Friedman   Zelda Nathanson Friedman

Contents

Es bewies mehr wie alles andere Platons, eines versuchten Mathematikers, philosophischen Geist, daß er über die große, den Verstand mit so viel herrlichen und unerwarteten Prinzipien in der Geometrie berührende reine Vernunft in eine solche Verwunderung versetzt werden konnte, die ihn bis zu dem schwärmerischen Gedanken fortriß, alle diese Kenntnisse nicht für neue Erwerbungen in unserm Erdenleben, sondern für bloße Wiederaufweckung weit früherer Ideen zu halten, die nichts geringeres, als Gemeinschaft mit dem göttlichen Verstande zum Grunde haben könnte. Einen bloßen Mathematiker würden diese Produkte seiner Vernunft wohl vielleicht bis zur Hekatombe erfreuet, aber die Möglichkeit derselben nicht in Verwunderung gesetzt haben, weil er nur über seinem Object brütete, und darüber das Subject, so fern es einer so tiefen Erkenntniß desselben fähig ist, zu betrachten und zu bewundern keinen Anlaß hatte. Ein bloßer Philosoph, wie Aristoteles, würde dagegen den himmelweiten Unterschied des reinen Vernunftvermögens, so fern es sich aus sich selbst erweitert, von dem, welches, von empirischen Prinzipien geleitet, durch Schlüsse zum allgemeinern fortschreitet, nicht genug bemerkt und daher auch eine solche Bewunderung nicht gefühlt, sondern, indem er die Metaphysik nur als eine zu höhern Stufen aufsteigende Physik ansahe, in der Anmaßung derselben, die sogar aufs Übersinnliche hinausgeht, nichts Befremdliches und Unbegreifliches gefunden haben, wozu den Schlüssel zu finden so schwer eben seyn sollte, wie es in der That ist.

Welches sind die wirklichen Fortschritte, die die Metaphysik seit Leibnitzens und Wolf's Zeiten in Deutschland gemacht hat? Beilage No. I

The philosophical spirit of Plato, an experienced mathematician, was demonstrated above all else by the fact that he could be transported into such a wonderment by the greatness of pure reason—which concerns the understanding with so many magnificent and unexpected principles in geometry—that he was carried away even to the fantastic thought that all this knowledge was not to be taken for a new acquisition in our earthly life, but rather for a mere reawakening of much earlier ideas which could have no less of a basis than community with the divine understanding. A mere mathematician would no doubt have delighted and perhaps even rejoiced in these products of his reason, but their possibility would not have set him into wonderment, because he was intent only on his object and had no cause to consider and to wonder about the subject—in so far as it is capable of such a deep cognition thereof. On the other hand, a mere philosopher, such as Aristotle, would not sufficiently have noted the vast difference of the *pure* faculty of reason—in so far as it amplifies itself from itself alone—from that which, guided by empirical principles, progresses through inferences to more general [principles]; and therefore he would also not have felt such a wonderment, but rather, in that he viewed metaphysics only as a physics ascending to a higher level, he would have found nothing strange and incomprehensible in its presumption, which aims even at the supersensible—the key to which must be precisely as difficult to find as it in fact is.

What Real Progress Has Metaphysics Made in Germany since the Time of Leibniz and Wolff? Supplement No. I

Preface

Immanuel Kant was deeply engaged with the science of his time—with the mathematical physics of Newton, in particular—during his entire philosophical career. His first published work, *Thoughts on the True Estimation of Living Forces* (1747), initiates a fundamental philosophical reconsideration of Newtonian physics which is then continued throughout the so-called pre-critical period: we here see Kant attempting to redefine the nature and method of metaphysics in light of the recent breathtaking advances in mathematics and mathematical physics. In the great period of the critical works (1781–1790) Kant achieves this metaphysical revolution by self-consciously following "the examples of mathematics and natural science, which through a suddenly accomplished revolution have become what they now are" (Bxv–xvi). Accordingly, in the *Prolegomena to Any Future Metaphysics* (1783) Kant explicitly addresses the question "How is metaphysics in general possible?" by way of the questions "How is pure mathematics possible?" and "How is pure natural science possible?" And it is no accident that the *Metaphysical Foundations of Natural Science* (1786), which presents Kant's most developed account of the foundations of Newtonian mathematical physics, is written at the height of the critical period. In Kant's post-critical reflections contained in his unpublished *Opus postumum* (1796–1803), moreover, we see a final reconsideration of the philosophical foundations of the sciences as Kant contemplates a new work, intended to complete his philosophy, entitled *Transition from the Metaphysical Foundations of Natural Science to Physics*.

Yet there has been a marked tendency to downplay and even to dismiss the philosophical relevance of Kant's engagement with contemporary science, particularly among twentieth-century English-language commentators. The reason for this is not far to seek: to read Kant in close connection with the specifics of the mathematics and physics of his time would seem inevitably to diminish his relevance to our current concerns. Much of twentieth-century philosophy finds its starting point and inspiration in the overthrow of Euclidean geometry and Newtonian physics by Einstein's theory of relativity. Hence from our present point of view geometry and physics certainly do not have the fixed, and indeed the synthetic a priori, status attributed to them by Kant. If Kant's philosophical achievement is still to have significance for us, therefore, it must be read in terms of more general epistemological principles that transcend the specifics of Euclidean-Newtonian science. Indeed, even commentators who have attempted to take seriously Kant's philosophy of science (and have thus not rested content with general epistemology) have also sought to extract a more general conception of science that would be as acceptable in the twentieth as in the eighteenth century.

Although this attempt to read Kant, as far as possible, in independence from the details of his scientific context is therefore understandable, I believe it is also profoundly mistaken. Kant's philosophical achievement consists precisely in the depth and acuity of his insight into the state of the mathematical exact sciences as he found them, and, although these sciences have since radically changed in ways that were entirely unforeseen (and unforeseeable) in the eighteenth century, this circumstance in no way diminishes Kant's achievement. For, on the one hand, Kant had an astonishing grasp of the philosophical foundations of the exact sciences of his time—a grasp that we, with all our increase in purely scientific sophistication, have hardly been able to match vis-à-vis twentieth-century exact science. So Kantian thought stands as a model of fruitful philosophical engagement with the sciences. On the other hand, our current philosophical predicament evolves directly from the breakdown of the Kantian philosophy in light of twentieth-century scientific developments (via the development of logical positivism and its aftermath): it is precisely because Kantian philosophy is so well adapted to eighteenth-century science—and not, therefore, to twentieth-century science—that our present philosophical situation has the specific shape it does. A better understanding of Kant's thought within its eighteenth-century context is therefore most relevant indeed to our twentieth-century problems.

This book is an attempt to provide detailed support for the above claims—especially concerning the depth and acuity of Kant's philosophical insight into the foundations of the sciences—by developing a reading of Kant's engagement with the exact sciences in all three periods: the pre-critical period of the *Thoughts on the True Estimation* through the *Inaugural Dissertation* (1770); the critical period of the *Prolegomena*, the *Metaphysical Foundations*, and the *Critique of Pure Reason* (1781, 1787); and finally the post-critical period of the *Opus postumum*. My aim throughout is to show that and how central aspects of the Kantian philosophy are shaped by—are responses to—the theoretical evolution and conceptual problems of contemporary mathematical science. I do not mean to suggest, however, that the Kantian philosophy can be thereby seen as wholly parasitic on the exact sciences—so that, for example, one can simply read off the content of that philosophy from the scientific developments in question. On the contrary, Kant's achievement consists rather in adapting and radically transforming independently given philosophical and metaphysical ideas—ideas stemming largely from the Leibnizean philosophical tradition he inherited—within the essentially new scientific context wrought by Newton. In this way, philosophy and the exact sciences are set into a fruitful interaction with one another that illuminates both the nature and function of the former and the conceptual foundations of the latter.

Kant's pre-critical period is the main subject of the Introduction. In these years Kant self-consciously seeks to refashion the Leibnizean (Leibnizean-Wolffian) tradition so as better to harmonize metaphysics with Newtonian natural philosophy. The notions of substance, the active force of substance, and the interaction of substances are the central metaphysical concepts at issue. The problem of interaction—as expressed in the well-known conflict among the systems of physical influx, occasionalism, and pre-established harmony—is Kant's philosophical starting point. In particular, Kant attempts throughout the pre-critical period to defend a modified version of physical influx—according to which distinct substances really do interact with one another via genuine "transuent forces"—explicitly modeled on the Newtonian theory of universal gravitation; and this effort then leads him to a radically new conception of the relationship between (phenomenal) space and the underlying (noumenal) realm of ultimate substances. The pre-critical period is also noteworthy for Kant's attempt to reformulate the proper method of metaphysics. In explicit agreement with the Newtonianism of Euler, Kant argues that metaphysics must begin with—must take as its "data"—the far more certain and secure results of the mathematical exact sciences. Indeed, it is only in this way that metaphysics can possibly aspire to a properly scientific status for itself.

The critical period is the subject of Part One. The great innovations of this period are explained as revolving around two central ideas. The first, explored in detail in Chapters 1 and 2, involves a sharp distinction of the faculties of the mind into a conceptual or intellectual faculty, on the one hand, and a sensible or intuitive faculty, on the other. This division of the faculties of understanding and sensibility is here traced back to Kant's extraordinary insight into the logic and proof-structure of Euclid's *Elements* (including the arithmetic of Books VII–IX and the theory of proportion of Book V). In Euclid existence assumptions are represented not via propositions formulated in modern quantificational logic but rather by constructive operations—generating lines, circles, and so on in geometry—which can then be iterated indefinitely. Such indefinite iteration of constructive operations takes the place, as it were, of our use of quantificational logic, and it is essentially different, moreover, from the inferential procedures of traditional syllogistic logic. From Kant's perspective, then, mathematical reasoning—which necessarily involves such indefinite iteration in geometry, in arithmetic, and in "algebra" as well—thereby must involve non-logical, and hence non-conceptual elements; and this appeal to "construction in pure intuition" underlies Kant's radical division of the faculties of the mind. Perhaps the most important result is the characteristically critical theory of space: space is no longer constituted by the interactions among intellectually conceived ultimate substances (as it was for the pre-critical Kant); rather, it is an autonomous "form of sensible intuition" governed by *sui generis* (non-intellectual) laws of its own—namely, the laws of Euclidean geometry.

The second great innovation of the critical period is explored in Chapters 3 and 4. Not only is the sensible faculty fundamentally distinct from the intellectual faculty, but the concepts and principles of the intellectual faculty themselves have "sense and meaning" only when applied to the sensible faculty and thus to space (and time). In particular, the intellectual concepts of substance, active force (or causality), and interaction (or community) now have genuine content only as applied to the spatio-temporal world of sense. This spatio-temporal "schematism" of the pure intellectual concepts or categories marks a profoundly new conception of their nature and function: such metaphysical concepts no longer characterize an underlying (noumenal) intellectual realm located somehow beneath or behind the (phenomenal) world of sensible experience; rather, their function now is precisely to constitute the conditions of possibility of sensible experience itself. From the present point of view, the spatio-temporal application of the pure concepts or categories is exemplified, above all, by Kant's penetrating analysis, developed in the *Metaphysical Foundations of Natural Science*, of Newton's law of universal gravitation. Here Kant starts from a rejection of Newtonian absolute space, but still attempts, nonetheless, to do justice to Newton's central distinction between "true" and "apparent" motion. Kant views the argument of Book III of the *Principia* as first defining or constructing a privileged frame of reference—an empirical counterpart of absolute space—and, at the same time, as constructing or deducing the law of universal gravitation from the "phenomena." This procedure then exemplifies how the categories discharge the function of "as it were prescribing laws to nature and even making nature possible" (B159).

Kant's post-critical reflections in his *Opus postumum* are the subject of

Part Two (Chapter 5). The problem underlying Kant's attempt to formulate the *Transition from the Metaphysical Foundations of Natural Science to Physics* is understood in the following way. In the critical period, as indicated above, Kant is able to show how the understanding "prescribes laws to nature," and thus makes experience possible, in the case of the Newtonian theory of universal gravitation. But the rest of the phenomena of nature—chemical phenomena especially—remain entirely unaccounted for. Indeed, in the critical period Kant explicitly despairs of the properly scientific status of chemistry. If we are truly to see how experience—experience as a whole—is possible, therefore, some kind of extension of the argument of the *Metaphysical Foundations* is needed. The *Transition* project takes its inspiration from the new developments in the science of heat and in chemistry that constitute Lavoisier's chemical revolution, and this project is therefore based on Kant's appreciation of the emerging scientific status of chemistry. In particular, Kant wrestles with the central theoretical construct of Lavoisier's new chemistry—the imponderable caloric fluid or aether—and ultimately attempts to show that it has an a priori, not merely hypothetical, status. In the course of the *Transition* project Kant's conception of the faculties of the mind and their relation—this time, the faculties of understanding and judgement—is once again fundamentally transformed.

Earlier versions of some of the chapters were published previously, as detailed below. In the present book I have revised these earlier versions substantively. In particular, I have now attained greater clarity and consistency, I hope, on the very delicate question of the relationship between geometry and physics, between sensibility and understanding, and between "form of intuition" and "formal intuition." It is also my hope that the recurrence of some themes and texts, approached from different directions in different chapters, will both facilitate access to the book (which can be read starting with any chapter of special interest) and prove to be philosophically illuminating.

Several pages from the Introduction are reprinted from "Causal Laws and the Foundations of Natural Science," in P. Guyer, ed., *The Cambridge Companion to Kant* (Cambridge University Press, 1992). I am grateful to Paul Guyer and Cambridge University Press for permission to reproduce this material.

An earlier version of Chapter 1 appeared as "Kant's Theory of Geometry," *Philosophical Review* 94 (1985): 455–506. I am grateful to the editors of *The Philosophical Review* for permission to publish the present version.

An earlier version of Chapter 2 appeared as "Kant on Concepts and Intuitions in the Mathematical Sciences," *Synthese* 84 (1990): 213–257,

© 1990 Kluwer Academic Publishers, and is reprinted by permission of Kluwer Academic Publishers.

An earlier version of Chapter 3 appeared as "The Metaphysical Foundations of Newtonian Science," in R. Butts, ed., *Kant's Philosophy of Physical Science* (D. Reidel, 1986). I am grateful to Robert Butts and D. Reidel for permission to publish the present version.

An earlier version of Chapter 4 appeared as "Kant on Space, the Understanding, and the Law of Gravitation: *Prolegomena* §38," *Monist* 72 (1989): 236–284. I am grateful to the editor of *The Monist* for permission to publish the present version.

My intellectual debts incurred in writing this book are many and various.

Reading Gerd Buchdahl's *Metaphysics and the Philosophy of Science* (1969) in 1980 first awoke my interest in Kant's philosophy of science. Before this my interest in Kant had taken a more traditional direction, as I had assumed (as have many others) that Kant's commitment to the science of his time presented a formidable obstacle to the twentieth-century relevance of his scientific thought. In the meantime I had pursued independently studies of the conceptual foundations of space-time physics. Buchdahl's rich and suggestive treatment of Kant in the context of the development of seventeenth- and eighteenth-century science first showed me how to combine my interests in Kant and in the foundations of physics and formed the starting point for my ensuing investigations.

During this same period I benefited greatly from the encouragement and advice of Thomas Ricketts, who urged me to forge links between my study of Kant and my study of the philosophy of space and geometry. In attempting to do so I have since profited repeatedly from Ricketts's philosophical insights.

My first efforts focused on Kant's philosophy of mathematics and geometry. Here I drew inspiration from the work of Jaakko Hintikka, Charles Parsons, and Manley Thompson. I also benefited from the comments and advice of Parsons and Thompson—and later, in this connection especially, from extensive conversations with William Tait. I am further indebted to Thompson for his fundamental criticisms of my 1985 version of Chapter 1.

In the Spring of 1984 I held a Canada Council Visiting Foreign Scholars Fellowship at the University of Western Ontario. This invaluable opportunity allowed me to begin to work out my ideas on the *Metaphysical Foundations of Natural Science,* which I presented in a series of seminars. I am indebted to the participants, particularly Richard Arthur, Robert Binkley, Robert Butts, Philip Catton, William Demopoulos, Malcolm Forster, William Harper, Clifford Hooker, Thomas Lennon, Ausonio Marras, John Nicholas, Kathleen Okruhlik, and Graham Solomon. Since then I have returned yearly to Western Ontario and have continued to present my evolving ideas there. I am grateful for the support and encouragement of Butts, Demopoulos, Lennon, and Marras, who have made this continuing relationship possible.

In attempting to come to terms with Kant's analysis of Newton's remarkably subtle argument for the law of universal gravitation I am indebted, above all, to the writings and advice of Howard Stein. Without the benefit of Stein's understanding of the intricacies of the Newtonian argument I simply would not have been able to pursue Kant's analysis as far as I have. As comments from Stein, and also especially from Robert DiSalle, have made clear, my account still needs to be developed further.

My work on the *Opus postumum* has benefited particularly from the writings of Eckart Förster and Burkhard Tuschling. I am also indebted to Förster for his comments on an earlier draft of Chapter 5.

I developed the ideas on the pre-critical period articulated in the Introduction while working with Alison Laywine on her dissertation on this topic. There is no doubt that I learned as much from her as she did from me.

For comments on the penultimate draft I am indebted to an anonymous referee for Harvard University Press and also, once again, to William Harper.

In addition to those already named, I am indebted for comments, conversations, and criticisms over the years to Henry Allison, Karl Ameriks, Gordon Brittan, John Carriero, Richard Cartwright, Alberto Coffa, Joshua Cohen, Graciela De Pierris, Burton Dreben, Hannah Ginsborg, Warren Goldfarb, Anil Gupta, Paul Guyer, Peter Hylton, Philip Kitcher, Thomas Kuhn, Ernan McMullin, Ralf Meerbote, Carl Posy, John Rawls, Roberto Torretti, Daniel Warren, Scott Weinstein, Margaret Wilson, and Mark Wilson. I regret that I became acquainted with the *scharfsinnig* work of Martin Carrier on Kant's scientific thought only as the present book was already in press—I hope to benefit from it in future work.

For support of my work on this book I am indebted, in addition, to the Institute for the Humanities at the University of Illinois at Chicago for a Fellowship in the academic year 1984–1985 (I am also grateful to Gene Ruoff, the Director of the Institute, for technical assistance in the preparation of the manuscript), to the Department of Philosophy at Harvard University for a George Santayana Fellowship in the Fall of 1986, to the University of Illinois for a Senior University Scholars Award from 1987 to 1990, to the National Science Foundation for a Grant (SES 86-19813) in 1988, and to the John Simon Guggenheim Memorial Foundation for a Fellowship in the academic year 1988–1989.

Kant and the Exact Sciences

· INTRODUCTION ·

Metaphysics and Exact Science in the Evolution of Kant's Thought

Kant began his philosophical career as an enthusiastic student of Leibnizean-Wolffian metaphysics and Newtonian natural philosophy. As is well known, Kant's interest in both systems of thought was inspired and nurtured by his teacher Martin Knutzen at Königsberg, a moderate Wolffian revisionist who was one of the first in Germany to accept Newtonian attraction.[1] Kant himself accepts Newtonian attraction as an immediate action-at-a-distance throughout his career, and, in fact, he consistently takes the law of universal gravitation as his paradigm of a well-established physical law.[2] Yet Kant also consistently holds that, whereas the Newtonian natural philosophy is correct as far as it goes, it does not go far enough: a true natural science requires a grounding in metaphysics—a metaphysics based on the prior notions of *substance* and *active force*. The following passage from the *Physical Monadology* of 1756 is typical in this regard:

1. See Erdmann [23]; Tonelli [110]; Cassirer [18]; Beck [5], p. 430. Knutzen's revisionism consisted primarily in a defense of physical influx as opposed to the Leibnizean system of pre-established harmony, and it is probable that Knutzen influenced Kant decisively in this regard (Erdmann, for example, argues on p. 143 of [23] that Knutzen is the "certain clear-sighted author" to whom Kant alludes in §6 of *Thoughts on the True Estimation of Living Forces:* 1, 21.3–4). For Knutzen and Newtonian attraction see Tonelli [110], p. 67 and n. 200 thereto on p. 117.

2. Kant's consistent acceptance of Newtonian attraction as a true and immediate action-at-a-distance actually represented a rather extreme position, as compared with the position of Newton himself and the majority of continental Newtonians of Kant's day. Euler, for example, who was certainly no friend otherwise of Leibnizean-Wolffian metaphysics, notoriously rejected action-at-a-distance, and even such staunch Newtonians as Maupertuis and Voltaire expressed themselves extremely cautiously on the matter. Kant's position corresponds to that of the second generation English Newtonians, especially to that of John Keill. See Tonelli [110], pp. 66–69.

> Clear-headed philosophers have unanimously agreed that those who seriously undertake to investigate nature should be on guard lest anything made with a certain freedom of conjecture and without reason should find its way into natural science, and lest anything be undertaken in it without the support of experience and without geometrical interpretation. Certainly nothing can be thought more useful to philosophy, and sounder, than this counsel. Since hardly any mortal can steadily advance along the straight line of truth without here and there turning aside in one direction or another, those who have to a great extent obeyed this law in investigating the truth have little dared to commit themselves to the high sea but have considered it more advantageous to remain always close to the shore, and to admit nothing except what is immediately known through the testimony of the senses. Setting out in this sound way, we can expound the laws of nature but not the origin and causes of these laws. For those who only pursue the phenomena of nature are as far removed from the abstruse understanding of primary causes, and from attaining a science of the very nature of bodies, as those who persuade themselves that by ascending to the summits of higher and higher mountains they are at last about to touch the heavens with their hands.
>
> Therefore metaphysics, which many say can be properly avoided in the field of physics, is in fact its only support and what gives light. (1, 475.2–19)[3]

It is for this reason that the *Physical Monadology* then attempts to reconcile the natural philosophy of Newton with the basic principles of Leibnizean-Wolffian metaphysics: that is, with a monadology.

Yet such a reconciliation is no trivial matter, for the two systems of thought appear to be entirely incompatible. According to the Leibnizean-Wolffian system reality consists of non-spatial, non-temporal, unextended simple substances or monads. Moreover, these simple substances or mo-

3. Newton is clearly paradigmatic of the non-metaphysical investigators of nature referred to in the first paragraph: compare, e.g., the Introduction to the *Enquiry Concerning the Clarity of the Principles of Natural Theology and Ethics* (1764): "*Newton's* method in natural science changed the unrestrained freedom of physical hypotheses into a secure procedure in accordance with experience and geometry" (2, 275.8–11). Similar sentiments to those expressed in the above passage from the *Physical Monadology* are found, e.g., in *Thoughts on the True Estimation of Living Forces* (1747): "But we must connect the laws of metaphysics with the rules of mathematics in order to determine the true measure of force of nature; this will fill the gap and better effect the satisfaction of the ends of God's wisdom" (1, 107.27–30); and in the *Metaphysical Foundations of Natural Science* (1786): "Thus these mathematical physicists could certainly not avoid metaphysical principles, and among them certainly not such as make the concept of their proper object, namely matter, a priori suitable for application to outer experience: as the concepts of motion, the filling of space, inertia, etc. However, they rightly held that to let merely empirical principles govern these concepts would be absolutely inappropriate to the apodictic certainty they wished their laws of nature to possess; they therefore preferred to postulate such principles, without investigating them in accordance with their a priori sources" (4, 472.27–35).

nads do not interact with one another: the evolution of the states of each is completely determined by a purely internal principle of active force, and the appearance of interaction is explained by a pre-established harmony between the states of the diverse substances established originally at the creation by God. Finally, since metaphysical reality is thus essentially non-relational, neither space nor time is metaphysically real: both are ideal phenomena representing the pre-established harmony—the mirroring of the universe from various points of view—between the really non-interacting simple substances. In the Newtonian system, by contrast, we begin with absolute space and time existing prior to and independently of all material substances that are found therein. Material substances or bodies themselves are essentially spatio-temporal and therefore inherit all the geometrical properties—such as extension and continuity—possessed antecedently by space and time. Moreover, material substances or bodies are in essential interaction with one another: according to the law of inertia bodies cannot change their state unless acted upon by an external impressed force, and according to the law of universal gravitation each body in the universe is simultaneously in immediate interaction with every other body in the universe—no matter how distant.[4]

Indeed, the conflict between these two systems of thought was a fact of central importance in the intellectual life of the early and mid-eighteenth century. Two episodes in particular were centrally important to Kant. First, in the years 1725–1746 there was a confrontation between Wolffians and Newtonians at the St. Petersburg Academy of Sciences. This controversy—emerging publicly with a paper by Wolff himself in the official journal of the St. Petersburg Academy in 1728—concerned the *vis viva* dispute primarily, and it is this controversy that Kant is attempting to resolve in his first publication, *Thoughts on the True Estimation of Living Forces* (1747).[5] Second, and more important, there was a confron-

4. Thus a phenomenal physics based on elastic impact—that advocated by Huygens and Leibniz—harmonizes much better with the underlying principles of a monadology than does the physics of Newtonian attraction: whereas the change of state of a body in elastic impact can be conceived as determined by the body's own inherent elasticity (see, e.g., Leibniz [68], p. 521; [72], p. 33), the change of state of a body due to the immediate action-at-a-distance of Newtonian attraction can in no way be conceived as determined internally. It is for this reason that both Leibniz and Wolff vehemently oppose action-at-a-distance and seek to account for gravitational attraction as the effect of underlying impacts or pressure (of the aether).

5. For the confrontation at the St. Petersburg Academy see Calinger [14]; with special attention to Kant's role in the controversy, see Polonoff [98], pp. 24–62. The St. Petersburg Academy was founded by Peter the Great in 1724, as the result of earlier urgings by Leibniz, in order to provide Wolffian philosophy a refuge after Wolff's expulsion from Prussia (due to Pietist anti-rationalist agitation) by Frederick William I in 1723. Nevertheless, the Acad-

tation between Wolffians and Newtonians at the Berlin Academy of Sciences in the years 1740–1759. The Berlin Academy was led by the French Newtonian Pierre Maupertuis, who proceeded to orchestrate a concerted attack on Wolffianism—an attack which opened publicly with a Prize Question for the years 1745–1747 on monadology. It is this controversy which Kant is attempting to resolve in 1756 in the *Physical Monadology*.[6]

Thus, in the *Physical Monadology* Kant is perfectly clear about the apparent incompatibility between the Leibnizean-Wolffian and Newtonian systems of thought and perfectly clear, therefore, about the seemingly overwhelming obstacles standing in the way of his philosophical project (which is to demonstrate *The Use in Natural Philosophy of Metaphysics Combined with Geometry*):

> But how in this business can metaphysics be reconciled with geometry, where it appears easier to be able to unite griffins with horses than transcendental philosophy with geometry? The former precipitously denies that space is infinitely divisible, while the latter asserts this with its customary certitude. The latter contends that void space is necessary for free motion, the former denies it. The latter shows most exactly that attraction or universal gravitation is hardly to be explained by mechanical causes but rather from forces innate in bodies that are active at rest and at a distance, the former reckons this among the empty playthings of the imagination. (1, 475.22–476.2)

Much of Kant's philosophical development can be understood, I think, as a continuous attempt—an attempt faced with a succession of more and more fundamental problems—to construct just such an apparently paradoxical reconciliation of Newtonian and Leibnizean-Wolffian ideas, and to construct thereby a genuine metaphysical foundation for Newtonian natural philosophy.

---

emy was quickly embroiled in controversy via the Newtonian attacks on *vis viva* led by Daniel and Nicholas III Bernoulli (later joined quietly by Euler); the Leibnizean-Wolffian position was then defended by Georg Bilfinger and Jakob Hermann. Kant argues that the Newtonian position is correct mathematically and mechanically, but that there are nonetheless metaphysical grounds supporting *vis viva*. See below.

6. See Calinger [15]; Polonoff [98], pp. 77–89; Beck [5], pp. 314–319. The Berlin Academy was founded by Frederick the Great in 1740 in a deliberate attempt to introduce the French enlightenment into Prussia. To this end he invited Maupertuis (as president) and Euler, although Wolff was also invited on being recalled to Halle in 1740 (he declined). The attack on monadology was initiated by Euler in 1746 in his "Thoughts on the Elements of Bodies," in which he argued that monadology is incompatible with both the law of inertia and the infinite divisibility of space and matter. Samuel Formey then replied on behalf of the Wolffians in 1747: in effect, he repeated the standard Wolffian line that metaphysics grasps the real nature of things while geometry and mechanics deal only with creatures of our imagination by which we confusedly represent phenomenally the true monadic reality. The prize was awarded to an anti-Wolffian paper by J. Justi in 1747. Kant's position will be discussed below.

· I ·

Beginning with *Thoughts on the True Estimation of Living Forces* Kant attempts to revise the Leibnizean-Wolffian monadology in light of Newtonian physics. Reality consists of non-spatial, non-temporal, unextended simple substances; space, time, and motion are phenomena derivative from this underlying monadic realm. But Kant breaks away decisively from the Leibnizean-Wolffian conception of active force and interaction. The primary notion of active force is not that of an internal principle by which a substance determines the evolution of its own states, it is rather that of an action exerted by one substance on another substance whereby the first changes the inner state of the second (§4: 1, 19). Kant has thus imported Newton's second law of motion into the very heart of the monadology.

In the next few sections Kant's sense of liberation is palpable as he rapidly introduces a series of revolutionary metaphysical moves. Since the primary and general notion of active force is that by which one substance changes the inner state of a second substance, there is no difficulty in conceiving of an action of matter upon the soul or of the soul upon matter: there is no obstacle, therefore, to the triumph of physical influx over pre-established harmony (§§5, 6: 19–21).[7] Moreover, since substances are connected with one another only through a mutual exercise of active force (and not through pre-established harmony), a substance can exist along with others without actually being connected with them. Such a non-interactive substance is in fact not part of the same world constituted by the others, and it follows that more than one *actual* world can exist (§§7, 8: 21–23). Finally, since space depends upon the connection and order of simple substances, without an active force whereby substances act outside themselves there is no extension and no space. Indeed, the basic properties of space are entirely derivative from the fundamental law of interaction by which substances are connected into a single world, and Kant concludes with his famous speculation deriving the three-dimensionality of space from the inverse-square law of gravitational attraction (§§9–11: 23–25). Kant has thus imported universal gravitation into the monadology as well.

In the *New Exposition of the First Principles of Metaphysical Cognition*

7. In particular, Kant's notion of active force is not tied, as is the notion of *vis viva,* to the *motion* of the being that exercises it. The problem is no longer to explain how motion is communicated from one body to another but rather to explain in general how one substance can change the inner state of another (§§2–4: 18–19). Hence active force (like the Newtonian force of attraction which inspires it) can be exerted by a substance at rest, and there is no need to suppose either that the soul is in motion or that bodies act on the soul by producing motions.

of 1755 Kant announces two new metaphysical principles which make his Newtonian version of the monadology fully explicit. According to the first principle, the principle of succession, no change can occur in the internal states of a substance except in so far as it is united by mutual interaction with others. A substance considered in isolation can do no more than conserve or maintain the very same state, and the sufficient reason for an alteration must be sought in factors external to the substance itself (Prop. XII: 1, 410–411). Kant is perfectly clear that he is here decisively rejecting the notion of active force as an internal principle of change:

> Although this truth is sustained by such an easy and unmistakable chain of reasons, those who reckon themselves under the Wolffian philosophy have paid so little attention to it that they rather contend that a simple substance is subject to continuous changes from an internal principle of activity. (411.16–19)

Kant proceeds on this basis to argue against both subjective idealism (since the soul can experience no inner changes except in virtue of a connection with other substances external to it) and the Leibnizean pre-established harmony (411–412), and so it is once again evident that a metaphysical version of the law of inertia is a powerful weapon indeed.[8]

Kant's second principle, the principle of coexistence, further articulates this new conception of active force and mutual interaction. Since simple substances are in no way connected with one another by virtue of their

8. This is not to say that Kant is entirely clear about the law of inertia itself. Indeed, Kant here invokes a *vis insita* or innate force of bodies (the *vis inertiae* of Definition III of Book I of *Principia*) as a principle of resistance and reaction in impact by which the total quantity of force (action + reaction) is conserved (408.16–22). He thereby conceives the internal *vis inertiae* by which a body conserves its state as a force completely on a par with the external impressed force (impulse) by which one body is induced to change its state by another. But in *Thoughts on the True Estimation of Living Forces* the confusions on this matter are far worse. Kant argues that in addition to the external force recognized by mechanics there is also an internal force or *intension* by which a body seeks to conserve its own state of motion. This force is measured by the *velocity* of the body, and, moreover, it is only manifested in certain conditions (of so-called free motion) in which a body strives to conserve the motion imparted to it by an external force through its own internal force or intension. In such circumstances the total force of the body is given by the product of the intension by the externally communicated force; and, since the latter is always proportional to the velocity, the total force is here proportional to the square of the velocity (§§117–120: 1, 141–144). This conception, as Kant acknowledges, is completely incompatible with the law of inertia (§132: 155.13–22). Kant only achieves a fully correct understanding of inertia in the *New System of Motion and Rest* of 1758 (2, 19–21), although in the *Physical Monadology* he is already clear that *vis inertiae* is measured by mass, not velocity (Prop. XI: 1, 485.15–26). At the height of the critical period, in the *Metaphysical Foundations of Natural Science,* Kant emphatically rejects self-activity as a general characteristic of substance (4, 543–544, 550–551). It appears that Kant's evolving sophistication concerning the law of inertia arises directly out of a careful consideration of Euler's views: see Timerding [109].

existence alone (again due to the failure of pre-established harmony), their constituting a single world is therefore something quite distinct from their mere existence. Over and above the mere existence of substances we also require a general law or principle of universal mutual interaction, and it follows that God does not simply create the existence of substances *simpliciter* in creating a world, but also establishes a universal principle of interaction by what Kant calls a "schema of the divine intellect" (Prop. XIII: 412–414). The universal mutual interaction of substances constitutes position, situation, and space (414.10–20), and its phenomenal manifestation is universal gravitation:

> Further, since the determinations of substances are reciprocally related, i.e., substances diverse from one another act mutually (for one determines something in the other), the notion of space is accomplished by the interwoven actions of substances, with which reaction is always necessarily joined. If the external phenomenon of universal action and reaction through all of space, in which bodies relate to one another, is their mutual approach, it is called *attraction*—which, since it is effected by copresence alone, extends to arbitrary distances and is *Newtonian Attraction* or universal gravitation; and it is probable, therefore, that it is effected by the same nexus of substances as that which determines space, and that it is the most primitive law of nature to which matter is subject—but which perpetually endures only because God immediately supports it, according to the view of those who profess to be partisans of Newton. (415.5–16)

In this way the Leibnizean-Wolffian doctrine of space as a phenomenon derived from the order of non-spatial simple substances is transmuted into the Newtonian doctrine of divine omnipresence.[9]

This transformation is of central importance, for it implies that in calling space phenomenal Kant is not asserting its *ideality*. For Leibniz and Wolff space is ideal because relations between substances are ideal: each substance mirrors the entire universe internally due to its own inner principle, and space is an ideal representation of the underlying order of monads expressed in the pre-established harmony. Indeed, since each simple substance by itself already expresses completely the order of the entire universe, nothing but the mere existence of substances is necessary to consti-

9. The same doctrine is found in the *Universal Natural History and Theory of the Heavens* of 1755: "Attraction is without a doubt a property of matter extending precisely as far as the coexistence that constitutes space, in that it connects substances through mutual dependencies; or, to speak more properly, attraction is precisely this universal relation which unites the parts of nature in one space: it therefore extends itself to the entire extension of space into all distances of its infinity" (1, 308.27–34). In the course of the same discussion Kant speaks of "empty space, this infinite extent of divine presence" (306.32–33) and "the infinite space of divine presence" (312.36, 313.27–28). Compare also the Scholium to §22 of the *Inaugural Dissertation* (2, 409.28–410.6).

tute phenomenal space.[10] For Kant, by contrast, relations of interaction between substances are in no way ideal: a universal principle of mutual interaction is a distinct reality over and above the mere existence of substances (requiring a distinguishable divine action going beyond the creation of the existence of substances *simpliciter*), and the universal principle of mutual interaction constitutes space. In this context calling space an external phenomenon means only that it is *derivative* from or constituted by the underlying non-spatial reality of simple substances. But it is as metaphysically real as is the universal principle of mutual interaction itself.[11]

Kant, in the *Physical Monadology*, can therefore not escape the dilemma posed by infinite divisibility, as did the Wolffians, by invoking the ideality of space and the imaginary nature of geometrical concepts (see note 6 above). On the contrary, although he continues to hold that space is "entirely lacking in substantiality and is a phenomenon of the external relations of united monads" (1, 479.25–26), Kant is just as clear that the dilemma cannot be evaded by impugning the metaphysical reality of geometrical space:

> There is certainly no thought in a disquisition on the elements which more obstructs the marriage of geometry with metaphysics than the preconception—an opinion never sufficiently examined—that the divisibility of the space which an element occupies proves the division of the element itself into substantial parts. This is so commonly asserted as to be placed beyond all doubt that those who defend the infinite division of real space shrink from monads *toto coelo,* while those who subscribe to monads have thought that they had to maintain that the properties of geometrical space are as good as imaginary. Since, however, it is clearly evident from the above demonstrations that neither is geometry deceived nor does the thought of the metaphysician deviate from the truth, the opinion which divides them—that an element absolutely simple with respect to its substance cannot fill a space while saving its simplicity—is necessarily deceptive. (480.14–25)

Thus, one of the previous demonstrations (Prop. III: 478–479) is a standard geometrical proof of infinite divisibility, which Kant has "accommo-

10. "As for my own opinion, I have said more than once, that I hold space to be something purely relative, as time is; that I hold it to be an order of coexistences, as time is an order of successions. For space denotes, in terms of possibility, an order of things which exist at the same time, considered as existing together; without enquiring into their manner of existing" (§4 of Leibniz's Third Paper in *The Leibniz-Clarke Correspondence:* [67], vol. 7, p. 363; [70], pp. 25–26).

11. Thus, in #2 of Proposition XI of the *New Exposition* (409–410) Kant rejects the principle of the identity of indiscernibles on the grounds that spatial location is a genuine determination of substances sufficient for their individuation—something which is of course quite impossible on a Leibnizean conception of the phenomenality of space.

dated to physical space, in order that those who make use of a general distinction between geometrical and natural space cannot somehow escape through an exception" (478.36–479.1). Kant himself is maintaining precisely the infinite divisibility of *real* space.[12]

The desired reconciliation between geometry and metaphysics is then effected, not by attacking the metaphysical reality of geometrical space, but rather by arguing that a simple substance fills a space by the "sphere of activity" of its *repulsive force*—where the latter is conceived as radiating continuously outwards from a central point (Props. VI, VII: 480–482). In dividing the space thus occupied by a monad we are therefore dividing only the external relations of activity by which one monad acts upon others, we are not dividing the substance of the monad itself:

> Since, however, space is not a substance, but a phenomenon of certain external relations of substances, that a relation of one and the same substance can be divided in two does not contradict the simplicity, or, if you prefer, the unity of substance. For what is found on both sides of the divided line is not something that can be separated from the substance and which would retain its own existence without it—which is certainly required in the case of a real division that destroys simplicity—rather, it is an action or relation exercised by one and the same substance on both sides such that to find some plurality in it is not to separate the substance itself into parts. (480.27–35)

Hence, although the monad fills a space and thereby has a spatial location (namely, the central point of its sphere of activity), it nevertheless remains essentially non-spatial and unextended: the substance or ultimate subject of activity is certainly not in space.[13] Kant has in this way preserved the monadology from the threat of infinite divisibility by once again ingeniously importing the Newtonian doctrine of central forces.[14]

12. Thus I cannot follow Vuillemin in his interpretation of the *Physical Monadology*, according to which Kant is there maintaining a Leibnizean conception of the *ideality* of space: [117], pp. 122–123. I agree rather with Adickes, who in [1], vol. 1, §68, depicts the *Physical Monadology* as maintaining a Newtonian view of the *reality* of space together with a Leibnizean view of the *relational character* of space—derivative from the external relations of monads. (Curiously, however, both Adickes—[1], p. 147—and Vuillemin—[117], pp. 121–122—misread the passage from the Preface at 1, 475.22–476.2, quoted above where Kant contrasts "metaphysics" and "geometry": both misread the final contrast in such a way that the former accepts and the latter rejects Newtonian attraction as a true action-at-a-distance.)

13. "But in addition to the external presence, i.e., relational determinations of substance, there are other internal determinations, and if the latter were not the former would have no subject in which to inhere. But the internal determinations are not in space precisely because they are internal. Nor are these themselves divided by the division of external determinations; no more than the subject itself or the substance is in this way divided" (481.27–32).

14. Kant's key move is to conceive impenetrability or the filling of space as the manifestation of a repulsive central force—understood as precisely analogous to the central force of

Just as Kant's conception of the phenomenality of space diverges fundamentally from the Leibnizean-Wolffian conception of the ideality of space, so his conception of how the universal harmony of substances depends on God diverges fundamentally from the pre-established harmony. Kant articulates this latter divergence in the following passage from the *New Exposition:*

> However, since no substance has the power of determining others different from itself through what belongs to it internally (as demonstrated), but only in virtue of the nexus in which they are united by the idea of an infinite being, whatever determinations and mutations are found in any substances are always certainly externally relative, but a physical influx properly so-called is excluded, and there is a universal *harmony* of things. Yet there does not result thereby the *pre-established* harmony of *Leibniz,* which properly introduces a *consensus,* not a mutual *dependence* of substances; for God does not use the craftsmen's artifice [*artificiorum technis*] in a series of causes fitted together and adapted to effect the agreement of substances, nor is there established an always special influx—i.e., an interaction [*commercium*] of substances through the *occasional causes* of Malebranche. For the same indivisible action which confers existence on substances and conserves them also brings together their mutual and universal dependency, so that the divine action is not determined this way or that according to circumstances; but there is a real action of substances mutually among themselves, or an interaction [*commercium*] through truly efficient causes, because the same principle which establishes the existence of things also makes them such as to be bound by this law, and thereby a mutual interaction [*commercium*] is established through those determinations which pertain to the origin of their existence. (415.20–36)

Thus, although Kant explicitly distances himself from the vulgar system of physical influx which posits interaction without explanation, it is clear that his conception of how the universal harmony of things depends on God is intended to be closer to real physical influx than to pre-established harmony.[15]

---

gravitational attraction introduced later (Prop. X: 483–485). Kant first introduces the force of repulsion alongside the force of gravitational attraction in the *Universal Natural History and Theory of the Heavens,* where he asserts that the two forces are "both borrowed from the Newtonian philosophy" (1, 234.31–32). In the *Physical Monadology* he calls both forces "innate forces [*virium insitarum*]" (485.5)—a term customarily used (and previously used by Kant himself) for internal forces (such as *vis inertiae*) *as opposed to* Newtonian external impressed forces (see note 8 above). Kant thereby saves the Leibnizean-Wolffian monadology from the infinite divisibility of space only by fundamentally reinterpreting the central notion of active force in a way that would be entirely unacceptable to Leibniz and Wolff.

15. Compare the parallel passage in §22 of the *Inaugural Dissertation:* "Thus all the interaction [*commercium*] of substances in the world is *established externally* (by the common cause of all), and is either established generally through a (corrected) physical influx,

In order fully to appreciate what is at stake here it is necessary to consider Kant's views on physico-theology, as first expressed in the *Universal Natural History and Theory of the Heavens* of 1755 and later, in a more complete and developed form, in *The Only Possible Basis for a Proof of the Existence of God* of 1763. Kant's basic idea is that the order and harmony manifest in the material universe arise neither from special divine intervention nor from special divine arrangement, whereby God adapts the world to useful ends through his providential wisdom. Rather, the order and harmony of the material universe can be completely explained by the fundamental laws of material interaction—first and foremost, by the law of universal gravitation—which determine an orderly *evolution* of the structure of the universe out of a primordial chaos. Yet this purely mechanistic explanation is itself the best proof of a divine origin of the universe; for it is God, and God alone, who has established these fundamental laws of interaction (laws which have such harmonious consequences of necessity) as belonging to the very essence of matter:

> The matter which is the primitive material of all things is therefore bound by certain laws such that if these laws are allowed to operate freely matter must necessarily bring forth beautiful combinations. Matter has no freedom to deviate from this plan of perfection. Since it is thus subject to a supremely wise purpose, it must have necessarily been placed in such harmonious relations through a first cause ruling over it; and *there is a God precisely because nature itself even in chaos cannot proceed otherwise than in a rule-governed and orderly way.* (1, 228.3–11)[16]

Kant is therefore appealing to the origin of the universal laws of interaction in the schema of the divine intellect (here a divine "plan of perfection") discussed above.

---

or is negotiated for their states individually; in this last case it is founded *originally* through the first constitution of any substance, or is impressed on the *occasion* of any change: the former is called *pre-established harmony,* the latter *occasionalism.* If, therefore, through the sustaining of all substances by a single one the *conjuction* of all, whereby they consitute a unity, is *necessary,* the universal interaction [*commercium*] of substances is by *physical influx* , and the world is a real whole" (2, 409.14–23).

16. This basic idea of the *Theory of the Heavens* corresponds to Maupertuis's polemic against Wolffian physico-theology of 1748. Because Wolff adhered to the mechanical natural philosophy and rejected Newtonian attraction (see note 4 above), he could envisage no *single* mechanical cause controlling the evolution of the material universe. He therefore resorted to particular teleological explanations (such as the utility of the arrangement of the stars for navigation, and the like) in explaining why God has disposed the celestial bodies and the state of the earth in the harmonious way in which we find them. Maupertuis, by contrast, argued for the possibility of an evolution of all astronomical and physical phenomena from the fundamental laws of simple masses and maintained that precisely this proves the existence of divine wisdom far better than any collection of particular teleological explanations. See Tonelli [110], pp. 54–56.

In *The Only Possible Basis* Kant develops this idea by distinguishing two types of natural order to which physico-theology may be applied: contingent order of nature and necessary order of nature (Third Consideration: 2, 106–108). The former pertains properly to the *organic* realm, where inherently distinct structures are assembled together in an organism in a wonderfully adapted fashion that has no single material cause—as it were, by artifice.[17] The latter pertains properly to the *inorganic* realm, where harmonious order flows necessarily from the action of a single cause—namely, from the fundamental laws of material interaction:

> On the other hand, there are not different causes which fashion the earth in a spherical shape, still others which retain bodies on the earth against its rotation, and still another that preserves the moon in its orbit; rather, a single gravity is the cause, which suffices in a necessary fashion for all of these. Now it is no doubt a perfection that grounds are found in nature for all these effects, and if the same ground that determines one is also sufficient for the others then so much more unity thereby accrues to the whole. But this unity, and with it the perfection, is in these cases necessary and adheres to the essence of the thing; and all adaptedness, fruitfulness, and beauty that results from this depends on God through the essential order of nature, or through that which is necessary in the order of nature. (2, 106.32–107.10)

The result of the orderly evolution of the cosmos via the action of universal gravitation sketched in the *Theory of the Heavens* (and recapitulated here in the Seventh Consideration: 137–151) is thus paradigmatic of necessary order of nature.

Kant then proceeds to argue that a physico-theology based on necessary order of nature is superior from a philosophical point of view to that based on contingent order of nature. In the latter case we view the order in question as independent of the fundamental laws of material interaction and as therefore resulting from an artificial arrangement of matter due to a special divine purpose.[18] But this customary approach to physico-theology

17. "The creatures of the plant and animal kingdoms provide throughout the most remarkable examples of contingent unity, yet one still harmonizing with great wisdom. Vessels that draw sap, vessels that draw air, those which work up the sap, and those which exhale the air, etc.—a great manifold in which each single part has no suitability for the actions of the others, and where the uniting of these together into a total perfection is artificial [künstlich], so that the plants themselves with their relations to such diverse ends constitute a contingent and optional unity" (2, 107.14–22).

18. Kant maintains that the organic realm is independent of the fundamental mechanical laws of nature in precisely this way, so that teleological explanations are perfectly appropriate there: "Now, because the forces of nature and their laws of action contain the ground of an order of nature, which, in so far as it comprises manifold harmonies in a necessary unity, brings it about that the connection of many perfections in *one* ground becomes a law,

suffers from three main defects. First, it tends to view *all* perfection, harmony, and beauty of nature as an arrangement of providential wisdom—whereas much of this harmony in fact follows with necessary unity from the essential laws of matter (118–119). Second, it thereby engenders an inclination to set limits to the investigation of nature and to abandon the search for physical causes too soon (119–122). Third, and perhaps most important, this approach can only serve to prove a creator of the connections and artifical arrangements of matter, but not of the matter itself and the ultimate constituents of the universe:

> This considerable failing must keep all those who use it alone in danger of that error one calls the more refined atheism, and according to which God in the proper understanding is viewed as a master craftsman [Werkmeister] and not as a creator of the world, who, to be sure, has ordered and formed matter, but has not brought it into being and created it. (122.35–123.3)

By contrast, a physico-theology based on necessary order of nature derives the harmonious arrangements of matter from the necessary laws of interaction constituting the very essence of matter itself, and it thereby proves the existence of a wise *creator* of the world.[19]

We are now in a position to appreciate the full force of Kant's divergence from the system of pre-established harmony. For Kant there is a necessary order of nature: the original laws of mutual interaction without which the various existing simple substances would not constitute a single world. These laws comprise the fundamental laws of physics—governing Newtonian forces of attraction and repulsion—and they alone make posi-

---

one has to consider diverse natural effects with respect to their beauty and utility under the essential order of nature and through this under God. On the other hand, since there are also many perfections that are not possible in a whole through the fruitfulness of a single ground [viz., organic wholes] but rather require various grounds optionally united for this purpose, many an artificial [künstlich] arrangement will be in turn the cause of a law, and the actions that take place in accordance with it stand under the contingent and artificial [künstlich] order of nature—but through this under God" (107.32–108.7).

19. "However, because this unity is nonetheless grounded in the very possibilities of things, there must be a wise being without which these natural things would not even be possible, and in which, as a great ground, the essences of so many natural things are united into such rule-governed relations. It is then clear, however, that not only the mode of connection but the things themselves are possible only through this being: that is, they can only exist as his effects—which circumstance alone sufficiently makes known the complete dependence of nature on God" (125.19–27). It is in this way that Kant's preferred approach to physico-theology harmonizes with his own version of the ontological argument for the existence of God, which conceives God as the supreme ground of all *possibility:* 77–92; compare also the *New Exposition,* Prop. VII: 1, 395–396. It is in this way also that Kant's approach to physico-theology gives priority to the divine *intellect* over the divine *will:* compare the footnote at 2, 125.33–126.35, with that at 2, 109.29–37.

tion, situation, space, and time first possible. The contingent order of nature, on the other hand, is in no way necessary to bind the ultimate simple substances into a single world order: on the contrary, it consists in precisely the circumstance that the matter which is already given in a single world order is, in addition, also formed into special adventitious arrangements independent of the necessary order of nature. The matter forming a living organism, for example, would still be matter—and would still belong to a single world—even if no living organisms existed; but if the necessary order of nature is itself suspended then we no longer have a single material world at all. The necessary order of nature, then, flows directly from that act by which God creates the ultimate simple substances *as belonging to a single nature or world,* and, in this way, it is a direct result of the creative schema of the divine intellect.

In the system of pre-established harmony, however, the idea of a necessary order of nature is missing—more precisely, there is no distinction between necessary and contingent order of nature. Each simple substance by itself already expresses or represents a complete world order from its own point of view, and the pre-established harmony is then only needed to insure that all simple substances thereby express or represent the *same* world order. Hence, the pre-established harmony is either fully in effect or it is not in effect at all, and there is no room for Kant's distinction between necessary and contingent order: between inorganic nature and organic nature. There is no room, that is, for a distinction between the fundamental laws of material interaction without which there would not even be a world in the first place and more particular, adventitious arrangements of already existing matter constituting artifical natural products. On the contrary, in the system of pre-established harmony *all* order and harmony is necessarily on the same footing. From Kant's point of view, then, since what is missing from the system of pre-established harmony is precisely the idea that the fundamental laws of physics are those laws without which the coexisting simple substances would no longer constitute a single world via their mutual interactions, this system appears to reduce all order of nature to contingent order. The system of pre-established harmony thereby appears to reduce God to the role of one who "use[s] the craftsmen's artifice in a series of causes fitted together and adapted to effect the agreement of substances."[20]

20. Compare the passage from the *Inaugural Dissertation* quoted in note 15 above: "If, therefore, through the sustaining of all substances by a single one the *conjunction* of all, whereby they consitute a unity, is *necessary,* the universal interaction [*commercium*] of substances is by *physical influx*, and the world is a real whole." The passage continues: "But if not, the interaction [*commercium*] is sympathetic (i.e., harmony without true interaction [*commercio*]), and the world is no more than an ideal whole" (2, 409.23–25).

· II ·

Alongside of Kant's attempt to transform the content of Leibnizean-Wolffian metaphysics in light of the Newtonian natural philosophy is an equally serious attempt to rethink the form and method of metaphysics in light of Newtonian methodology. The problem of the relationship between the method of metaphysics and the mathematical method of the exact sciences was much discussed in the eighteenth century—particularly in connection with the Wolffian system, and it presented itself to Kant in the following way. On the one hand, the Wolffian philosophy had attempted to develop metaphysics in the form of a strictly deductive system, *more geometrico:* metaphysics was to begin with the most abstract and general concepts (such as being, essence, attribute, and the like) and to construct a chain of definitions of successively more concrete and particular concepts (such as body, state, motion, and the like). At the same time, one began with the most abstract and general principles (the highest of these being the principle of contradiction) and built up a chain of syllogisms leading from there to the most concrete principles: at this point the entire structure could be confronted with experience. On the other hand, however, despite this philosophical imitation of the form of the mathematical method, the Wolffian philosophy remained suspicious of the content of mathematics itself, especially of geometry. This is because space, according to Wolff, is a *confused* sensible representation of the purely intellectual reality described by metaphysics; therefore geometrical (and hence mechanical) concepts are in essence purely imaginary. And it is in the latter way, in fact, that the Wolffians responded to the problem of reconciling simplicity of substance with the infinite divisibility of geometrical space discussed above.[21]

Now such a conception of the nature of metaphysics and of the relationship between metaphysics and the exact sciences is of course completely unacceptable from the point of view of the Newtonian natural philosophy. In the first place, the great achievement of Newtonian methodology was to demonstrate how one could begin with the most certain and uncontroversial experience—that is, with *phenomena*—and ascend from there by means of evident mathematical reasoning to a knowledge of the first principles of natural bodies. And this procedure of "deduction from the phenomena"—as exemplified, above all, in the argument for universal gravitation—was held to be infinitely more secure than all merely "hypothetical" attempts to begin with first principles and work one's way down to phenomena. In the second place, moreover, since "deduction from the

21. See note 6 above. See Tonelli [111]; and Polonoff [98], who on p. 86 helpfully distinguishes between Wolff's adherence to the *form* of the mathematical method and his suspicion of the *content* of mathematics.

phenomena" appears to be thus privileged over all purely a priori speculation, there can be no question of using metaphysical arguments in any way to limit the established results of geometry and mechanics. On the contrary, if metaphysics is admissible at all, it is necessarily subordinate to and regulated by the more certain results of the exact sciences.

Such a broadly Newtonian attitude is perfectly expressed by Euler in his famous paper on space and time presented to the Berlin Academy in 1748:[22]

> ... since it is metaphysics which is concerned in investigating the nature and properties of bodies, the knowledge of these truths [of mechanics] is capable of serving as a guide in these intricate researches [of metaphysics]. For one would be right in rejecting in this science all the reasons and all the ideas, however well founded they may otherwise be, which lead to conclusions contrary to these truths [of mechanics]; and one would be warranted in not admitting any principles which cannot agree with these same truths. The first ideas which we form for ourselves of things, which are found outside ourselves, are ordinarily so obscure and so indefinite that it is extremely unsafe to draw from them conclusions of which one can be certain. Thus it is always a great step in advance when one already knows some conclusions from some other source, at which the first principles of metaphysics ought finally to arrive: and it will be by these conclusions, that the principal ideas of metaphysics will be necessarily regulated and determined. ([27], pp. 376–377; [28], pp. 116–117)

And it is for precisely this reason that Euler responds to the problem of simple substance versus infinite divisibility by totally rejecting the monadology.[23]

In his first published work on the *True Estimation of Living Forces* Kant exhibits a certain amount of ambivalence on the question of the relationship between metaphysics and the sciences of mathematics and mechanics. The work consists of three parts. The first part briefly outlines the metaphysical conception discussed above. The second part takes the point of view of mathematics, and argues that from this point of view *vis viva* is definitely excluded as a proper measure of force: the only forces considered by mechanics are the external or impressed forces, and from this (clearly Newtonian) point of view force must be measured in propor-

22. I say "broadly Newtonian" because Euler of course staunchly resisted Newtonian attraction as a genuine action-at-a-distance (although he naturally accepted Newton's argument for the *law* of universal gravitation).

23. See again note 6 above. In the section of the essay on space and time which follows that quoted above Euler criticizes the "metaphysicians" in that they "upbraid the mathematicians, because they [the mathematicians] apply these principles [of mechanics] inappropriately to ideas of space and time which are taken to be only imaginary and destitute of all reality": [27], p. 377; [28], p. 117.

tion to the simple velocity (compare §115: 1, 140). The third part then attempts to find a place for *vis viva* after all on the basis of a metaphysical consideration of the internal force of bodies by which they strive to preserve the motion imparted to them by an external force. And, in the course of this metaphysical consideration, Kant introduces a distinction between *mathematical* and *natural* bodies intended to limit the validity of mathematics in a manner clearly reminiscent of the Wolffians: "laws which are found to be false in mathematics can still occur in nature" (§§114, 115: 139–140).[24]

Nevertheless, at the end of the first or properly metaphysical part of the work Kant is very hesitant about the claims of metaphysics and appears, in fact, to express an attitude more in line with the above quotation from Euler:

> I may not promise to achieve something decisive and consistent in a consideration that is merely metaphysical; therefore I now turn to the following chapter, which through the application of mathematics can perhaps make more claim to conviction. Our metaphysics is, like many other sciences, in fact only on the threshold of a correctly well-grounded [gründlich] cognition; God knows when one will see it step over that threshold. It is not difficult to see its weaknesses in many of the things it undertakes. One very often finds prejudice as the greatest force of its proofs. Nothing is more to be blamed here than the overpowering inclination of those who seek to increase the extent of human cognition. They would like to have a great philosophy, but it is to be hoped that it will also be a well-grounded one. (§19: 30.28–31.4)[25]

24. See note 8 above. The crux of Kant's argument is this. Mathematical quantities necessarily obey the principle of continuity in a very strong form: whatever is true of any arbitrary finite quantity of a certain kind is also true for an infinitely small quantity of the same kind. In the mathematical defense of *vis viva,* however, the finitude of (actual) motion is thought to be the mark distinguishing the measure of force appropriate here from that appropriate to the "dead force" manifested in infinitely small (potential) motion—which latter is unanimously agreed to be proportional to the simple velocity. This conception of *vis viva* is therefore inconsistent with a purely mathematical consideration. From a metaphysical point of view, by contrast, we can introduce a discontinuity: we can ask whether or not the moving body contains a sufficient reason for continuing in motion within itself. Only in cases where this sufficient reason is present—which, for Kant, occurs only at certain *finite* minimum velocities—is *vis viva* the appropriate measure of force. Once again, it is clear that this argument, although not without a certain ingenuity, is wholly incompatible with the law of inertia.

25. It is likely that the criticism here is directed precisely at the Wolffians. Compare §106, where Kant explicitly names Wolff as the source of the mistaken conception of *vis viva* he seeks to reject, and then rather contemptuously continues: "Anyone who will read his treatise in the indicated work of the [St. Petersburg] Academy will find that it is very difficult to find anything in it that constitutes a correct proof, so much is all made diffuse and incomprehensible in virtue of the analytical inclination which is so very prominent therein" (113.30–34).

And it is clear, moreover, that the third part of the work is intended not so much to supplant the second or properly mathematical part as to complement it. As noted above (note 3) Kant's aim there is to found a new dynamics that will "connect the laws of metaphysics with the rules of mathematics in order to determine the true measure of force of nature." Thus, although Kant here certainly intends to limit the claims of mathematics so as to provide room for metaphysics, this metaphysical treatment itself must still be capable of consistent and fruitful cohabitation with mathematics.[26]

Forging such a union between metaphysics and mathematics is, as we have seen, the explicit aim of the *Physical Monadology*. Moreover, as we have also seen, Kant appears to be considerably more confident about the prospects for metaphysics there, for he argues that Newtonian physics by itself cannot achieve a properly scientific status without a metaphysical foundation—which "is in fact [physics's] only support and what gives light."[27] Yet there is no remaining trace of the idea that the content of mathematical exact science can be limited or denied full reality from the point of view of metaphysics. Metaphysics must accept the infinite divisibility of space as a given fact or datum from geometry, and it is only by proceeding from this datum that a proper understanding of impenetrability, the filling of space, and the composition of bodies can be attained. Kant therefore insists, in particular, that infinite divisibility characterizes "real space," so that "those who make use of a general distinction between geometrical and natural space cannot somehow escape through an exception." In sharp contrast to the argument in *Thoughts on the True Estimation,* then, Kant here explicitly refrains from invoking a distinction between "mathematical" and "natural" on behalf of metaphysics.

Similarly, in the *Theory of the Heavens* and *The Only Possible Basis* Kant takes Newtonian attraction as a given datum and attempts to construct an improved approach to cosmology and physico-theology on the

26. It is therefore significant that Kant provides mathematical-empirical confirmation for his metaphysical defense of *vis viva:* "in order to please those for whom anything is suspicious that has only the appearance of a metaphysics, and require throughout an experiment [Erfahrung] on which to ground its consequences, so I will indicate a method according to which these considerations can be brought to their better satisfaction. Namely, towards the end of this chapter I will verify from an experiment [Erfahrung] with mathematical rigor: that in nature actual forces are to be found that have the square of the velocity" (150.2–9). The reference is to the discussion of the experiments of Ricciolus, s'Gravesande, Poleni, and Muschenbroek on the penetration of falling bodies into soft materials in §§157–162 (176–180).

27. The *Theory of the Heavens* and *The Only Possible Basis* have made it clear what this "light" consists in: grounding Newtonian physics in Kant's revised version of monadological metaphysics precisely elucidates the total dependence of all of nature on the divine intellect.

basis of this datum. And he clearly has no patience at all for metaphysical doubts about Newtonian attraction:

> I shall be hereby presupposing the universal gravitation of matter according to *Newton* or his followers. Those, who through some definition of metaphysics according to their taste believe that they can overthrow the conclusions of clear-sighted men from observation and mathematical inference, can pass over the following propositions, as something which has besides only a distant kinship with the main purpose of this work. (2, 139.17–23)[28]

Given the overriding importance of Newtonian attraction in Kant's new physico-theological argument—indeed, given the fundamental role we have seen it play in Kant's central conception of the dependence of the world on the schema of the divine intellect—the hint of tolerance voiced here must be considered as an expression of sarcasm.[29]

In the *Attempt to Introduce the Concept of Negative Magnitudes into Philosophy* of 1763 Kant definitively aligns himself with the point of view of Euler. After distinguishing two uses one might make in philosophy of mathematics—"the imitation of its method, or the real application of its propositions to the objects of philosophy" (2, 167.3–5)—Kant decisively rejects the first and adopts the second:

> Metaphysics seeks, for example, the nature of space and to find the highest ground through which its possibility can be understood. Now certainly nothing can be more helpful here than being able in some way to borrow data proved to be trustworthy in order to base one's consideration upon it. Geometry provides one with some such data concerning the most general properties of space: e.g., that space absolutely does not consist of simple parts. But one ignores this and places one's trust solely in the ambiguous consciousness of this concept, in that one thinks it in an entirely abstract manner. If then speculation in accordance with this procedure does not harmonize with the propositions of mathematics, one seeks to save one's artificial concept by reproaching this science, as if the concepts lying at its basis did not derive from the true nature of space but were rather optionally fabricated. The mathematical consideration of motion, together with the cognition of space, yields in the same fashion much data for keeping the metaphysical consideration of time on the path of truth. The famous *Euler* has among other things given a few indications here,* but it seems more comfortable to hold onto

28. The "definition of metaphysics" referred to in this passage from *The Only Possible Basis* is most likely the definition of *contact* criticized in the *Enquiry Concerning the Clarity of the Principles of Natural Theology and Ethics* (1764) at 2, 287.36–288.35.

29. See 148.14–21, where Kant consigns the celestial vortices—the only real competitor to Newtonian attraction—to "Milton's limbo of vanity." Moreover, it is clear from Kant's remarks in the Preface at 68.22–69.9 that his gravitational cosmology has an intimate kinship indeed with the main purpose of the work.

obscure and difficult to test abstractions than to enter into an association with a science that is involved only with comprehensible and evident insights. (168.3–24)[30]

And there is no doubt that those who are here characterized as mistakenly attempting to pursue metaphysics in isolation from the exact sciences—even to the point of casting doubt on the latter for the sake of the former—are precisely the Wolffians.[31]

But Kant's most extensive discussion of the relationship between metaphysical and mathematical method in this period is of course the *Enquiry Concerning the Clarity of the Principles of Natural Theology and Ethics* of 1764. This essay was explicitly addressed to the Prize Question of the Berlin Academy for 1763 on the comparative certainty of metaphysics (especially natural theology and ethics) and geometry,[32] and it fully articulates the Newtonian-Eulerian conception of metaphysical method. The crucial point is that mathematical method is *synthetic* whereas metaphysical method is necessarily *analytic.* Mathematics begins with a few simple and evident primitive concepts and a small number of equally evident primitive propositions governing these concepts. We then proceed by building up further concepts by "optional connection" (2, 276.8) of these primitive concepts—that is, by synthesis—and by similarly deductively developing all further propositions from the primitive propositions. In this way synthetic definitions of concepts are the basis of our science. In metaphysics, on the other hand, our problem is that we are dealing with obscure concepts (such as body, time, freedom, and so on) which we do not yet know how properly to define. Definitions are therefore the least secure part of our science, as it were, and our task is rather to begin with some trustworthy knowledge involving the concepts in question—on the

30. The reference is to Euler's paper on space and time cited above. For the consideration of time in conjunction with the mathematical theory of motion (Newton's), see the *Inaugural Dissertation* §14, 5, at 2, 400.31–401.7 (including the footnote thereto at 401.28–38). Kant there appears to follow §§20, 21 of Euler's paper in rejecting the idea that the equality of times can be rendered intelligible independently of the law of inertia—by the mere order or quantity of successive changes, for example. Instead, Kant appears to hold that the law of inertia *defines* the equality of times: equal times are those during which an inertially moving body traverses equal distances. I will return to this matter below.

31. "As for metaphysics, this science, instead of making use of some of the concepts or doctrines of mathematics, has rather more often taken up arms against it, and, where [metaphysics] could have perhaps borrowed secure principles on which to ground its considerations, one sees it concerned to make the concepts of mathematics into nothing but artful fictions that have little of truth intrinsically outside of their field. One can easily guess which side has the advantage in the struggle between two sciences in which one surpasses all others together in certainty and clarity while the other is first striving to attain this" (167.18–168.2).

32. As is well known, the prize was awarded to Mendelssohn and Kant's essay took second place. For the details of the prize competition and its background see Tonelli [111].

basis of which we can hope finally to ascend to proper definitions "through *abstraction* from that cognition which is made clear through analysis" (276.8–9). In this way definitions come rather at the end than at the beginning of metaphysical inquiry.[33]

It is clear, once again, that the incorrect metaphysical method attempting to proceed *more geometrico* is that of Wolff,[34] and Kant is perfectly explicit that his preferred model for the correct metaphysical method is the Newtonian method:

> The genuine method of metaphysics is in principle identical with that which *Newton* introduced into natural science and which was there of such useful consequences. One should, it is said there, seek the rules according to which given appearances of nature proceed through secure experience, and in every case with the help of geometry. If one does not comprehend the first ground thereof in bodies, it is nevertheless certain that they act in accordance with this law; and one explains the natural events involved if one clearly shows how they may be contained under these well-established rules. Precisely so in metaphysics: seek through secure inner experience—i.e., an immediately evident consciousness—those characteristics that certainly lie in the concept of some or another general condition; and, although you are not acquainted with the entire essence of the thing, you can still make secure use of it and derive therefrom a great deal about the thing. (286.8–21)

Just as the Newtonian method in natural science does not begin with

33. There is no question here, however, of a distinction between synthetic and analytic *judgements*. We are merely distinguishing two different routes by which one may arrive at definitions of concepts: either by conjoining initial primitive concepts so as subsequently to build up complex concepts or by beginning with given complex concepts for which the desired primitive components are then to be sought via analysis. A particularly clear discussion of this point is found in Menzel [78], pp. 172–184 in particular. Indeed, so far is Kant from the critical distinction between analytic and synthetic judgements that he here asserts: "Metaphysics therefore has no formal or material ground of certainty that would be of another kind from that of geometry. In both the formula of judgement occurs in accordance with the propositions of identity and contradiction" (295.36–296.2). Menzel also emphasizes—as does Tonelli [111], p. 65—that Kant envisages an ideal future state of metaphysics in which all of its concepts are finally clarified and in which it can then proceed synthetically as well: see 290.22–26.

34. See 277.19–35, 282.34–283.9, 283.13–24, and especially 289.6–26. It is true that Kant himself had previously proceeded *more geometrico* in both the *New Exposition* and the *Physical Monadology* (and also in the little dissertation on fire of 1755). However, Kant's criticism here really pertains only to the external form of these works: the content of the main theological argument of the *New Exposition* is essentially identical to that of *The Only Possible Basis,* where Kant expresses the same scruples against proceeding mathematically by beginning with definitions as in the *Enquiry* (2, 66.12–22)—moreover, the very same argument is offered as an example of certainty in natural theology in §1 of the Fourth Consideration in the *Enquiry* (2, 296–297); the content of the main argument of the *Physical Monadology* is presented as the primary illustration of the correct metaphysical method in the Second Consideration of the *Enquiry* (286.22–288.35).

direct insight into the ultimate principles governing the behavior of bodies but rather infers these principles regressively or analytically from more mundane and less controversial facts of experience ("deduction from the phenomena"), so Kant's improved method of metaphysics does not begin with direct and complete insight into the meanings of the concepts in question (that is, with definitions) but rather hopes to derive such knowledge of essences regressively or analytically from more certain and less controversial judgements that simply record evident knowledge we already possess involving these concepts.[35]

In thus maintaining that metaphysics must renounce a strictly deductive or synthetic method in which all subsequent propositions are derived from a small number of initial principles, and must instead adopt a quasi-inductive or regressive method which takes as its basis a large number of heterogeneous unprovable propositions as data, Kant is explicitly following the example of Crusius—as he makes very clear in §3 of the Third Consideration. Kant also follows Crusius in carefully distinguishing the *formal* principles of cognition, such as the principles of identity and contradiction, from the *material* principles of cognition: namely, the unprovable data just mentioned. The primary error of past (namely, Wolffian) metaphysics was to place exclusive reliance on the former:

> Now in philosophy there are many unprovable propositions, as were also introduced above. These certainly stand under the formal first principles, but immediately; however, in so far as they also contain grounds of other cognitions, they are the first material principles of human reason. E.g., *a body is composite* is an unprovable proposition, in so far as the predicate can only be thought as an immediate and first characteristic in the concept of body. Such material principles constitute, as *Crusius* correctly asserts, the foundation and solidity of human reason. For as we have mentioned above, they are the material for explanation and the data from which one can securely infer even if one has no explanation.
>
> And here *Crusius* is correct to reproach other schools of philosophy for ignoring these material principles and having held onto merely the formal ones. (295.1–15)[36]

35. For the idea that such judgements must be *uncontroversial*, see the criticism of Crusius in §3 of the Third Consideration: "But one can never concede the value of highest material principles to some propositions if they are not evident for every human understanding" (295.19–21).

36. For the relationship between Kant and Crusius see Tonelli [110], especially chap. III and chap. IV, §D. As Tonelli points out, this same Crusian doctrine is present in §6 of *The False Subtilty of the Four Syllogistic Figures* (1762). Kant there makes it clear that the *immediacy* of such unprovable material principles consists simply in the circumstance that they are not known to be true on the basis of a definition: "All judgements that immediately stand under the propositions of identity or contradiction—i.e., in which neither the identity

The crucial question, however, now concerns the nature and status of these "first material principles of human reason." What is the precise character of such data and how is their certainty secured? Kant criticizes Crusius's own appeal to "the feeling of conviction" here as excessively subjective (295.32–35); but how then does he himself propose to provide a more objective ground?

There is no better way to answer this question than to consider the example Kant presents of "the only certain method of metaphysics in the cognition of bodies" in the Second Consideration (286–287). This example recapitulates the central argument of the *Physical Monadology*. Bodies are composite and thus consist of simple parts. However, "I can verify by means of infallible proofs of geometry that space does not consist of simple parts" (287.1–2). Therefore the simple parts of bodies must occupy a space, not by a plurality of substantial parts in turn (which would contradict their simplicity), but rather through a *repulsive force* of impenetrability. The simple parts of bodies are therefore unextended (having no multiplicity of parts external to one another) and instead fill a space through the "multiplicity in their external action" (287.32–33).

If we now compare this example with the passage from the essay on *Negative Magnitudes* quoted above it is clear that prominent among the material propositions Kant intends as data here are mathematical propositions from the exact sciences. Thus the geometrical proposition that space is infinitely divisible is absolutely crucial to Kant's example, for if space itself consisted of simple parts no inference to the existence of a repulsive force would in fact be possible.[37] Kant's example turns precisely on combining a metaphysical proposition—which he presumably understands as entirely uncontroversial and "evident for every human understanding" (see note 35 above)—with an "evident" and "infallible" proposition of

nor the conflict is discerned through an intermediate characteristic (hence not by means of the analysis of concepts) but which is rather immediately discerned—are unprovable judgements; those which can be mediately cognized are provable. Human cognition is full of such unprovable judgements. Some of these precede every definition, as soon as one, in order to arrive at [the definition], represents that as a characteristic of a thing which one previously and immediately cognizes in the thing. Those philosophers are mistaken who proceed as if there were no unprovable fundamental truths except a single one" (2, 60.27–61.7). Note that there is again no question here of *synthetic* judgements in the critical sense, for *all* judgements stand under the principles of identity and contradiction (compare the passage from the *Enquiry* at 2, 295.36–296.2, quoted in note 33 above).

37. Moreover, as is clear from the immediately following discussion at 287.36–288.35, this explanation of impenetrability via a repulsive *force* is itself absolutely central to Kant's defense of action-at-a-distance against metaphysical doubts (compare note 28 above), and Kant employs precisely the same strategy in this regard from the *Physical Monadology* through the *Metaphysical Foundations of Natural Science*.

geometry.[38] We can therefore infer that the data for Kant's Newtonian method of metaphysics consists in uncontroversial metaphysical propositions (not definitions!) acceptable to all competing schools, on the one hand, and established results of the mathematical exact sciences, on the other.[39] By placing both types of data side by side and allowing them to interact it is possible to regulate—and if need be to correct—metaphysical reasoning by means of "a science that is involved only with comprehensible and evident insights," and it is possible to hope that a well-grounded science of metaphysics will thereby be someday achieved. Moreover, as we have seen, it is by means of just this kind of interaction with the exact sciences that Kant has constructed his own distinctive metaphysical position.

38. The uncontroversial metaphysical proposition in question is of course the judgement that bodies are composite. The argument from this to the existence of simple parts is presented in §2 of the First Consideration at 279.11–25. Note that Kant presents this argument as a paradigm of purely philosophical, purely abstract reasoning—where the universal is considered only *in abstracto,* and he explicitly contrasts it here with his standard argument for the infinite divisibility of space derived from a figure (see *Physical Monadology,* Prop. III: 1, 478)—where the univeral is considered under signs *in concreto* in that the geometer "cognizes in this sign with the greatest certainty that the division must proceed without end" (279.10–11). In the above passage from the essay on *Negative Magnitudes* Kant criticizes the metaphysician for ignoring this latter demonstration and instead placing his trust "solely in the ambiguous consciousness of this concept [of space], in that one thinks it in an entirely abstract manner." As Kant explains in the Second Consideration of the *Enquiry* (284.29–285.2), ambiguity is one of the main pitfalls of purely abstract philosophical reasoning as opposed to concrete and intuitive mathematical reasoning.

39. In light of Kant's comments in the Introduction, where he says that in accordance with the Newtonian method the content of his treatise consists of "secure propositions of experience and the consequences immediately drawn therefrom" (275.17–18), there is some temptation to view the data for Kant's method as *empirical judgements* and, accordingly, to view the *Enquiry* as defending a version of Newtonian empiricism: for such a reading see especially de Vleeschauwer [116], pp. 33–37. As we have seen, however, the "propositions of experience" in question here are nothing but the results of "secure inner experience—i.e., an immediately evident consciousness." And in §6 of *The False Subtilty* Kant defines the faculty of judgement in general as "nothing other than the faculty of inner sense, i.e., to make one's own representations into objects of one's thought" (2, 60.13–15). Moreover, this faculty of judgement is the basis of "the entire higher faculty of cognition" (60.18), which is customarily divided into "*understanding* and *reason*" (59.14). In particular, then, there is no suggestion at all that what will later be called *sensibility* plays any essential role in the "secure inner experience" that furnishes Kant's data. On the contrary, all Kant's examples are *a priori* judgements of conceptual connection—either uncontroversial metaphysical judgements or evident mathematical judgements. See also Tonelli [110], where Kant's notion of "inner sense" here is compared with the Crusian conception thereof and also with the traditional notions of "good sense," "right reason," "the natural light of reason," and so on. Where Kant goes beyond Crusius here is precisely by conjoining (following Euler) mathematically established results of the exact sciences with purely metaphysical results of "right reason."

· III ·

The metaphysical position we have sketched in §I above is not without a certain brilliance and power—and even, with respect to Kant's new approach to physico-theology, a certain sublimity. Nevertheless, this revised version of a monadology is also faced with insurmountable problems which, in the end, undermine it decisively. These problems all center around Kant's new conception of the nature of space and how, in particular, space is grounded in the fundamental law of interaction governing the external relations of non-spatial unextended monads. The key idea, it will be recalled, is that the most basic laws of dynamics—as codified in the Newtonian theory of universal gravitation—express the schema of the divine intellect by which God first brings it about that the original simple substances not only exist but are present to one another in a single world. Moreover, in thus first constituting the copresence of simple substances in a single world God simultaneously constitutes space. The properties of space are therefore entirely derivative from the properties of the fundamental law of interaction contained in the schema of the divine intellect.

The first problem lies in understanding precisely how the properties of space can be so derived from the external relations governing a non-spatial realm of unextended simple substances. Let us grant, for example, that Kant has shown that the absolute simplicity of the monads is consistent with the infinite divisibility of space. This does not yet show how the infinite divisibility of space can actually be *derived* from the properties of the underlying monadic realm. Indeed, in view of Kant's insistence that bodies not only consist of simple substances, but consist of a definite finite number of such simple substances (*Physical Monadology,* Corollary to Proposition IV: 1, 479.35–36), it is particularly difficult to see how such a derivation is possible.[40] Thus, the idea of a sphere of activity by which the monad fills a space by its repulsive force presupposes rather than explains the continuity of space. Similarly, Kant's sketch of how the three-dimensionality of space can be derived from the inverse-square law of gravitational attraction in §§9–11 of the *True Estimation* presupposes that the fundamental law of interaction is expressed in terms of a continuous function of (Euclidean) distance (and therefore that distance is itself a continuous quantity). Given this the argument then explains why space has three rather than some other number of dimensions. No explanation has been offered, however, for the fact that space is a continuous extended (Euclidean) manifold in the first place.

From a modern point of view, there is of course no difficulty in principle

40. In this respect Kant's views contrast sharply with those of Leibniz, who maintains that an actual infinity of monads are contained in every material body no matter how small: see, e.g., [68], p. 522; [72], p. 34.

in envisioning a derivation of the properties of space from the properties of an underlying realm of entities that are not themselves intrinsically spatial. What is required is an axiomatic characterization of an abstract system of relations so that all the properties of three-dimensional Euclidean space, for example, can be derived therefrom. And the essence of this modern point of view, of course, is that the abstract system of relations in question is otherwise completely undetermined: we do *not* need to begin with intrinsically spatial elements. In Kant's time, however, the logical resources by means of which alone this kind of characterization can be developed are entirely lacking. In particular, the logical resources necessary for a purely axiomatic characterization of continuity (or even of denseness or infinite divisibility, which in Kant's time is of course not clearly distinguished from continuity) simply do not exist until the latter part of the nineteenth century. And this means that *Kant's* "relationalist" project is not completable in principle.[41]

Suppose, however, that Kant's version of a "relational" theory of space could be fully implemented and that the essential properties of space could after all be completely derived from the fundamental law of interaction governing simple substances. This would create even greater problems, for it would then follow that our knowledge of the essential properties of space had precisely the same status as our knowledge of the fundamental law of interaction. But the latter is expressed phenomenally as the law of universal gravitation—which is itself known only empirically on the basis of Newton's "deduction from the phenomena" presented in Book III of *Principia*.[42] It follows that if Kant's version of a "relational" theory of space were correct, our knowledge of the essential properties of space such as three-dimensionality and continuity should be equally empirical. And it would therefore be entirely incomprehensible how we could determine these properties a priori in pure geometry antecedent to any consideration whatsoever of the fundamental laws of dynamics. From the point of view of Kant's version of "relationalism," in other words, our knowl-

41. Leibniz's version of "relationalism," by contrast, is not so clearly subject to these difficulties. On the one hand, Leibniz envisions an actual infinity of monads in every small region of space (see note 40 above), and, on the other hand, space for Leibniz is something ideal or imaginary. Indeed, problems involving continuity are evidence for precisely this ideal or imaginary status: see, e.g., [68], pp. 522–523; [72], p. 34—and compare [67], vol. 4, p. 436; [72], p. 44. (The relationship between continuity or infinite divisibility and Kant's impoverished conception of logic is discussed in detail in Chapter 1 below.)

42. Kant himself presents a lucid summary of this "deduction from the phenomena" in a short introductory section to the First Part of the *Theory of the Heavens* at 1, 243–246. Moreover, as we have seen, he consistently characterizes the Newtonian method as proceeding regressively based upon experience and geometry. For Kant, then, there is no doubt that our knowledge of the law of universal gravitation is based on precisely an inference from "phenomena."

edge of the essential properties of space should be entirely derivative from our (empirical!) knowledge of the laws of dynamics.[43]

To appreciate the full force of the problem here one should recall that space is in no way dependent on the laws of dynamics for Leibniz. On the contrary, for Leibniz space arises simply from the (co-)existence of the underlying simple substances considered in itself, and no more specific rules or laws of interaction are involved.[44] Space thus represents what is common to all possible systems of dynamical laws, and Leibniz accordingly distinguishes explicitly between the necessary status of the laws of geometry and the contingent status of the laws of dynamics: the former are true in all possible worlds and thus depend only on the principle of identity, the latter result from God's choice of the actual world as the best of all possible worlds and thus depend essentially on the principle of sufficient reason.[45] For Leibniz, then, there can be no question of an empirical foundation for the laws of geometry. Hence, it is precisely by replacing Leibniz's conception of the ideality of space with his own conception of the fundamentally dynamical character of space that Kant himself has first exposed geometry to the threat of empirical disconfirmation.

Finally, there is a serious problem concerning how the interaction between material substances such as bodies and immaterial substances such as the soul is to be conceived. Kant is convinced, as we have seen, that material substances and immaterial substances do really interact and therefore that they are present to one another in a single world. But copresence and the possibility of real interaction are grounded, in the first instance, precisely by the fundamental laws of physical dynamics. It is these laws, and these laws alone, which first constitute both the copresence

43. In *Thoughts on the True Estimation* Kant argues that the inverse-square law—and therefore the three-dimensionality of space—rests wholely on God's arbitrary choice and is thus contingent: God could have chosen a different law and a different number of dimensions. The impossiblity of our imagining a space of other than three dimensions is equally contingent and depends simply on the fact that our perceptions are subject to the very same arbitrary law (§10: 1, 24.19–25.2). In *The Only Possible Basis*, on the other hand, Kant maintains that both the properties of space and the basic laws of motion are necessary, in that both flow from the divine intellect in which all possibilities are grounded (First Consideration of Part Two: 2, 93–100). Nevertheless, the properties of space and the basic laws of motion are necessary in the very same sense, and there is no suggestion whatever of any essential difference in *epistemic* status.

44. See note 10 above, and note that Leibniz explicitly abstracts there from the "manner of existing" of the simple substances.

45. Sections 345–351 of the *Theodicy* ([67], vol. 6, pp. 319–323; [69], pp. 332–336) articulate this distinction with particular clarity. Kant could hardly have been unaware of this fundamental difference between Leibniz's conception of space and his own, for he explicitly refers to §351 of the *Theodicy* in the course of presenting his own argument for the dependency of the three-dimensionality of space on the inverse-square law in *Thoughts on the True Estimation* (§9: 1, 23.13–16).

of all simple substances and space. Hence, both material and immaterial substances are in space, and both, moreover, are subject to the very same fundamental laws of physical dynamics. There is thus a very real danger of collapsing the distinction between material and immaterial substances completely—a danger which Kant explicitly acknowledges in §2 of the Third Consideration of the *Enquiry:*

> I admit that the proof we possess demonstrating that the soul is not matter is good. But be on your guard in inferring therefrom that the soul is not of material nature. For by this everyone understands not only that the soul is not matter, but that it also is not such a simple substance as could be an element of matter. This requires a special proof: namely, that this thinking being is not in space—as is a corporeal element—through impenetrability, and that it could not constitute together with others something extended and massive [ein Ausgedehntes und einen Klumpen]; whereof no such proof has actually been given, which, if it were discovered, would intimate the inconceivable manner in which a spirit is present in space. (2, 293.7–18)

Once again, it is Kant's own particular view of copresence and space which has made it virtually impossible to conceive how a soul or spirit could be present to matter without actually being itself "of material nature."[46]

The solution to all of these problems is to acknowledge the autonomy of space (and time) and eventually the autonomy of the material or phenomenal world of bodies in space (and time). Kant takes the first step in the essay *On the First Grounds of the Distinction of Regions in Space* of 1768. He there argues *"that absolute space has its own reality independently of the existence of all matter and even as the first ground of the possibility of its composition"* (2, 378.9–11), so that "the determinations of space are not consequences of the positions of the parts of matter relative to one another, rather, the latter are consequences of the former" (383.13–15). Kant is particularly careful to separate the question of the independent reality of space from the laws of dynamics and, in this connection, explicitly to distance himself from Euler's argument in the latter's 1748 essay on space and time

46. For this point I am especially indebted to Alison Laywine. In [65] she argues, in particular, that precisely this problem underlies Kant's preoccupation, in the *Dreams of a Spirit-seer, Interpreted through Dreams of Metaphysics* of 1766, with Swedenborg's fantastic stories of intercourse with spirits. The problem, in other words, is that Kant's own metaphysics has itself placed the material and the immaterial worlds in uncomfortably close proximity. See, for example, Kant's remarks at 2, 321.34–38: "On the other hand, in the case of spiritual substances that are to be united with matter, as, e.g., the human soul, the difficulty is manifest that I am to think a mutual connection of such with corporeal beings into a whole, and yet I am nevertheless supposed to annul the only known manner of connection, which occurs among material beings."

> . . . which, however, did not quite achieve its end, because it only indicates the difficulties in giving the most universal laws of motion a determinate meaning if one assumes no other concept of space than that arising from abstraction from the relation of real things, but leaves untouched the no less significant difficulties that remain in the application of the aforesaid laws if one wants to represent them *in concreto* in accordance with the concept of absolute space. The proof I seek here is intended to provide, not the students of mechanics [Mechanikern]—as was the intention of *Euler*—but even the geometers [Meßkünstlern] a convincing ground for being able to assert the actuality of their absolute space with their customary evidence. (378.20–30)

Kant's aim is thus to vindicate the autonomous reality of "universal absolute space, *as it is thought by geometers* [die Meßkünstler]" (381.17–18: my italics).

It is important to note that Kant is here leaving the manner in which space thus has autonomous reality entirely open. In particular, he is not here endorsing a *Newtonian* conception of the autonomous reality of "absolute space." This is clear from the above passage, where Kant deliberately refrains from endorsing the Newtonian conception—adopted by Euler—of *absolute motion.*[47] Moreover, towards the end of the essay Kant says that "absolute space" is "no object of outer sensation, but is rather a fundamental concept [Grundbegriff], which makes all of these first possible" (383.19–20). At the very least, then, the later doctrine of the *Inaugural Dissertation*—that space has autonomous reality, not as a self-subsistent independent object in the manner of Newton, but rather as an autonomous form of sensible intuition—is in no way excluded here. What is definitely excluded is only an explicitly "relationalist" conception—in either Leibniz's original form or Kant's revised version—according to which the essential properties of space are entirely derivative from those of the ultimate constituents of matter.

In any case, the characteristic Kantian doctrine of space and time as autonomous forms of sensible intuition is of course first articulated in the *Inaugural Dissertation* of 1770. Kant there combines his new insight of the essay on *Regions in Space* into the necessarily autonomous status of geometrical space with his ideas from the *Enquiry* concerning the essentially sensible and intuitive character of mathematical reasoning,[48] and he

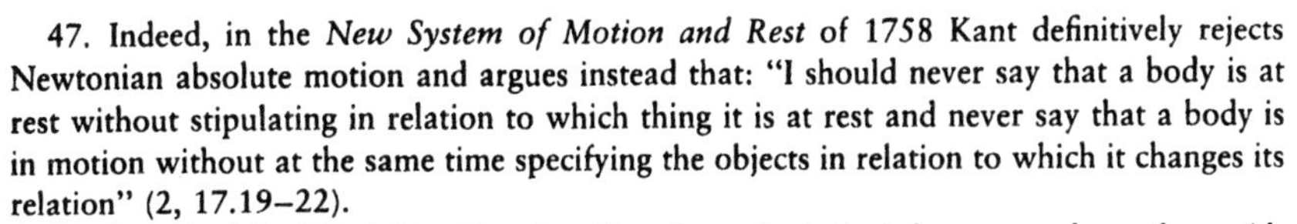

47. Indeed, in the *New System of Motion and Rest* of 1758 Kant definitively rejects Newtonian absolute motion and argues instead that: "I should never say that a body is at rest without stipulating in relation to which thing it is at rest and never say that a body is in motion without at the same time specifying the objects in relation to which it changes its relation" (2, 17.19–22).

48. See §1 of the Third Consideration: "mathematics in its inferences and proofs considers its universal cognition under signs *in concreto,* but philosophy does this alongside the signs yet always *in abstracto.* This constitutes a notable difference in the manner in which

finally arrives at the idea that there is a distinct faculty of sensible cognition in addition to the faculty of intellectual cognition: it is through this distinct faculty—and through this faculty alone—that we first represent space and time in pure mathematics, and thus *"there is given a science of the sensible"* (2, 398.4) independently of all purely intellectual cognition.[49] Space and time are no longer realities derivative from or constituted by the underlying monadic realm of non-spatio-temporal simple substances; rather, they are autonomous forms of pure sensible intuition through which this underlying monadic realm manifests itself or appears to creatures with our particular faculty of sensibility.[50] The essential properties of space and time can therefore in no way be derived from our cognition of the underlying monadic realm—which latter is necessarily purely intellectual—but constitute instead the autonomous subject matter of the mathematical exact sciences.[51]

It follows that space and time are both phenomenal and ideal in a radically new sense. Human cognition has two distinct principles: the

---

each arrives at certainty. For, since the signs of mathematics are sensible means of cognition, one can, with the same confidence with which one is assured of what one sees with one's eyes, also know that one has left no concept out of attention, that each individual comparison occurs according to easy rules, etc. . . . the intuitive nature of this cognition, in so far as correctness is concerned, is greater in mathematics than in philosophy: since in the former the object is considered in sensible signs *in concreto* but in the latter always only in universal abstracted concepts, whose clear impression can be nowhere near as great as in the former. In geometry, where the signs have a similarity with the designated things, moreover, this evidence is still greater—although in algebra the certainty is precisely as trustworthy" (2, 291.24–292.17). Compare note 38 above.

49. Kant's reliance on both the essay on *Regions in Space* and the *Enquiry* is clear in §15.C, where he first recapitulates the example of incongruent counterparts from the former (403.1–10), and then continues: "Hence geometry uses principles which are not only indubitable and discursive, but fall under the gaze of the mind, and the *evidence* in demonstrations (which is the clarity of certain cognition in so far as it is assimilated to the sensible) is not only the greatest possible, but is the unique case given in pure science and is the *exemplar* and medium of all *evidence* in other sciences. . . . For the rest, geometry does not demonstrate its universal propositions by thinking the object through a universal concept, as happens with rational things, but by subjecting it to the eye through a singular intuition, as happens with sensitive things" (403.11–22).

50. In particular, space is no longer a reality constituted by the fundamental law uniting substances into a single world, but is rather precisely the *appearance* of this fundamental law to our faculty of sensibility: see §16, where Kant speaks of the *"principle upon which this relation of all substances rests, which when viewed intuitively is called space"* (407.6–7).

51. Compare §12: "Hence pure mathematics, which exhibits the form of all our sensitive cognitions, is the organon of each and every intuitive and distinct cognition; and, since its objects themselves are not only the formal principles of all intuitions, but are themselves *original intuitions,* it provides the truest cognition and at the same time the exemplar of the highest evidence in other cases" (397.33–398.4); and the Corollary to Section III: "Therefore all primitive affections of these concepts [of space and time] are outside the boundaries of reason and can therefore not be explained intellectually in any way" (405.22–24).

intellectual faculty of understanding or reason by which we represent the underlying monadic realm as it is in itself, and the new faculty of sensibility by which we represent this underlying reality as it appears to creatures like ourselves. Not all rational beings have a faculty of sensibility constituted like our own, and so it is not the case that reality appears spatio-temporally to all rational beings. Kant's radical division of the human cognitive faculties has thus introduced a radically new element of *subjectivity* into the representations of space and time: space and time are "*subjective* and ideal and proceed from the nature of the mind by a constant law, as it were as a schema for coordinating with one another absolutely all external sensibles" (403.24–26). And it is precisely in virtue of Kant's radical division of the cognitive faculties that this new kind of ideality goes far beyond that attributed to space and time by Leibniz and Wolff.[52] At the same time, however, because Kant's delineation of the new faculty of sensibility is grounded entirely in the mathematical exact sciences, sensible cognition can no longer be characterized as either "confused" or "imaginary" in comparison with intellectual cognition: on the contrary, the former represents the very paradigm of clear, evident, and well-grounded human knowledge (§7: 394–395).

Kant's articulation of this radically new conception of space and time here is clearly intended to solve the three serious problems afflicting his earlier conception of space and time discussed above. In the first place, the essential properties of space and time expressed in pure mathematics are no longer derivative from those of an underlying realm of non-spatio-temporal entities. Continuity, in particular, is now a primitive *intuitive* property of space (403.30–37), which, in the end, depends on the *intuitive* continuity of time—for it is only in time that the human mind can represent a potentially infinite manifold "in so far as, by the successive addition of part to part, the concept is possible genetically, i.e., by SYNTHESIS" (397.18–20).[53] Thus:

52. Leibniz, for example, asserts that extension and motion "are not substances but true phenomena, like rainbows and parhelia." ([68], p. 523; [72], p. 34). Since Leibniz has no truly distinct faculty of sensibility, however, there is no reason on his conception that substances should not appear spatio-temporally to all rational beings (or, at least, to all finite rational beings). (Immediately before the above quotation Leibniz deleted: "Space, time, extension, and motion are not things, but modes of consideration having a foundation.")

53. Kant also holds that the *actual* or *completed* infinite is not representable intuitively—yet is not therefore impossible (388.6–40). This does not mean, however, that the actual infinite is *intellectually* representable for us: "*measurability* here denotes only the relation to the scale of the human intellect, through which one can arrive *at the definite concept of a multitude* only by successively adding unit to unit and can arrive at the *complete* concept, which is called *number,* only by terminating this progression in a finite time . . . for there could be given an intellect, although certainly not a human one, which, without the successive application of a measure, would clearly determine a multitude at one glance" (388.33–40).

> *Time is a continuous quantity* and is the principle of the law of continuity in the alterations of the universe. For the continuum is a quantity that is not composed of simples. But since through time nothing but relations are thought, without any entities being given in relation to one another, in time as a quantity there is composition such that if all of it is imagined to be annulled nothing at all remains. But that composite of which nothing at all remains when all composition is annulled does not consist of simple parts. (399.21–27)

Continuity is in this way a most essential feature of the *sensible* world, and "it is therefore unsuitable to wish to arm reason against the first postulates of pure time, e.g., continuity" (402.2–3).

In the second place, however, since the essential properties of space and time are no longer derivative from the fundamental laws of dynamical interaction, there is no longer a danger of subjecting the laws of pure mathematics—especially the laws of geometry—to empirical disconfirmation. Thus in §15.D Kant contrasts his new conception of space with both the "absolute" conception of the Newtonians and the "relational" conception of those who follow Leibniz. He reserves his harshest criticism for the latter philosophers, who

> . . . contend that [space] is the relation *itself* of existing things, which vanishes entirely if the things are annulled and can only be thought in actualities, as, after Leibniz, most of us [Germans] maintain. . . . [These philosophers] are in open conflict with the phenomena themselves and the most faithful interpreter of all phenomena: geometry. For—without bringing into view the evident circularity in the definition of space in which they are necessarily entangled—they dislodge geometry from the apex of certainty and cast it down to the rank of those sciences whose principles are empirical. For if all the affections of space are only borrowed through experience from external relations, then no universality enters into the axioms of geometry except comparative universality, as is acquired through induction (i.e., extending as far as is observed), and no necessity, except according to established laws of nature, and no precision, except that which is arbitrarily fabricated; and there is hope, as happens with the empirical, of sometime discovering a space endowed with other primitive affections, and perhaps even a two-sided rectilinear figure. (403.28–404.20)[54]

Hence the only way the *human* intellect can represent infinity at all is potentially and in time. (The connections between continuity or infinite divisibility and Kant's conception of the intuitive character of space and time are extensively discussed below, especially in Chapter 1.)

54. In the first edition of the *Critique of Pure Reason* Kant illustrates the danger of an empirical foundation for geometry with the property of three-dimensionality: "What is borrowed from experience has therefore only comparative universality, namely through induction. One would thus only be able to say that, as far as has been observed so far, no space has been found that has more than three dimensions" (A24). As we have seen, Kant

As we have seen, this criticism applies to the views of no other philosopher so well as it does to the earlier conception of Kant himself.[55]

Finally, since the sensible world—that is, the appearance to our particular faculty of sensible intuition of the underlying purely intelligible realm—has now been definitively separated from this intelligible world, there is no longer a danger of bringing material and immaterial substances into uncomfortably close proximity. On the contrary, Kant's new method in metaphysics emphasizes above all that "one should anxiously be on guard *that the principles of sensitive cognition do not transgress their domestic boundaries and affect the intellectual*" (411.29–31). In particular, then, there is no danger that immaterial substances such as the soul are governed by the fundamental laws of physics, for these laws—together with the geometrical space within which they are formulated—apply only to sensible or material things:

> But there is a virtual presence of immaterial [substances] in the corporeal world, not a local presence (although the latter is improperly customarily asserted); space, however, contains the conditions of no other possible mutual actions except those of matter; but what truly constitutes the relations of external forces for immaterial substances—both among one another and with bodies—entirely escapes the human intellect, as also the most perspicacious Euler, otherwise a great investigator and judge of phenomena, acutely noted (in letters to a certain German princess). (414.8–14)[56]

The interaction between soul and body does not take place in accordance with the fundamental laws of physical dynamics, but only in accordance with the purely intellectual idea of a conjunction of substances through a common origin in the divine creator (§20: 408). "Certainly the human

argues in *Thoughts on the True Estimation* that the three-dimensionality of space is contingent and depends on God's arbitrary choice of the form of a law of interaction—he even speculates that there may be other *actual* worlds (not in connection with our world) that have spaces of more than three dimensions (§11: 1, 25).

55. If the blame appears disingenuously to be placed rather with Leibniz here, despite the fact that, as we have seen, Leibniz himself carefully distinguishes the status of geometry from that of the laws of dynamics (a fact of which Kant could hardly fail to have been aware: see note 45 above), it is perhaps because Kant is reasoning in the following way. The original Leibnizean monadology is in open conflict with the phenomena, for it is entirely incompatible with the Newtonian doctrine of universal attraction. A Leibnizean has no other option, then, but to modify the monadology into something like Kant's earlier view. And once this is done geometry then necessarily acquires the same empirical status as is possessed by the fundamental laws of dynamical interaction.

56. The reference appears to be to the Letter of November 29, 1760, "On the Nature of Spirits" ([25], p. 319; [26], p. 270): "This union of the soul with the body undoubtedly is, and ever will be, the greatest mystery of the divine Omnipotence—a mystery which we shall never be able to unfold."

mind is not affected externally, and the world is not open to its vision to infinity, except *in so far as it is sustained with all others by the same infinite force of a single being*" (409.31–34).[57]

It is important to note, however, that, although Kant has indeed radically revised his conception of the relationship of space and time (and thus the phenomenal world in space and time) to the underlying monadic realm, he has in no way rejected his metaphysical conception of this underlying realm itself. Indeed, the metaphysical treatment of this realm—which now falls under the exclusive purview of the intellectual faculty—is taken over from Kant's earlier writings virtually unchanged. The intelligible world consists of two distinguishable aspects: matter, that is, the simple substances which originally populate the world (§2: 389–390); and form, that is, the *real* coordination of substances into a single whole in virtue of a "nexus"—which "constituting the *essential* form of a world, is viewed as a principle of the *possible influxes* of the substances constituting the world" (390.18–20). This nexus, as we just observed, is itself only possible as a consequence of the dependence of all simple substances on a single divine being (§§16–21: 406–408), and, since some such nexus is in fact essential to the world, the system of physical influx is still to be preferred to the system of pre-established harmony (§22: 409). Moreover, even though space and time are no longer constituted by such an essential nexus of simple substances, but are rather merely appearances thereof, which "bear witness to some such common principle of a universal nexus without exhibiting it" (391.8–9), we also find a suggestion of the doctrine of divine omnipresence: "space, which is the sensibly cognized universal and necessary condition of the copresence of all, can be called PHENOMENAL OMNIPRESENCE" (410.1–3). In this way, the sensible world studied by the mathematical exact sciences is no longer constituted by the underlying—purely intelligible—monadic reality studied by metaphysics, but it is still somehow a reflection of this metaphysical reality.

· IV ·

Just as the defects in Kant's earler metaphysics clearly stand out from the point of view of the *Inaugural Dissertation,* so the defects in the latter clearly stand out from the point of view of the critical philosophy. The main problem, of course, is that, by definitively securing the autonomy of space, time, and the phenomenal world in space and time from the underlying metaphysical reality of simple substances, Kant has also entirely isolated this underlying monadic realm from our epistemic purview. In

57. Again, this aspect of the *Inaugural Dissertation* is especially emphasized by Alison Laywine: see note 46 above.

particular, since space, time, and the phenomenal world are no longer derivative from or constituted by the external relations of the original monads, it is no longer clear how they are at all connected with the metaphysical reality which they are still supposed somehow to express or reflect. Hence, since it appears that objects are given to us only through our sensible faculty of cognition, it is entirely unclear how our intellectual faculty of cognition—which is supposed to represent the fundamental monadic reality as it is in itself—has any access at all to its objects.[58]

This general difficulty can also be understood in terms of the Eulerian-Newtonian conception of metaphysical method described in §II above. The crucial idea of this conception is that metaphysics must be regulated and guided by the clearer and more secure results of the mathematical exact sciences: only so can metaphysics aspire to a properly scientific status itself. Now, however, metaphysics and the exact sciences have been definitively split apart: the latter describe the spatio-temporal phenomenal world and are products of the sensible faculty of cognition, the former describes the non-spatio-temporal monadic realm of simple substances and is a product of the entirely distinct intellectual faculty of cognition. It appears, then, that metaphysics and the exact sciences are now positively excluded from the fruitful interaction envisioned by the Eulerian-Newtonian conception of metaphysical method, and, if this is the case, it is now entirely unclear how metaphysics as a science is possible at all.

Indeed, in a letter to J. H. Lambert of September 2, 1770, concerning the *Inaugural Dissertation* (which he has just sent to Lambert under separate cover), Kant outlines a division of philosophy into two distinct parts: general phenomenology, which determines the principles and limits of sensibility, and metaphysics proper, which describes the non-sensible world known by the pure intellect:

> The most universal laws of sensibility mistakenly play a great role in metaphysics, where merely concepts and principles of pure reason are nevertheless in question. It appears that an entirely special, although purely negative science (general phenomenology), must precede metaphysics, in which the validity and limits of the principles of sensibility are determined, in order that they do not confuse the judgements about object of pure reason, as has almost always happened until now. For space and time and the axioms [Axio-

58. This is of course the problem Kant himself raises in his letter to Marcus Herz of February 21, 1772: "In my dissertation I had contented myself with expressing the nature of intellectual representations merely negatively: namely, that they were not modifications of the soul by means of the object. But how then a representation is possible that could otherwise relate to an object without being in any way affected by it I passed over in silence. I had said: the sensible representations represent the things as they appear, the intellectual as they are. But through what are these things then given to us, if they are not [so given] through the mode in which we are affected . . . " (10, 130.33–131.5).

> men] which consider all things under the relations of [space and time] are very real in consideration of empirical cognitions and all objects of the senses, and actually contain the conditions of all appearances and empirical judgements. But if something is thought absolutely not as an object of the senses, but rather through a universal and pure concept of reason—as a thing or substance in general, etc.—then very mistaken results are forthcoming if one wants to subject it to the above fundamental concepts of sensibility. It also appears to me—and perhaps I will be fortunate to gain your agreement in this through this still very inadequate attempt—that such a propaedeutic discipline, which would preserve metaphysics proper from all such admixture of the sensible, could be easily brought to a useful [state of] fullness of detail and evidence. (10, 98.17–36)

Metaphysics and the exact sciences could hardly be further apart.[59]

Kant did not gain Lambert's agreement in this, as is clear from the latter's reply of October 13, 1770:

> I will certainly allow it if one views time and space as mere images [Bilder] and appearances [Erscheinungen]. For, besides the fact that constant appearance [Schein] is truth for us—where that which lies at the basis is either never discovered or only in the future—it is useful in ontology also to take up the concepts secured from appearance [Schein], *because its theory must still finally be again applied to the phenomena.* For so also the astronomer begins with the phenomenon, derives the theory of the world structure [Weltbau] therefrom, and applies it in his Ephemerides again to the phenomena and their prediction. In metaphysics, where the difficulty of appearance [Schein] is so central, the method of the astronomer will certainly be the safest. The metaphysician can admit everything as appearance [Schein], separate the empty from the real, and infer from the real to the true. (108.23–36)

Thus Lambert not only reemphasizes the crucial importance of the Eulerian-Newtonian conception of metaphysical method here, but he also in some measure anticipates the radical new conception of metaphysics Kant himself is to articulate in the *Critique of Pure Reason.*

According to this new conception, as we know, metaphysics no longer describes an underlying monadic realm of simple substances at all—whether this realm be conceived of as constituting the phenomenal world in space and time or as that which lies beyond the phenomenal world

59. When Kant first writes to Lambert on December 31, 1765, there is no suggestion at all of such a separation. On the contrary, Kant there discusses his work on the proper method of metaphysics and envisions two examples *in concreto* which are to illustrate this method: *Metaphysical Foundations of Natural Philosophy* and *Metaphysical Foundations of Practical Philosophy* (10, 56). It appears, then, that Kant's conception of metaphysical method is still that of the *Enquiry* here, for it is hard to see how such a discipline as "metaphysical foundations of natural philosophy" is possible at all on the conception of metaphysics of the *Inaugural Dissertation.*

in space and time as reality relates to appearance. Instead, metaphysical concepts of the pure intellect (now the pure understanding)—such as possibility, existence, necessity, substance, cause, etc.[60]—are to be *applied to* the phenomenal world in space and time as the conditions of the possibility of experience. In this way metaphysics functions precisely to ground empirical knowledge and, at the same time, to explain how the mathematical exact sciences are themselves possible. For, as Kant has already intimated in the above letter to Lambert, the mathematical exact sciences—which delimit the form and principles of the spatio-temporal phenomenal world—constitute the conditions of the possibility of empirical knowledge, and what Kant now sees is that just this circumstance requires a further, and very elaborate explanation. The explanation in question is to show in detail how the pure concepts of the understanding are in fact applied to sensibility so as to make empirical knowledge possible through the exact sciences, and metaphysics is thereby absorbed into what Kant has earlier called general phenomenology.[61]

Kant calls this process of applying pure intellectual concepts to the phenomena the *schematism* of the pure concepts of the understanding, and he now holds that, considered apart from such schematization, pure intellectual concepts have no relation whatsoever to any object:

> In fact, however, a meaning does remain for the pure concepts of the understanding even after separation from all sensible conditions, but only the logical meaning of mere unity of representations, for which no object and therefore no meaning that could yield a concept of an object is given. Thus, e.g., substance, if one leaves aside the sensible determination of permanence, signifies nothing further than a something that can be thought as a subject (without being a predicate of something else). But I can make nothing of this representation, in that it absolutely does not indicate to me which determination the thing has which is to hold as such a first subject. Therefore, the categories, without schemata, are only functions of the understanding for concepts but represent no object. This meaning comes to them from sensibility, which realizes the understanding in that it simultaneously restricts it. (A147/B186–187)

The relationship between the pure intellectual concepts of metaphysics and the world of phenomena has thus been reinterpreted in a profoundly radical fashion. Pure intellectual concepts no longer characterize an underlying reality situated at a deeper and more fundamental level than the

60. These concepts are given as examples of the "real use" of the pure intellect in §8 of the *Inaugural Dissertation:* 2, 395.24.

61. Of course I am here considering metaphysics only in its *theoretical* aspect. Metaphysics considered as a *practical* science is still quite distinct from general phenomenology. Indeed, Kant of course still wants to limit the validity of the principles of sensibility—this time, however, to make room for the *metaphysics of morals.*

phenomena themselves; on the contrary, such concepts can acquire a relation to an object in the first place only by being realized or schematized at the phenomenal level. And it is this radical reinterpretation of the relationship between metaphysics and the phenomena—even more than the division of the cognitive faculties into sensibility and understanding—that constitutes Kant's truly decisive break with the Leibnizean-Wolffian tradition.

As the above passage suggests, the radical character of Kant's reinterpretation of the relationship between metaphysics and the phenomena is expressed most clearly, perhaps, in his reinterpretation of the concept of substance. Substance is now "*substantia phaenomenon* in space" (A265/B321): that is, *matter* (B278, B291). And phenomenal substance is characterized by permanence or conservation, in that its total quantity in nature is neither increased nor diminished (B224). Precisely as such, however, substance is no longer in any way *simple*. This becomes especially clear in the *Metaphysical Foundations of Natural Science* of 1786, where Kant argues that material substance is infinitely divisible (4, 502–504) and explicitly points out the incompatiblity of this with his own earlier *Physical Monadology* in particular (504–505) and with Leibnizean metaphysics in general (505–508); he then argues for the conservation of the total quantity of matter (541–542) and points out that this is true *only* in virtue of matter's divisibility into spatial parts external to one another (542–543)—the monadist, for whom the quantity of matter must be conceived as intensive rather than extensive, simply cannot make intelligible the conservation of matter (539.32–540.4).[62] Kant, in importing the Newtonian conception of force into the monadology, has already rejected the Leibnizean doctrine of the self-activity of substance (compare note 8 above); now, in explicitly rejecting the simplicity of substance, he has broken with the monadology completely.[63]

An analogous result holds for the interaction of substances. This concept too must be schematized spatio-temporally, in terms of the simultaneous copresence or community of material substances in space (B256). Interaction can therefore in no way be rendered intelligible from the point

62. Similarly, since the soul as mere object of inner sense has no extensive quantity, permanence or conservation can in no way be proved for it (543.3–14 and compare B413–415). Thus spiritual or immaterial beings do not fall under the schema of substantiality, and it follows that there are no spiritual or immaterial substances at all in Kant's new ontology—although there are living or thinking *material* substances (544.7–10): viz., human beings (B415). And it is in this way that the vexing problem of the interaction between immaterial and material substances is finally (dis-)solved.

63. The radical character of Kant's reinterpretation of the notion of substance is rightly stressed by Vuillemin [117], chap. X.

of view of the pure intellect alone independently of sensibility:

> Finally, the category of *community* is absolutely not conceivable according to its possibility through mere reason, and thus the objective reality of this concept cannot possibly be comprehended without intuition—and in fact outer [intuition] in space. For how is one to think the possibility that, when several substances exist, one can infer something (as effect) mutually from the existence of one to the existence of another—and thus, because something exists in the first, something must therefore exist also in the other, which cannot be understood from the existence of the latter alone? . . . But we can perfectly well make conceivable the possibility of community (of substances as appearances) if we represent it in space and thus in outer intuition. For the latter already contains a priori in itself formal outer relations as conditions of the possibility of real [outer relations] (in action and reaction, hence community). (B292–293)[64]

In this way the possibility of real interaction between substances is representable only by means of the mathematical exact sciences themselves and, in particular, through the Newtonian theory of universal interaction—which of course continues to be paradigmatic here.[65] Kant thus comes to see that *if* one attempts to represent community of substances purely intellectually via the pure understanding, one has no option but the system of pre-established harmony—true physical influx is impossible *here* (A274–275/B330–331). Kant's earlier attempt to establish physical influx in an intellectually conceived monadic realm of simple substances must be viewed, from the present perspective, as necessarily a failure.

But how exactly is the understanding so applied to sensibility? How does the intellectual faculty ground or make possible the empirical knowledge represented by means of the sensible faculty? Section 24 of the second edition transcendental deduction is entitled "On the application of the categories to objects of the senses in general." Kant there introduces the notion of *figurative synthesis* or *transcendental synthesis of the imagination,* which is "an action of the understanding on sensibility and the first

64. The problem "How I am to understand *that, because something is, something else should be?*" is first raised by Kant in the essay on *Negative Magnitudes:* 2, 202.20–21. Kant now sees that the possibility of such a relation of "real ground" is conceivable in virtue of—and only in virtue of—the mathematical consideration of space and time founded upon the faculty of sensibility.

65. In the *Metaphysical Foundations of Natural Science* Newton's third law of motion—the equality of action and reaction—instantiates the category of community (4, 544–551). And it is clear from Kant's discussion in this work, as well as from the examples of community provided in the Third Analogy (B257: the earth-moon system; A213/B260: the heavenly bodies), that it is universal gravitation which establishes dynamical community in the phenomenal world *in concreto.* Compare the passage from the *Theory of the Heavens* at 1, 308.27–34, quoted in note 9 above.

application of the understanding (at the same time the ground of all the rest) to objects of our possible intuition" (B151–152). He then illustrates this notion as follows:

> This we also always observe in ourselves. We can think no line without *drawing* it in thought, no circle without *describing* it. We can absolutely not represent the three dimensions of space without *setting* three lines at right-angles to one another from the same point. And even time we cannot represent without attending in the *drawing* of a straight line (which is to be the outer figurative representation of time) merely to the action of synthesis of the manifold, by which we successively determine inner sense—and thereby attend to the succession of this determination in it. Motion, as action of the subject (not as determination of an object*), and thus the synthesis of the manifold in space—if we abstract from the latter and attend merely to the action by which we determine *inner* sense according to its form—even produces the concept of succession in the first place. (B154–155)

As Kant explains in the footnote: "motion, as the *describing* of a space, is a pure act of successive synthesis of the manifold in outer intuition in general through the productive imagination and belongs not only to geometry but even to transcendental philosophy" (B155n). It is therefore motion considered purely mathematically—the mere motion of a mathematical point[66]—which illustrates the application of the understanding to sensibility here.

According to the above passage the motion under consideration has two distinguishable aspects. In the first place, it underlies the constructive procedures of pure geometry, which are implemented—in accordance with Euclid's first three postulates—by *drawing* lines and *describing* circles. Geometrical construction is thus a spatio-*temporal* process.[67] This conception of geometry represents a marked advance in sophistication

66. That the describing of a space = the motion of a mathematical point follows from the Observation to Definition 5 of the first chapter or Phoronomy of the *Metaphysical Foundations of Natural Science:* 4, 489.6–11.

67. "I can represent to myself no line, however small, without drawing it in thought, i.e., generating from a point all its parts continuously and thereby first recording this intuition. . . . On this successive synthesis of the productive imagination in the generation of figures is based the mathematics of extension (geometry), together with its axioms, which express the conditions of a priori sensible intuition under which alone the schema of a pure concept of outer appearance can arise" (A162–163/B203–204). That such spatio-temporal construction depends on the action of the *understanding* is explicitly stated at B137–138: "Thus the mere form of sensible intuition, space, is still absolutely no cognition; it yields only the manifold of a priori intuition for a possible cognition. To cognize anything at all in space, e.g., a line, I must *draw* it, and thus synthetically achieve a determinate combination of the given manifold, in such a way that the unity of this act is simultaneously the unity of consciousness (in the concept of a line), and thereby an object (a determinate space) is first cognized."

over that of the *Inaugural Dissertation,* which tends to view the role of intuition in geometry in terms of a static, quasi-perceptual capacity for *seeing* the truths of geometry, as it were in the concrete particular.[68] The crucial new idea is that geometrical construction depends on universal *schemata* rather than on particular *images* (A140–142/B179-181): that is, on general procedures for constructing each and every figure of a particular kind (lines, circles, triangles, and so on) rather than on the particular figures themselves. And it is in this way possible—and in this way alone—to reconcile the singularity and non-discursive character of spatio-temporal intuition with the generality and conceptual character required by geometrical proof.[69]

Geometrical construction thus illustrates the notion of figurative synthesis or transcendental synthesis of the imagination in its purely mathematical aspect. In the second place, however, the above passage suggests that such synthesis has also a *dynamical* aspect:[70] for the concept of *succession* (as exhibited in the motion of a mathematical point) is of course intimately involved with the dynamical categories.[71] Motion, considered as "the *drawing* of a straight line," therefore "belongs not only to geometry but even to transcendental philosophy." And this comes about, I suggest, be-

68. "That there are not given in space more than three dimensions, that between two points there is only one line, and that from a given point in a plane surface with a given line a circle can be described, etc.—these cannot be concluded from some universal notion of space, but can only be *seen,* as it were, *in concreto* in [space] itself. . . . For the rest, geometry does not demonstrate its universal propositions by thinking the object through a universal concept, as happens with rational things, but by subjecting it to the eye through a singular intuition, as happens with sensitive things" (2, 402.32–403.22). And, although Kant does briefly mention "construction" here (at 402.31), it is clear that the *activity* of geometrical construction plays no essential role in his account.

69. These ideas are elaborated in Chapters 1 and 2 below.

70. See B201n for the distinction between *mathematical* synthesis (composition [Zusammensetzung] of the homogeneous) and *dynamical* synthesis (connection [Verknüpfung] of the inhomogeneous).

71. The schema of causality "consists in the succession of the manifold, in so far as it is subject to a rule" (A144/B183). According to B291: "In order to present *alteration,* as the intuition corresponding to the concept of *causality,* we must take motion as alteration in space as example." Moreover, the schema of possibility "is the agreement of the synthesis of different representations with the conditions of time in general (e.g., where contraries can be represented in a thing not simultaneously but only after one another), and thus the determination of the representation of a thing to some or another time" (A144/B184); and, according to B292: "How it is now possible that from a given state a contrary state of the same thing follows, can not only not be made conceivable by reason without example but can never be made understandable without intuition, and this intuition is the motion of a point in space, whose existence in different places (as a sequence of contrary determinations) alone first makes alteration intuitive for us." (And compare B48–49.) The motion of a point exhibited in the drawing of a straight line is therefore intimately connected with the schema of the category of possibility as well.

cause motion so considered represents the law of inertia—the privileged state of force-free motion fundamental to all of modern physics.[72] In particular, then, motion so considered first makes it possible "to determine inner sense according to its form" or to represent time itself as a magnitude: equal temporal intervals are to be *defined* as those during which an inertially moving point traverses equal distances, and the representation of time as a magnitude is thus parasitic on the representation of space.[73] In other words, we do not define the inertial motions as those traversing equal distances in equal times, rather, we first pick out a class of inertial motions (whose spatial projections must of course be straight lines) and then define from them the equality of times by means of the notion of the equality of spaces (taken from geometry).[74]

Now our consideration of motion so far belongs to pure kinematics—which Kant calls pure phoronomy (in the first chapter of the *Metaphysical Foundations*), pure mechanics (in §12 of the *Inaugural Dissertation*), or the (pure) general doctrine of motion (in §5 of the first *Critique*). We are so far considering only the motion of a mere mathematical point in pure intuition (in geometrical space). In order to use this representation in grounding a science of the sensible world, however, we must also know how to apply it to *empirical* intuition (in physical space). But the problem of applying the pure representation of inertial motion to our actual experience of nature is highly non-trivial—for it forces us explicitly to confront the classical problem of absolute versus relative motion. On the one hand, the law of inertia is itself entirely incompatible with a purely relativistic approach to motion: no state of motion is uniform and rectilinear in *every* frame of reference, and so the law of inertia requires the existence of a privileged frame of reference.[75] On the other hand, in our experience of

72. In the *Metaphysical Foundations* the law of inertia is explicitly taken to realize or instantiate the category of causality (4, 543). Compare also the footnote to A207/B252 appended to a discussion of causality, action, and force: "One should note well: that I do not speak of the alteration of certain relations in general but of alteration of state. Thus, if a body moves uniformly, it absolutely does not alter its state (of motion), but only if its motion increases and diminishes."

73. See B293: "Precisely so it can easily be verified that the possibility of things as *quantities*, and hence the objective reality of the category of quantity, can also only be exhibited in outer intuition, and by means of the latter alone can it then be applied to inner sense." This point is elaborated especially in Chapter 2 below.

74. This, I suggest, is why Kant says that we represent time by the drawing of a straight line—but he does not add that the drawing must be uniform: uniformity of temporal duration only becomes comprehensible in the first place in virtue of precisely the representation in question. Compare also note 30 above.

75. More precisely, the law of inertia requires a *class* of privileged frames of reference—what we now call inertial frames of reference—all of which move rectilinearly and uniformly relative to one another (and to Newtonian absolute space: see *Principia,* Corollary V to the

nature we are not directly given the privileged frame of reference in question: Relative to *what* are inertial motions uniform and rectilinear? In elucidating the conditions for applying the representation of motion to experience Kant is therefore forced once again to confront the problem of Newtonian absolute space.[76]

As we observed above, Kant explicitly refrains from confronting this problem when he first argues for the autonomy of space in the essay on *Regions in Space*—nor does he confront it in the *Inaugural Dissertation*. As we also noted (note 47 above), however, he does consider Newtonian absolute space in the *New System of Motion and Rest* of 1758. He there argues that motion must always be considered relative to some or another physically defined frame of reference but, at the same time, that there is always a privileged such frame for considering certain interactions. In particular, in the case of two-body collision problems the privileged frame of reference is always the *center of mass frame* of the interaction, relative to which both bodies in question are necessarily moving with reference to one another, and it is for this reason, in fact, that action and reaction are necessarily equal in such cases (2, 17–19, 23–25).[77] This solution is correct as far as it goes (for the center of mass frame of a sufficiently isolated system is always inertial: see note 75 above), but it is obviously not general enough. The only interactions considered are two-body collisions, and there is no indication of how a single, all-embracing privileged frame of reference can be constructed.

Kant generalizes the approach suggested in the *New System of Motion and Rest* in the *Metaphysical Foundations of Natural Science*, which is devoted to an exposition of "pure natural science" or "the pure doctrine of nature." The principles of pure natural science are expounded in four chapters, corresponding to the four headings of the table of categories from the first *Critique*.[78] Of particular importance are the principles of

---

Laws of Motion). Yet Kant understands the problem, following Newton, as that of determining a *single* such privileged frame. This matter is further discussed in Chapter 3 below.

76. In emphasizing the fundamental importance of the pure representation of motion in grounding the possibility of a pure science of nature (and, indeed, its central importance to the critical philosophy as a whole) I am in agreement with Gloy [38], especially chap. III.B. Yet Gloy overlooks the fundamental problems involved in *applying* this representation to experience and thereby underestimates, I think, the extent to which properly empirical factors must necessarily enter into both Kant's concept of motion and his concept of matter (as the movable in space).

77. And it is for this reason that Kant there finally rejects the concept of *vis inertia*, conceived of as that force with which a body *at rest* resists motion: 2, 19–21. Compare note 8 above.

78. Compare Kant's observations on the table of categories in §11 at B110, together with the footnote thereto referring to the *Metaphysical Foundations*. This general correspondence is discussed in great detail in Vuillemin [117].

pure natural science expounded in the third chapter or Mechanics, which thus correspond to and instantiate the relational categories of substance, causality, and community. These principles, parallel to the three analogies of experience, are given by Kant as the three "laws of mechanics": (1) the principle of the conservation of mass or quantity of matter, (2) the law of inertia ("Every body persists in its state of rest or motion, in the same direction and with the same speed, if it is not necessitated through an external cause to leave this state," 4, 543.16–20), (3) the principle of the equality of action and reaction. And it is clear, moreover, that Kant views these as synthetic a priori principles—very closely related to the transcendental relational principles themselves.[79]

Of even greater importance, from the present point of view, is the fourth chapter—which is appropriately entitled "Phenomenology." This chapter thus corresponds to and realizes the modal categories of possibility, actuality, and necessity, and it has as its aim the transformation of *appearance* [*Erscheinung*] into *experience* [*Erfahrung*]. More specifically, its aim is to transform *apparent motions* into *true motions*. Here it appears that Kant is following the lead of Book III of Newton's *Principia,* which applies the laws of motion to the observable, so far merely relative or apparent motions in the solar system so as to derive therefrom the law of universal gravitation and, at the same time, to establish a privileged frame of reference (the center of mass frame of the solar system) relative to which the notion of true (or absolute) motion is first empirically defined.[80] In particular, Kant outlines a procedure for applying the laws of mechanics expounded in the previous chapter so as to subject the given appearances (apparent motions) to the modal categories in three steps or stages.[81]

In the first stage, we record the observed relative motions in the solar system of satellites with respect to their primary bodies and the fixed stars: the orbits of the moons of Jupiter and Saturn, the orbits of the planets with respect to the sun, and the orbit of the earth's moon. We begin, then, with precisely the empirical "Phenomena" that initiate Newton's argument for universal gravitation. We note that all such observed relative motions are described by Kepler's laws, and we subsume these so far merely apparent motions under the category of possibility.

79. Compare the discussion of the synthetic a priori principles of pure natural science at B17–18, B20n, and §15 of the *Prolegomena* (4, 294–295).

80. Thus, for example, it is only after establishing the center of mass frame of the solar system in Proposition XI of Book III that Newton can settle the issue of heliocentrism in Proposition XII. Compare Kant's remarks on this at Bxxii,n.

81. This reading of the Phenomenology is elaborated and defended in Chapters 3 and 4 below. Something like this interpretation of Kant's distinction between true and apparent motion is also suggested by Timerding [109], especially §11, pp. 42–45.

In the second stage, we assume that the above relative motions approximate to true motions (from a modern point of view, that the above-mentioned frames of reference approximate, for the purpose of describing these motions, to inertial frames of reference), and we thence can apply Kant's law of inertia (Newton's first and second laws of motion) to infer that the relative accelerations in question manifest an "external cause" or *impressed force* directed towards the center of each primary body. Moreover, it now follows purely mathematically from Kepler's laws that these given forces—together with the true accelerations engendered thereby—satisfy the inverse-square law. Accordingly, we now subsume these true orbital motions (inverse-square accelerations) under the category of actuality.

In the third and final stage, we apply the equality of action and reaction (Newton's third law of motion) to conclude that the above true accelerations are *mutual*—equal and opposite—and also to conclude that gravitational acceleration is directly proportional to mass. To infer the latter result from the equality of action and reaction we need to assume, in addition, that *all* bodies in the solar system—not merely the satellites in question—experience inverse-square accelerations towards each primary body (and thus, in effect, that gravitational attraction is *universal*), and we also need to apply the third law of motion directly to these mutual interactions of the primary bodies (and thus, in effect, to assume that gravitational attraction acts *immediately* at a distance).[82] Given these assumptions and our previous results the law of universal gravitation now follows deductively: each body experiences an inverse-square acceleration towards each other body, which, in addition, is directly proportional, at a given distance, to the mass of the body towards which it accelerates. Moreover, we are now—and only now—in a position rigorously to estimate the masses of the various primary bodies in the solar system so as rigorously to determine the center of mass frame of the solar system, and, in this way, the true motions can now be explained precisely as motions relative to this privileged frame of reference. The inverse-square accelerations resulting thereby—which are universal, everywhere mutual, and directly proportional to mass—are subsumed under the category of necessity.

Kant has thereby outlined a procedure for "reduc[ing] all motion and rest to absolute space" (4, 560.5–6). This procedure begins with the observed, so far merely relative or apparent motions in the solar system, and its product is an empirically defined privileged frame of reference. The key

82. The crucial importance of these two additional assumptions of *universality* and *immediacy* is further discussed in Chapters 3 and 4 below.

point, however, is that Kant does not view this procedure as finding or discovering the true or absolute motions in the solar system, but rather as that procedure by which the notion of true or absolute motion is first *objectively defined.* It is not that we have discovered that the center of mass of the solar system is at rest relative to a pre-existing absolute space (compare Proposition XI of *Principia,* Book III); rather, it is only subsequent to the Newtonian argument itself that the notions of true motion and rest—along with that of "absolute space"—have any objective meaning. And it is for precisely this reason that Kant views the Newtonian laws of motion—or equivalently his own three laws of mechanics[83]—as true a priori. These laws of motion do not state facts, as it were, about a notion of true or absolute motion that is antecedently well defined; rather, they alone make this notion possible in the first place. Moreover, since in applying the representation of motion to experience we are thereby applying that representation through which the understanding first "acts" on sensibility so as to make experience possible, it follows that the laws of motion—together with the analogies of experience which they instantiate and realize—are conditions of the possibility of objective experience.

The *Metaphysical Foundations,* which thus elucidates the conditions for applying the pure representation of motion to our given experience of nature, illuminates the precise sense in which the categories of the understanding—the dynamical categories in particular—are conditions of possibility of the *sensible* world. These categories first make it possible to apply the representation of motion therein and thereby explain the possibility of our best example of a mathematical science of nature: Newton's theory of universal gravitation. Antecedent to such metaphysical explanation, by contrast, the possibility of this mathematical science of nature—precisely because of its apparent dependence on an impossible and nonsensible notion of "absolute space"—must remain entirely unclear. Moreover, since geometry, as we have seen, also depends on the very same "act" by which the understanding first so determines sensibility, we have simultaneously explained the possibility of applying pure geometry to *physical* space.[84] It is in this way, finally, that metaphysics is necessary as

83. Kant's first law—the conservation of mass—simply expresses one aspect of the conservation of momentum (Corollary Three to the Laws of Motion), which of course follows from Kant's third law—the equality of action and reaction. Interestingly enough, in his early writings Kant tends to state his conservation principle precisely in terms of momentum: e.g., in the *New Exposition* (1, 407–408) and in the essay on *Negative Magnitudes* (2, 194–197).

84. Compare A165–166/B206: "The synthesis of spaces and times, as the essential form of all intuition, is that which at the same time makes possible all apprehension of appearance, and therefore every outer experience; and what mathematics proves of the former in its pure employment, necessarily holds also of the latter"; and also A224/B271: "precisely the same image-forming [bildene] synthesis, by which we construct a triangle in the imagination, is

a ground for the mathematical exact sciences—not as somehow providing these sciences with an otherwise missing justification from some higher and more certain standpoint, but rather as first securing the possibility of metaphysics itself:

> Pure mathematics and pure natural science had, *for the sake of their own security* and certainty, no need for the kind of deduction that we have made of both. For the former rests on its own evidence; and the latter, however, although originated from pure sources of the understanding, nevertheless rests on experience and its thoroughgoing confirmation—which latter testimony it cannot wholly renounce and dispense with, because, despite all of its certainty, as philosophy it can never imitate mathematics. Both sciences had need of the investigation in question, not for themselves, but for another science, namely metaphysics. (*Prolegomena,* 40: §4, 327.5–14)

Metaphysics and the exact sciences have finally achieved a union that is fruitful indeed.

· V ·

Kant, as we have seen, decisively rejects Newtonian absolute space. In its place he puts an empirical procedure for determining a privileged frame of reference, a procedure he derives from the argument of *Principia,* Book III. Newton's argument there culminates in the determination of the center of mass frame of the solar system, relative to which neither the earth nor the sun can be taken to be exactly at rest. For Kant, however, this empirical procedure cannot, strictly speaking, terminate here; for the center of mass of the solar system itself experiences a slow rotation with respect to the center of mass of the Milky Way galaxy. Nor can even this last point furnish us with a privileged state of rest; for, according to Kant, the Milky Way galaxy also experiences a rotation around a common center of galaxies; and so on.[85] For Kant, in other words, the Newtonian procedure for determining a privileged frame of reference is necessarily nonterminating: it aims ultimately at the "common center of gravity of all matter" (*Metaphysical Foundations:* 4, 563.4–5), and this point lies forever beyond our reach. In place of Newton's absolute space we are therefore left with no object at all, but only with a procedure for determining

entirely identical with that which we exercise in the apprehension of an appearance, so as to make an empirical concept thereof." This matter is intimately connected with Kant's doctrine of space as not only a form of intuition but also a "formal intuition"—as expressed, e.g., at B160n—and is further discussed in Chapters 2 and 4 below.

85. This conception of an ever-expanding sequence of rotating heavenly systems is articulated in the *Theory of the Heavens,* Second Part, Chapter 7: 1, 306–322. Compare also the *New System* at 2, 16–17—where, in particular, Kant describes Bradley's observations of the motion of the solar system.

better and better *approximations* to a privileged frame of reference (from a modern point of view, a privileged inertial state). This is why Kant calls absolute space "a necessary concept of reason, thus nothing other than a mere *idea*" (559.8–9); for absolute space can be thought of only as the ideal end-point towards which the Newtonian procedure for determining the center of mass of a rotating system is converging, as it were.

The procedure for "reducing all motion and rest to absolute space" thus provides Kant with a model for what he calls in the first *Critique* the regulative use of the ideas of reason. For the ideas of reason (God, the soul, the world as a totality) can never have an object corresponding to them actually given in experience, but function rather precisely to guide our investigation into nature asymptotically towards an ideal but forever unreachable end-point—a *focus imaginarius* (A644–645/B672–673)—in which science is absolutely complete. Foremost among these ideas is the idea of God conceived of as wise author of the world or as highest intelligence. We can in no way demonstrate or assume the existence of an object corresponding to this idea, but it serves nonetheless as an indispensible heuristic device according to which we pursue our empirical investigations of nature under the presupposition of systematicity and intelligibility:

> We declare, for example, that the things in the world must be so considered *as if* they had their existence from a highest intelligence. In such a way the idea is properly only a heuristic and not an ostensive concept, and it indicates, not how an object is constituted, but rather how we, under its guidance, are *to seek* the constitution and connection of the objects of experience in general. (A670–671/B698–699)

And it is in this way that physico-theology is now radically reinterpreted.

The doctrine of the regulative use of the ideas of reason becomes the doctrine of reflective judgement in the third *Critique*. Kant there argues that the foundation for the laws of natural science in the concepts and principles of the understanding articulated in the first *Critique*—and in the *Metaphysical Foundations*[86]—is necessarily incomplete, because it does not yet show us how to achieve a system of the totality of empirical laws of nature:

> We have seen in the *Critique of Pure Reason* that the whole of nature as the totality of all objects of experience constitutes a system according to transcendental laws, namely such that the understanding itself provides a priori (for appearances, in so far as they are to constitute an experience, bound together in one consciousness). For precisely this reason, experience must also constitute a system of possible empirical cognitions, in accor-

86. The grounding of the law of universal gravitation via the same procedure by which the center of mass frame of the solar system is determined is paradigmatic here: see especially Chapter 4 below.

> dance with universal as well as particular laws, so far as it is in general possible objectively considered (in the idea). For this is required by the unity of nature according to a principle of the thoroughgoing combination of all that is contained in this totality of all appearances. So far, then, experience in general is to be viewed as a system according to transcendental laws of the understanding and not as a mere aggregate.
>
> But it does not follow therefrom that nature is also a system *comprehensible* to the human faculty of cognition in accordance with *empirical laws,* and that the thoroughgoing systematic coherence of its appearances in an experience—and thus experience as a system—is possible for men. For the manifoldness and inhomogeneity of the empirical laws could be so great, that it would certainly be possible in a partial manner to connect perceptions into an experience in accordance with particular laws discovered opportunely, but it would never be possible to bring these empirical laws themselves to unity of affinity under a common principle—if, namely, as is still possible in itself (at least so far as the understanding can constitute a priori), the manifoldness and inhomogeneity of these laws, together with the corresponding natural forms, were so infinitely great and presented to us, in this respect, a crude chaotic aggregate and not the least trace of a system, although we equally had to presuppose such a system in accordance with transcendental laws. (First Introduction to the *Critique of Judgement,* §IV: 20, 208.22–209.19)

The principle of reflective judgement stipulates that such systematicity of empirical laws cannot be achieved from the top down, as it were—by directly specifying further the principles of pure understanding in the manner of the *Metaphysical Foundations*—but only from the bottom up: that is, by starting with the lowest level empirical laws and successively attempting to unify them under ever higher and more general empirical laws, according to the presupposition that "*nature specifies its universal laws to empirical* [laws], *in accordance with the form of a logical system, on behalf of the faculty of judgement*" (216.1–3).

In §§84–91 of the *Critique of Judgement* Kant then articulates his mature conception of physico-theology by bringing it into relation with his moral conception of divinity deriving from pure *practical* reason. The key point is that the idea of God supplied by *theoretical* reason—the idea of a wise author of the world or highest intelligence—is still not an adequate idea of divinity. To achieve such an adequate idea we require the additional attributes of omniscience, omnipotence, omnipresence, and omnibenevolence, which can only flow from the conception of the *highest good* to be produced in the world—a conception which can itself only be based on our idea of an ideal moral order. In this way, physico-theology necessarily requires supplementation by ethico-theology:

> From this so determined principle of the causality of the original being, we must think it not merely as intelligence and as law-giving for nature, but

> rather also as law-giving in general in a moral realm of ends. . . . In such a fashion *moral* theology makes up for the deficiency of *physical* [theology]: for the latter, if it did not tacitly borrow from the former but rather were to proceed consistently, could ground nothing but a *demonology* by itself alone, which is capable of no determinate concept. (5, 444.12–32)

Physico-theology is now seen as doubly defective: not only does it supply a mere idea of reason for which no actual object can be given, but the idea it supplies is too indeterminate for expressing the entire divine nature.[87] For this idea yields no determinate conception of the final purpose of the world, which conception can only be provided by pure practical reason. Kant's mature conception of physico-*ethico*-theology therefore secures the unity of theoretical and practical reason.

Kant is nevertheless able thereby to reinterpret a fundamental idea of his earlier physico-theology: namely, the idea of phenomenal omnipresence. This comes about through the new connections we have just been considering among space, gravitation, and divinity, and the idea is most clearly expressed in a remarkable footnote to the General Observation to the Third Part of *Religion within the Limits of Reason Alone* (1793):

> The *cause* of the universal gravity of all matter in the world is thus unknown to us, so much so that one can nevertheless comprehend that it could never be known by us: because the concept thereof already presupposes a first and unconditionally inherent moving force. But it is still no mystery; rather, it can be made manifest to all, because its *law* is sufficiently known. If *Newton* represents it, as it were, as divine omnipresence in the appearance (phenomenal omnipresence), this is no attempt to explain it (for the existence of God in space contains a contradiction), but is still a sublime analogy considered merely in the unification of corporeal beings into a world-whole, in that one bases it upon an incorporeal cause; and thus also would be the result from the attempt to comprehend the self-sufficient principle of the unification of rational beings in an ethical state and to explain the latter

87. Nevertheless, Kant retains the idea—which we saw above to be so essential to his earlier approach to physico-theology—that biological phenomena are not comprehensible purely mechanically. Compare the famous remark in §75 of the *Critique of Judgement:* "Namely it is entirely certain that we will never sufficiently come to know organized beings and their inner possibility in accordance with merely mechanical principles of nature, much less explain them; and it is indeed so certain that one can boldly say: it is unsuitable for men even to make such a plan or to hope that perhaps someday a Newton could still arise who could make comprehensible even the generation of a blade of grass in accordance with natural laws ordered by no purpose; rather, one must absolutely deny this insight to men" (5, 400.13–21), with the Preface to the *Theory of the Heavens:* "the formation of all heavenly bodies, the cause of their motions, in short the entire present constitution of the universe will be able to be comprehended before the generation of a single herb or a caterpillar will be clearly and completely made known from mechanical grounds" (1, 230.21–26). See also notes 17 and 18 above.

> therefrom. We know only the duty that draws us towards this; the possibility of the effect held in view if we obey the former lies entirely beyond the limits of our insight. (6, 138.25–139.28)

This reinterpretation of the doctrine of phenomenal omnipresence is striking indeed. Space is no longer a manifestation of the fundamental law of interaction binding together elementary substances, but is rather the ideal end-point towards which our empirical procedure for "reducing all motion and rest to absolute space" is asymptotically converging. Universal gravitation is no longer the phenomenal manifestation of the fundamental law of interaction governing the monadic realm, but is rather an original principle of matter (phenomenal substance) which is presupposed by—and whose mathematical law is known only in virtue of—this same empirical procedure. And the divinity is no longer the ultimate ground of both the elementary simple substances and their union in virtue of a schema of the divine intellect, but rather expresses the final ideal end-point—the *focus imaginarius*—towards which all human activities, both theoretical and practical, ideally converge.

In any case, however, the juxtaposition of the regulative, asymptotically converging procedure of reason and reflective judgement alongside of the constitutive, schematizing procedure of the principles of pure understanding and the *Metaphysical Foundations* was destined to create a new fundamental problem for Kant, which he soon came to characterize as the possiblity of a "gap in the critical system" in the *Opus postumum* (1796–1803).[88] For, as we have seen, the regulative procedure of reflective judgement and the constitutive procedure of the understanding move, as it were, in two contrary directions. The latter proceeds from the top down, as it were, by schematizing the pure concepts of the understanding in terms of sensibility so as to provide the basis for a mathematical science of nature: in particular, the Newtonian theory of universal gravitation. The former proceeds from the bottom up, by systematizing lower-level empirical concepts and laws under higher-level empirical concepts and laws so as to approximate thereby to an ideal complete science of nature. But what guarantee is there that the science towards which we are converging in proceeding from the empirical to the a priori will include that mathematical exact science we have already constituted in proceeding from the a priori to the empirical? Does the Newtonian theory of universal gravitation, in other words, have any *necessary* connection with the regulative procedure of reflective judgement and the systematization of empirical laws? In the absence of such a guarantee we have no assurance whatsoever that the two aspects of the critical system will harmonize with one an-

88. I am indebted to Förster [31] for stimulating my interest in this problem—although Förster and I do not in the end understand the "gap" in the same way. See Chapter 5 below.

other, and we are thus faced with the possibility of precisely a "gap" in the critical system.

Kant therefore undertakes an entirely new project intended to bridge this "gap," conceived of as "the transition from the metaphysical foundations of natural science to [empirical] physics." This project, as its name suggests, is to extend the constitutive procedure of the *Metaphysical Foundations* so as to bring it into contact with the more empirical branches of natural science: with chemistry and the theory of heat in particular. In this way, Kant hopes also to come to terms with the important scientific developments that have taken place since the critical period: in particular, with Black's discovery of latent and specific heats and with Lavoisier's fundamental reorganization of chemistry. Moreover, in attempting to carry out this project Kant is led radically to reconsider the status of space and its relation to matter once again; for a central part of his "transition" project turns out to be an a priori proof of the existence of an everywhere distributed, space-filling matter or aether.[89] Hence space—now conceived of as "perceptible space" or "space realized"—continues to be central to Kant's life-long attempt to grapple philosophically with the exact sciences to the end. But with this brief glimpse of things to come we must finally bring our Introduction to a close.

89. The details of this "aether-deduction" depend upon attributing to the space-filling aether many of the same properties that God possessed in the pre-critical proofs of divine existence found in the *New Exposition* and *The Only Possible Basis:* namely, the properties of the *ens realissimum* discussed in the first *Critique* in the chapter The Ideal of Pure Reason. See Förster [32], and also Chapter 5 below.

· PART ONE ·

The Critical Period

· CHAPTER 1 ·

Tractatus 4.0412: For the same reason the idealist's appeal to 'spatial spectacles' is inadequate to explain the seeing of spatial relations, because it cannot explain the multiplicity of these relations.

Geometry

Since the important work of early twentieth-century philosophers of geometry such as Russell, Carnap, Schlick and Reichenbach, Kant's critical theory of geometry has not looked very attractive. After their work and the work of Riemann, Hilbert, and Einstein from which they drew their inspiration, Kant's conception is liable to seem quaint at best and silly at worst. His picture of geometry as somehow grounded in our intuition of space and time appears thoroughly wrong; and there is a consequent tendency to view the Transcendental Aesthetic as an unfortunate embarrassment that one has simply to rush through on the way to the more relevant and enduring insights of the Analytic.[1]

The standard modern complaint against Kant runs as follows. Kant fails to make the crucial distinction between *pure* and *applied* geometry. Pure geometry is the study of the formal or logical relations between propositions in a particular axiomatic system, an axiomatic system for Euclidean geometry, say. As such it is indeed a priori and certain (as a priori and certain as logic is, anyway), but it involves no appeal to spatial intuition or any other kind of experience. Applied geometry, on the other hand, concerns the truth or falsity of such a system of axioms under a particular interpretation in the real world. And, in this connection, it matters little whether our axioms are interpreted in the physical world—in terms of light rays, stretched strings, or whatever, or in the psychological realm—in terms of "looks" or "appearances" or other phenomenological entities. In either case the truth (or approximate truth) of any particular axiom system is neither a priori nor certain but, rather, a matter for empirical investigation, in either physics or psychology. This modern attitude is epitomized in Einstein's famous dictum (in which he has geometry

1. One finds this attitude in even as sympathetic and sensitive a commentary as Kemp Smith [57], for example, pp. 40–41.

especially in mind): "As far as the laws of mathematics refer to reality, they are not certain; and as far as they are certain, they do not refer to reality" ([21], p. 28). From this point of view, then, Kant misconstrues the problem from the very beginning, and, accordingly, his teaching is hopelessly confused.

Yet this modern complaint is quite fundamentally unfair to Kant; for Kant's conception of *logic* is certainly not our modern conception. Our distinction between pure and applied geometry goes hand in hand with our understanding of logic, and this understanding simply did not exist before 1879, when Frege's *Begriffsschrift* appeared. The importance of relating Kant's understanding of logic to his philosophy of mathematics has been stressed by several recent commentators, notably, by Hintikka and Parsons.[2] In reference to geometry in particular, however, I think that no one has been as close to the truth as Russell, who habitually blamed all the traditional obscurities surrounding space and geometry—including Kant's views, of course—on ignorance of the modern theory of relations and uncritical reliance on Aristotelian subject-predicate logic.[3] I think Russell is exactly right, but I would like to turn his polemic on its head. Instead of using our modern conception of logic to disparage and dismiss earlier theories of space, we should use it as a tool for interpreting and explaining these theories, for deepening our understanding of the difficult logical problems with which they were struggling. This, in any case, is what I propose to undertake in reference to Kant's theory in what follows.

· I ·

What is most striking to me about Kant's theory, as it was to Russell, is the claim that geometrical *reasoning* cannot proceed "analytically according to concepts"—that is, purely logically—but requires a further activity called "construction in pure intuition." The claim is expressed most clearly in the Discipline of Pure Reason in Its Dogmatic Employment, where Kant contrasts philosophical with mathematical reasoning:

> Philosophy confines itself to general concepts; mathematics can achieve nothing by concepts alone but hastens at once to intuition, in which it considers

2. For Hintikka see [50]; "Kant's 'New Method of Thought' and His Theory of Mathematics" (1965) and "Kant on the Mathematical Method" (1967), both reprinted in [53]; and [52]. For Parsons see "Infinity and Kant's Conception of the 'Possibility of Experience' " (1964) and "Kant's Philosophy of Arithmetic" (1969), both reprinted in [92]; [90]; and [91]. See, in addition, Beth [6], which inspired Hintikka, and Thompson [108]. This last is oriented around the role of intuition in empirical knowledge, but it also contains a very important discussion of mathematics, logic, and the relationship between them.

3. See especially §434 of Russell [102], entitled "Mathematical reasoning requires no extra-logical element."

> the concept *in concreto*, although still not empirically, but only in an intuition which it presents a priori, that is, which it has constructed, and in which whatever follows from the general conditions of the construction must hold, in general, for the object [Objekte] of the concept thus constructed.
>
> Suppose a philosopher be given the concept of a triangle and he be left to find out, in his own way, what relation the sum of its angles bears to a right angle. He has nothing but the concept of a figure enclosed by three straight lines, along with the concept of just as many angles. However long he meditates on these concepts, he will never produce anything new. He can analyse and clarify the concept of a straight line or of an angle or of the number three, but he can never arrive at any properties not already contained in these concepts. Now let the geometer take up this question. He at once begins by constructing a triangle. Since he knows that the sum of two right angles is exactly equal to the sum of all the adjacent angles which can be constructed from a single point on a straight line, he prolongs one side of the triangle and obtains two adjacent angles which together equal two right angles. He then divides the external angle by drawing a line parallel to the opposite side of the triangle, and observes that he has thus obtained an external adjacent angle which is equal to an internal angle—and so on. In this fashion, through a chain of inferences guided throughout by intuition, he arrives at a solution of the problem that is simultaneously fully evident [einleuchtend] and general. (A715–717/B743–745)

Kant is here outlining the standard Euclidean proof of the proposition that the sum of the angles of a triangle = 180° = two right angles ([46]: Book I, Prop. 32). Given a triangle *ABC*, one prolongs the side *BC* to *D* and then draws *CE* parallel to *AB* (see Figure 1). One then notes that $\alpha = \alpha'$ and $\beta = \beta'$, so $\alpha + \beta + \gamma = \alpha' + \beta' + \gamma = 180°$. Q.E.D.

In contending that construction in pure intuition is essential to this proof, Kant is making two claims that strike us as quite outlandish today. First, he is claiming that (an idealized version of) the figure we have drawn is necessary to the proof. The lines *AB, BD, CE,* and so on are indispensable constituents; without them the proof simply could not proceed. So geometrical proofs are themselves spatial objects. Second, it is equally important to Kant that the lines in question are actually drawn or continu-

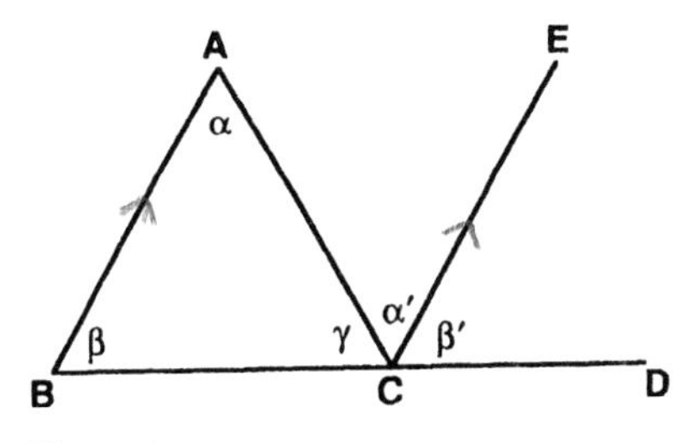

Figure 1

ously generated, as it were. Proofs are not only spatial objects, they are spatio-temporal objects as well. Thus, in an important passage in the Axioms of Intuition Kant says:

> I cannot represent to myself a line, however small, without drawing it in thought, that is gradually generating [nach und nach zu erzeugen] all its parts from a point. Only in this way can the intuition be obtained. . . . The mathematics of extension (geometry), together with its axioms, is based upon this successive synthesis of the productive imagination in the generation of figures [Gestalten]. (A162–163/B203–204)

That construction in pure intuition involves not only spatial objects, but also spatio-temporal objects (the motions of points), explains why intuition is able to supply a priori knowledge of (the pure part of) physics:

> . . . thus our idea of time [Zeitbegriff] explains the possibility of as much a priori cognition as is exhibited in the general doctrine of motion, and which is by no means unfruitful. (B49)

In other words, it is the spatio-temporal character of construction in pure intuition that enables Kant to give a philosophical foundation for both Euclidean geometry and Newtonian dynamics.

Kant's conception of geometrical proof is of course anathema to us. Spatial figures, however produced, are not essential constituents of proofs, but, at best, aids (and very possibly misleading ones) to the intuitive comprehension of proofs. Whatever the intended interpretation of the axioms or premises of a geometrical proof may be, the proof itself is a purely "formal" or "conceptual" object: ideally, a string of expressions in a given formal language. In particular, then, all that could possibly be missing from a purely "conceptual" or "analytic" derivation of ⌜*x*'s angles sum to 180°⌝ from ⌜*x* is a triangle⌝ are the *axioms* of Euclidean geometry. For us, the conjunction of ⌜*x* is a triangle⌝ with these axioms does of course imply ⌜*x*'s angles sum to 180°⌝ by logic alone; and no spatio-temporal activity of construction in pure intuition is necessary. To be sure, spatial objects may be needed to supply a particular interpretation of our axioms, but this is quite a different matter.

Is Kant simply forgetting about the axioms of Euclidean geometry here? This is most implausible, especially since the proof he sketches is Euclid's. No, his claim must be that even the conjunction of ⌜*x* is a triangle⌝ with these axioms does not imply ⌜*x*'s angles sum to 180°⌝ by logic alone: in other words, that Euclid's axioms do not imply Euclid's theorems by logic alone. Moreover, once we remember that Euclid's axioms are not the axioms used in modern formulations and, most important, that Kant's conception of logic is not our modern conception, it is easy to see that the claim in question is perfectly correct. For our logic, unlike Kant's, is

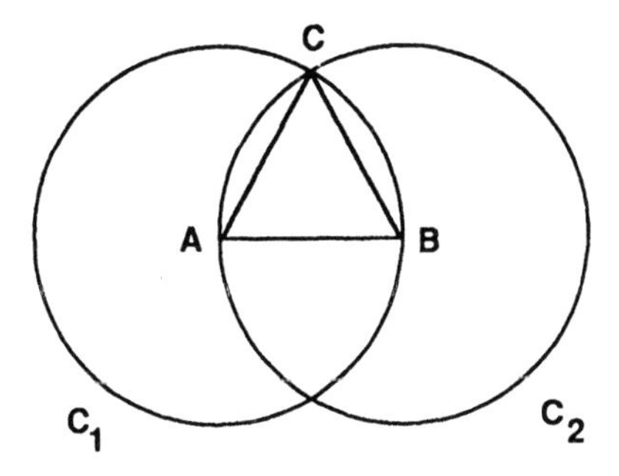

Figure 2

polyadic rather than monadic (syllogistic); and our axioms for Euclidean geometry[4] are strikingly different from Euclid's in containing an explicit, and essentially polyadic, *theory of order.*

The general point can be put as follows. A central difference between monadic logic and full polyadic logic is that the latter can generate an infinity of objects while the former cannot. More precisely, given any consistent set of monadic formulas involving k primitive predicates, we can find a model containing at most 2^k objects. In polyadic logic, on the other hand, we can easily construct formulas having only infinite models. Proof-theoretically, therefore, if we carry out deductions from a given theory using only monadic logic, we will be able to prove the existence of at most 2^k distinct objects: after a given finite point we will run out of "provably new" individual constants. Hence, monadic logic cannot serve as the basis for any serious mathematical theory, for any theory aiming to describe an infinity of objects (even "potentially").

This abstract and general point can be illustrated by Euclid's proof of the very first Proposition of Book I: that an equilateral triangle can be constructed with any given line segment as base. The proof runs as follows. Given line segment AB, construct (by Postulate 3) the circles C_1 and C_2 with AB as radius (see Figure 2). Let C be a point of intersection of C_1 and C_2, and draw lines AC and BC (by Postulate 1). Then, since (by the definition of a circle: Def. 15) $AC = AB = BC$, ABC is equilateral. Q.E.D.

There is a standard modern objection to this proof. Euclid has not proved the *existence* of point C; he has not shown that circles C_1 and C_2 actually intersect. Perhaps C_1 and C_2 somehow "slip through" one an-

4. These received their more-or-less definitive formulation in Hilbert [49]. I say "more-or-less" because there remains some confusion in Hilbert about the proper form of a continuity or completeness axiom.

other, and there is no point C. Moreover, in modern formulations of Euclidean geometry this "possibility" of non-intersection is explicitly excluded by a *continuity* axiom, an axiom which (apparently, anyway) does not appear in Euclid's list of Postulates and Common Notions. From this point of view, then, not only is Euclid's proof "defective," but so is his axiomatization: the existence of point C simply does not follow from Euclid's axioms.[5]

Why do we think that the existence of point C does not follow from Euclid's axioms? We might argue as follows. Cover the Euclidean plane with Cartesian coordinates in such a way that the midpoint of segment AB has coordinates $(0, 0)$, point A has coordinates $(-1/2, 0)$, and point B has coordinates $(1/2, 0)$. Then the desired point of intersection C has coordinates $(0, \sqrt{3}/2)$. Now throw away all points with irrational coordinates: the result is a model in Q^2, where Q is the rational numbers. This model appears to satisfy all Euclid's axioms, but, of course, point C does not exist in the model. So our model gives concrete form to the "possibility" of non-intersection, a "possibility" which therefore needs to be excluded by a continuity axiom.

But perhaps Euclid's formulation does contain such a continuity axiom, if only implicitly. After all, Postulate 2 states that straight line segments can be produced "continuously [κατὰ τὸ συνεξὲς]," while in our model straight lines are *dense* but not truly *continuous*. So one might think that our model is ruled out by Postulate 2. This attempt to "save" Euclid misses the central point. First, the intuitive notion of "continuity" figuring in Postulate 2 is not our notion of continuity: in particular, it is not explicitly distinguished from mere denseness. This distinction was not even articulated until late in the nineteenth century; before Dedekind mathematicians would commonly give what we call the definition of denseness when explaining what they meant by "continuity": namely, "for every element there is a smaller" or "between every two elements there is a third." Second, and more important, the notion of "continuity" in Postulate 2 is not logically analyzed: it appears as a simple (one-place) predicate. Therefore, whatever the intuitive meaning of "continuous" may be, there is certainly no valid *syllogistic inference* of the form:

$$\begin{array}{ll} & C_1 \text{ is continuous} \\ & C_2 \text{ is continuous} \\ \hline \therefore & C \text{ exists} \end{array}$$

5. A nice introductory account of such "defects" in Euclid is found in Eves [29], §8.1. Heath [46], vol. 1, pp. 234–240, provides a very detailed discussion of the "intersection" problem from a modern point of view. As far as I know the above criticism of Proposition I.1 was first made by Pasch [96], §6.

To get a valid inference of this form we need to analyze the notion of continuity in the modern style and to make an essential (and, as we shall see, rather strong) use of polyadic quantification theory.

Furthermore, once we start playing the game from a modern point of view we can generate trivial "counter-examples" to Euclid that do not depend on sophisticated considerations like continuity. Thus, in Figure 2, throw away all points of the plane except the two points *A, B*. Let the "line" *AB* be just the pair {*A*, *B*}, let the "circle" C_1 be the singleton {*B*}, and let the "circle" C_2 be the singleton {*A*}. Does "line" *AB* satisfy Postulate 2? Can it be "produced continuously"? Note again that neither "can be produced" nor "continuous" is logically analyzed: both appear as simple (one-place) predicates. So, from a strictly logical point of view, we can give them both any interpretation we like: let them both mean "has two elements," for example. Then Postulate 2 is obviously satisfied, and so are the other axioms. Hence, Euclid's axiomatization does not even imply the existence of more than two points.

Does this last "counter-example" show that Euclid's axiomatization is hopelessly "defective"? I think not. Rather, it underscores the fact that Euclid's system is not an axiomatic theory in our sense at all. Specifically, the existence of the necessary points is not logically deduced from appropriate existential axioms. Since the set of such points is of course infinite, this procedure cannot possibly work in a monadic (syllogistic) context. Instead, Euclid *generates* the necessary points by a definite process of construction: the procedure of construction with straight-edge and compass. We start with three basic operations: (i) drawing a line segment connecting any two given points (to avoid complete triviality we assume two distinct points to begin with), (ii) extending a line segment by any given line segment, (iii) drawing a circle with any given point as center and any given line segment as radius. We are then allowed to iterate operations (i), (ii), and (iii) any finite number of times. Euclid's Postulates 1–3 give the rules for this iterative procedure, and the points in our "model" are just the points that can be so constructed. In particular, then, the infinity of this set of points is guaranteed by the infinite iterability of our process of construction.[6]

More precisely, it is straightforward to show that the points generated by straight-edge and compass constructions (and, therefore, the points required for Euclidean geometry) comprise a Cartesian space (set of pairs) based on the so-called square-root (or "Euclidean") extension Q^* of the

6. See Eves [29], chap. IV, for a discussion of the mathematics of Euclidean constructions. Compare also the very helpful contrast between the Euclidean approach to existence and the modern approach typified by Hilbert in Mueller [81], pp. 11–15. I am indebted to William Tait for emphasizing the importance of straight-edge and compass constructions to me, and for helping me to get clearer about their essential properties.

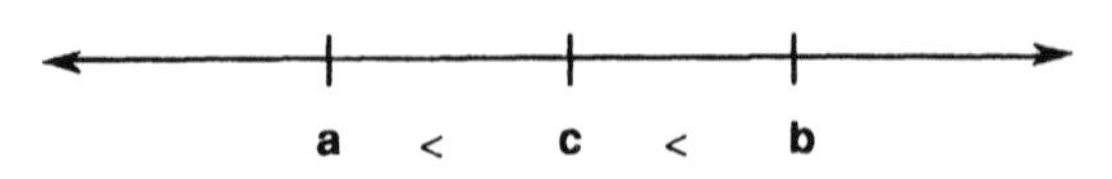

Figure 3

rationals, where Q* results from closing the rationals under the operation of taking real square-roots (see Eves [29], §9.3). In particular, then, the underlying set (Q* × Q*) is only a small fragment of the full Cartesian plane $\mathbf{R}^2$, where **R** is the real numbers. The former, unlike the latter, is a denumerable set, and each element is determined by a finite sequence of elementary operations. In this sense, there is no need in Euclidean geometry for anything as strong as a continuity axiom.

Compare Euclid's approach to the existence of points—in particular, to the existence of an infinity of points—with that taken by modern axiomatizations. The basis of the modern approach, beginning with Pasch in 1882 and culminating in Hilbert's *Foundations of Geometry* (1899), is to include an explicit *theory of order:* a theory of the order structure (and cardinality) of the points on a line. Thus, imagine the points on any line to be ordered by a two-place relation < of "being-to-the-left-of" (see Figure 3). Governing < is the theory of dense linear order without endpoints:

1.	$\neg(a < a)$	(irreflexivity)
2.	$a < c \,\&\, c < b \;\rightarrow\; a < b$	(transitivity)
3.	$a < b \;\vee\; b < a \;\vee\; a = b$	(connectedness)
4.	$\forall a\, \exists b\, (a < b)$	
5.	$\forall b\, \exists a\, (a < b)$	(no endpoints)
6.	$\forall a\, \forall b\, \exists c\, (a < b \rightarrow (a < c < b))$	(denseness)

The presence of some such axioms as 1–6 is the chief difference between Hilbert's axiomatization and Euclid's.[7]

7. I have left out the continuity axiom (which is of course second-order), so axioms 1–6 will be satisfied in the rationals Q. The resulting Cartesian space Q^2 will therefore be insufficient for Euclidean geometry. Nevertheless, as noted above, full continuity is certainly not required, and it suffices to supplement the Cartesian space based on axioms 1–6 with an axiom of intersection for straight lines and circles (this, of course, is where the square-roots come in). See the excellent survey by A. Tarski, "What is Elementary Geometry?" (1959)—reprinted in Hintikka [51]. In particular, Tarski gives a set of axioms sufficient to generate a Cartesian space based on the square-root extension Q* (see *Theorem 6* governing system $\mathscr{E}_2''$: the circle/line axiom is A13′ on p. 174 of [51]; like the denseness condition it has the logical form $\forall.\,.\forall\exists$). Adding a (first-order) continuity *schema* extends our underlying set to a real closed field (see *Theorem 1* governing system $\mathscr{E}_2$: the continuity schema is A13 on p. 167 of [51]; it has the (minimal) logical form $\forall.\,.\forall\exists\exists\exists\forall\forall$). Finally, adding a (second-order) continuity *axiom* gives us a system essentially equivalent to Hilbert's: the underlying set is precisely $\mathbf{R}^2$.

Axioms 1–6 have only infinite models, and, of course, they make an essential use of modern polyadic logic. Note, however, that it is not merely the presence of two-place as opposed to one-place predicates that is crucial here. After all, axioms 1–3 alone certainly have finite models. Rather, the essential new element is the *quantifier-dependence* exhibited in 4–6: the logical form $\forall x\ \exists y$.[8] This kind of dependence of one quantifier on another cannot arise in monadic logic, where we can always "drive quantifiers in" so that each one-place matrix is governed by a single quantifier. (Thus, for example, $\forall x\ \exists y\ (Fx \rightarrow Gy)$ is equivalent to $\forall x\ Fx \rightarrow \exists y\ Gy$.) Moreover, it is the dependence of one quantifier on another—specifically, of existential quantifiers on universal quantifiers—that enables us to capture the intuitive idea of an iterative process formally: any value x of the universal quantifier generates a value y of the existential quantifier, y can then be substituted for x generating a new value y', and so on. Hence, the existence of an infinity of objects can be deduced explicitly by logic alone.

We can now begin to see what Kant is getting at in his doctrine of construction in pure intuition. For Kant logic is of course syllogistic logic or (a fragment of) what we call monadic logic.[9] Hence for Kant, one cannot represent or capture the idea of infinity formally or conceptually: one cannot represent the infinity of points on a line by a formal theory such as 1–6 above. If logic is monadic, one can only represent such infinity intuitively—by an iterative process of spatial construction:

> Space is represented as an infinite given magnitude. A general concept of space (which is common to both a foot and an ell alike) can determine nothing in regard to magnitude [Größe]. Were there no limitlessness in the

8. Thus, Euclid's Common Notions contain axioms governing an *equality* or *congruence* relation and axioms governing the *part-whole* relation. The point, however, is that such axioms are "essentially monadic" in exhibiting no quantifier-dependence (we could formulate them using universal free-variables as in axioms 1–3). Moreover, these Euclidean axioms have finite models: they do not say anything about the cardinality of our underlying set of points. Interestingly enough, Kant explicitly says that these axioms are analytic: cf. B16–17, A164/B203. (For more on the notion of "essentially monadic" and Kant's conception of analyticity see note 14 below.)

9. Kant's actual views on logic involve many subtleties which I here pass over. See, in particular, the interesting discussion in Thompson [108]. Thompson argues very convincingly that Kant indeed made one substantial advance in logic by replacing the traditional logic of *terms* with a "transcendental logic" of *objects* and *concepts:* "a logic in which the form of predication is '*Fx*' and not '*S* is *P*' " ([108], p. 342). Yet I cannot follow Thompson when he says: "The general logic required by Kant's transcendental logic is thus at least first-order quantificational logic plus identity" ([108], p. 334). If we do not limit ourselves to the logical forms of traditional syllogistic logic, Kant's Table of Judgements makes no sense. It is more plausible, I think, to equate Kant's conception of logic with, at most, *monadic* (or perhaps "essentially monadic"—see note 8 above) quantification theory plus identity (which, as far as I can see, is all Thompson requires in his fascinating discussion of Kant, Strawson, and Quine on singular terms and descriptions: [108], pp. 334–335—especially n. 15).

progression of intuition, no concept of relations could, by itself, supply a principle of their infinitude. (A25)

> . . . that one can require a line to be drawn to infinity *(in indefinitum)*, or that a series of changes (for example, spaces traversed by motion) shall be infinitely continued, presupposes a representation of space and time that can only depend on intuition, namely, in so far as it in itself is bounded by nothing; for from concepts alone it could never be inferred. (*Prolegomena* §12: 4, 285.1–7)

> Space is represented as an infinite given quantity [Größe]. Now one must certainly think every concept as a representation which is contained in an infinite aggregate [Menge] of different possible representations (as their common characteristic [Merkmal]), and it therefore contains these *under itself.* But no concept, as such, can be so thought as if it were to contain an infinite aggregate of representations *in itself.* Space is thought in precisely this way, however (for all parts of space *in infinitum* exist simultaneously). Therefore the original representation of space is an a priori *intuition,* and not a *concept.* (B40)

Kant's point is that (monadic) conceptual representation is quite inadequate for the representation of infinity: (monadic) concepts can never contain an infinity of objects in their very idea, as it were. In particular, then, since our idea of space does have this latter property, it cannot be a (monadic) concept.[10]

The notion of infinite divisibility or denseness, for example, cannot be represented by any such formula as 6: this logical form simply does not exist. Rather, denseness is represented by a definite fact about my intuitive capacities: namely, whenever I can represent (construct) two distinct points *a* and *b* on a line, I can represent (construct) a third point *c* between them. Pure intuition—specifically, the iterability of intuitive constructions[11]—provides a uniform method for instantiating the existential quantifiers we would use in formulas like 6; it therefore allows us to capture notions like denseness without actually using quantifier-dependence. Before the invention of polyadic quantification theory there simply is no alternative.

Thus, in Euclid's geometry there is no axiom corresponding to our denseness condition 6. Instead, we are given a uniform method for actually

10. I am indebted to Manley Thompson for correcting my earlier discussion of B40 in which I uncritically assimilated Kant's notion of the extension of a concept to our own (as well as for correcting my earlier mistranslation of the second sentence of B40).

11. For the centrality of *indefinite iterability* to Kant's conception of pure intuition, I am indebted, above all, to Parsons, "Kant's Philosophy of Arithmetic," especially §VII. But see also Parsons, "Infinity and Kant's Conception of the 'Possibility of Experience' " for doubts about the "psychological" or "empirical" reality of such truly *indefinite* iterability. Space prevents me from here giving these doubts the extended discussion they deserve.

constructing the point bisecting any given finite line segment: it suffices to join C in the Proof of Proposition I.1 with its "mirror image" below AB—the resulting straight line bisects AB (Prop. I.10). This operation, which is itself constructed by iterating the basic operations (i), (ii), and (iii), can then be iterated as many times as we wish, and infinite divisibility is thereby represented. So we do not derive new points between A and B from an existential axiom, we construct a bisection function from our basic operations and obtain the new points as the values of this function:[12] in short, we are given what modern logic calls a *Skolem function* for the existential quantifier in 6.[13] For Kant, this procedure of generating new points by the iterative application of constructive functions takes the place, as it were, of our use of intricate rules of quantification theory such as existential instantiation. Since the methods involved go far beyond the essentially monadic logic available to Kant, he views the inferences in question as synthetic rather than analytic.[14]

12. A simpler illustration of these ideas is provided by the theory of successor based on a constant 0 and a one-place function-sign $s(x)$. Instead of saying "Every number has a successor," we lay down the axioms:

$$0 \neq s(x)$$

$$s(x) = s(y) \rightarrow x = y.$$

These axioms have only infinite models, for we have "hidden" the quantifier-dependence in the function-sign $s(x)$: we *presuppose* that the corresponding function is well defined for all arguments.

13. A Skolem function for y in $\forall x\, \exists y\, R(x, y)$ is a function $f(x)$ such that $\forall x\, R(x, f(x))$; Skolem functions for y, w in $\forall x\, \exists y\, \forall z\, \exists w\, B(x, y, z, w)$ are functions $f(x)$, $g(x, z)$ such that $\forall x\, \forall z\, B(x, f(x), z, g(x, z))$; and so on. See, for example, Enderton [22], §4.2. (Here I follow a suggestion by Thomas Ricketts.)

14. These ideas have much in common with Hintikka's reconstruction in [52]. As in the present account, Hintikka argues that Kant's analytic/synthetic distinction is drawn *within* what we now call quantification theory, and Hintikka calls a quantificational argument *synthetic* when (roughly) "new individuals are introduced." Thus, synthetic arguments, for Hintikka, will correspond closely to those in which Skolem functions figure essentially, and analytic arguments will correspond to those we are calling "essentially monadic." Hintikka also notes the importance, in this connection, of the (often ignored) fact that Kant's logic is syllogistic or monadic ([52], pp. 189–190). (In this respect, Hintikka has indeed made an important advance over Beth. For Beth considers only the trivial procedure of conditional proof followed by universal generalization, and therefore puts forward a conception of the role of "intuition" in proof that applies equally well to monadic or syllogistic logic: cf. [6], §§5–7.) Yet Hintikka views the problem of quantificational rules like existential instantiation in rather the wrong light, I think—particularly when he attempts to conceive Kant's "transcendental method" as, in part, a *justification* of such rules (see chap. V of [52]). As I understand it, the whole point of pure intuition is to enable us *to avoid* rules of existential instantiation by actually constructing the desired instances: we do not derive our "new individuals" from existential premises but construct them from previously given individuals via Skolem functions.

Finally, we should note that our modern distinction between pure and applied geometry, between an uninterpreted formal system and an interpretation that makes such a system true, cannot be drawn here. In particular, the only way to represent the theory of linear order 1–6 is to provide, in effect, an interpretation that makes it true.[15] The idea of infinite divisibility or denseness is not capturable by a formula or sentence, but only by an intuitive procedure that is itself dense in the appropriate respect. By the same token, the sense in which geometry is a priori for Kant is also clarified. Thus, the proposition that space is infinitely divisible is a priori because its truth—the existence of an appropriate "model"—is a condition for its very possibility.[16] One simply cannot separate the idea or representation of infinite divisibility from what we would now call a model or realization of that idea; and our notion of pure (or formal) geometry would have no meaning whatsoever for Kant. (In a monadic context a pure or uninterpreted "geometry" cannot be a geometry at all, for it cannot represent even the *idea* of an infinity of points.)

· II ·

The above considerations make a certain amount of sense out of Kant's theory, but one might very well have doubts about attributing them to Kant. After all, Kant certainly had no knowledge of the distinction between monadic and polyadic logic, nor of quantifier-dependence, Skolem functions, and so on. So using such ideas to explicate his theory may appear wildly anachronistic, and my reading of the passages from A25, *Prolegomena* §12, and B40 may appear strained. Thus, whereas in all these passages Kant does clearly state that general concepts are inadequate for the representation of infinity and does contrast purely conceptual representation with the unlimited or indefinite iterability of pure intuition, it is not at all clear that this inadequacy, for Kant, rests on the limitations of monadic or syllogistic logic. Indeed, it is very hard to see how Kant could possibly have comprehended what we would now express as the inadequacy of monadic logic.

15. Similarly, the theory of successor of note 12 contains an infinite sequence of terms—the so-called numerals 0, s(0), s(s((0)), and so on—that is itself a model for that theory. Compare Parsons, "Kant's Philosophy of Arithmetic," §VII, and Thompson [108], pp. 337–342, where this feature of the numerals is connected with Kant's views on "symbolic construction."

16. I am indebted to Philip Kitcher for prompting me to make this last point explicit. The proposition that *physical* space is infinitely divisible is quite a different matter, however, whose a priori truth requires transcendental deduction: see note 32 and §IV below. For further discussion of the precise sense in which geometry is a priori for Kant see Chapter 2.

The key passage in this connection is B40, for it is only here that Kant does more than simply assert the inadequacy of general concepts (and hence the need for pure intuition) in the representation of infinity: it is only here that Kant attempts to explain precisely what it is about general concepts that is responsible for this inadequacy. Now B40 operates with Kant's particular notions of the extension and intension of a concept.[17] The extension of a concept is the totality of concepts relating to it as species, subspecies, and so on to a higher genus: the extension of *body* includes *animate body, inanimate body, animal animate body, rational animal animate body,* and so forth. In Kant's terminology these species, subspecies, and so on are all contained *under* the given concept. For Kant, moreover, there is no lowest *(infima)* species: our search for narrower and narrower specifications of any give (empirical) concept necessarily proceeds without end (A654–656/B682–684). In this sense, the extension of a concept is always unlimited. The intension of a concept, on the other hand, is the totality of constituent concepts (Teilbegriffe) or characteristics (Merkmale) that occur in its definition: the intension of *man* thus includes *rational, animal, animate,* and *body*. In Kant's terminology these constituent concepts or characteristics are all contained *in* the given concept.

Kant is therefore making two basic points in B40. First, the extension of a concept is always unlimited or potentially infinite: "one must certainly think every concept as a representation which is contained in an infinite aggregate of different possible representations (as their common characteristic), and it therefore contains these *under itself*." Nevertheless, however, no given concept can be conceived as the conjunction of an infinite number of constituent concepts. There is no bound to the number of elements in a concept's intension, but this number is always finite. In brief: whereas extensions are always unlimited, intensions can never be infinite—as, for example, the intension of a Leibnizean complete concept would be (compare Allison [2], p. 93).

Yet this way of expressing the matter makes it appear that the point of B40 has nothing whatever to do with the inadequacy of monadic logic for representing the idea of infinity. For the latter depends on the fact that no set of monadic formulas has only infinite models: any satisfiable set of monadic formulas is satisfiable in a domain consisting of only a finite number of objects. What is at issue here, then, is a fact about extensions of concepts *in the modern sense:* no concept definable by purely monadic

17. See the *Jäsche Logik,* Part I, Section One: 9, 91–100. See also the illuminating discussion of B40 in Allison [2], pp. 92–94. Again, I am especially indebted to Manley Thompson for emphasizing the importance of Kant's particular notions of extension and intension to me.

means can force its extension—that is, the set of objects falling under it—to be infinite. And not only is the modern notion of the extension of a concept completely foreign to Kant (Kant's notion involves a relation between a concept and other concepts—its species, subspecies, and so on—rather than a relation between a concept and the objects falling under it), but Kant explicitly states that extensions in his sense are potentially infinite. To be sure, he also explicitly states that intensions are necessarily finite; but what does this have to do with our modern fact about finite extensions?

It seems to me that there is nonetheless an intimate connection indeed between our modern fact about the limitations of monadic concepts and what Kant is saying in B40. For how do we establish that, for example, no set of monadic formulas containing k primitive predicates can determine a model with more than 2^k objects? We observe that k primitive predicates $P_1, P_2, \ldots, P_k$ can partition the domain into only 2^k maximally specific subclasses: the classes of objects that are P_1 & P_2 & . . . & P_k, $\neg P_1$ & P_2 & . . . & P_k, P_1 & $\neg P_2$ & . . . & P_k, and so on. The number of distinct objects we can assert to exist, then, is bounded by 2^k: we can say that there is an object in the first partition, there is an object in the second partition, and so on—and this is all. But now each such partition corresponds to a Kantian intension of a concept: the concept defined by the given conjunction of primitive predicates (and their negations). Moreover, each primitive predicate includes all such conjunctions in which it occurs in its Kantian extension (together with all less specific conjunctions in which it occurs, of course). What Kant is saying in B40 is that, whereas there is no limit to the number of (empirical) concepts we may eventually introduce, we are operating at any given time with only a finite number: we are therefore never in a position to form an infinite conjunction (an infinite intension). Similarly, on a modern understanding of monadic logic, since the number of primitive predicates is always finite (although it can be as large as one pleases), the number of partitions of the domain one can construct by conjoining such predicates (and their negations) is also always finite: we are therefore never in a position to assert the existence of an infinite number of objects. It seems to me, then, that the situation can be fairly described as follows: although there can of course be no question of Kant explicitly comprehending the logical fact we would now express as the inadequacy of monadic logic for representing an infinity of objects, he nonetheless comes as close to this as is possible given his own understanding of logic.

This reading of B40 also illuminates its position and role within the general argument of the Metaphysical Exposition of the Concept of Space. The passage at B40 is of course the concluding paragraph of the argument for the intuitive character of our representation of space—an argument which begins at A24–25/B39:

> Space is not a discursive or, as one says, general concept of relations of things as such, but a pure intuition. For, first, one can represent to oneself only a single space; and if one speaks of several spaces, one means thereby only parts of one and the same unique space.

Kant begins, then, by asserting that space is a singular individual rather than a general concept: the various particular spaces do not relate to space as instances to a general concept but as parts to an individual whole.

However, as Kant explicitly asserts at A25, there is indeed a "general concept of space (which is common to both a foot and an ell alike)," and this concept of space—which we might represent by ⌜x is a space⌝—does relate to parts of space or spaces as general concept to the instances thereof. In other words, there is both the general concept ⌜x is a space⌝, of which particular spaces are instances, and the singular individual *space,* of which particular spaces are parts (and which is in turn itself an instance of the general concept ⌜x is a space⌝). The question is: why should the latter have priority over the former? Why should our idea or representation of space be identified with the singular individual *space* rather than the general concept ⌜x is a space⌝?

Kant's answer is given in the following three sentences:

> Nor can these parts precede the single all-inclusive space, as being, as it were, its constituents (and making its composition possible); on the contrary, they can be thought only *in it.* Space is essentially singular: the manifold in it, and hence the general concept of spaces as such, rests purely on limitations. It follows therefrom that an a priori intuition (that is not empirical) underlies all concepts of space. (A25/B39)

One cannot arrive at the singular individual *space* by starting from the general concept ⌜x is a space⌝—which includes all parts of space among its instances—and, as it were, assembling the individual *space* from these diverse spaces or parts of space. On the contrary, the only way to arrive at the general concept ⌜x is a space⌝ is via the intuitive act of "cutting out" parts of space from the singular individual *space.* It is only the latter intuitive procedure of "limitation" that makes the general concept of space and of spaces possible in the first place.

But now the question becomes why should this be so: Why should the singular intuition thus precede the general concept? In the sentence immediately following Kant appeals to our knowledge of geometry:

> So too are all principles [Grundsätze] of geometry—for example, that in a triangle two sides together are greater than the third—derived: never from

general concepts of line and triangle, but only from intuition, and this indeed a priori, with apodictic certainty.

In the end, therefore, Kant's claim of priority for the singular intuition *space* rests on our knowledge of geometry.[18] Our cognitive grasp of the notion of space is manifested, above all, in our geometrical knowledge. Hence, if we can show that this knowledge is intuitive rather than conceptual, we will have shown the inadequacy of the general concept of space and the priority of the singular intuition.

Continuing our line of questions, then, we must ask why, at bottom, is conceptual knowledge inadequate to geometry: why must intuition play an essential role? Surely the mere assertion that geometrical principles cannot be derived "from general concepts . . . but only from intuition" is not expected to convince those who, like the Leibnizeans and Wolffians, maintain precisely the opposite. It is at this crucial point that Kant inserts the argument of B40. What is required for establishing the intuitive character of our representation of space is not simply the fact that space consists of parts, but rather—as geometry demonstrates—the fact that it consists of an *infinite number* of parts: "all parts of space *in infinitum* exist simultaneously." Thus, for example, geometry shows us that space is divisible into a potentially infinite sequence of smaller and smaller parts;[19] and, as the argument of B40 makes clear, no mere (monadic) concept can possibly capture this essential feature of our representation of space.

Once again, Kant's conception of infinity and infinite divisibility can be clarified by contrasting it with modern formulations. We, of course, can easily represent infinite divisibility by means of *(polyadic)* concepts—as we did above in the theory of dense linear order. In such a theory the points on a line are taken as primitive, and the line itself is built up from them in just the way Kant says it cannot be: the points relate to the line as "its constituents (and making its composition possible)." Yet what makes this representation itself possible is precisely the quantifier-dependence of modern polyadic logic: the logical form $\forall \ldots \forall\exists$. In the

18. But see also the very interesting account in Melnick [77], chap. 1.A, emphasizing the *individuating* role of the singular intuition. Thus, for example, two cubic feet of space are not distinguished by the general concept ⌜x is a space⌝ (or by any other general concept), but only by their "positions" in *space* ([77], pp. 9–14). In this connection, see also Thompson, [108]. Nevertheless, although Kant does emphasize this individuating role of the singular intuition in various places (particularly at A263–264/B319–320 and A271–272/B327–328), there is no hint of such a role at A24–25/B39. On the contrary, the emphasis *here* is entirely on the priority of the intuitive procedure of "limitation."

19. This fact of geometry plays a central role for Kant in his opposition to Leibnizean-Wolffian metaphysics: see, in particular, the *Physical Monadology*, especially Proposition III (1, 478–479). Compare also A165–166/B206–207. See the Introduction above.

absence of such logical forms—and in accordance with the actual procedure of Euclid's geometry—the natural alternative is to represent infinite divisibility by an intuitive constructive procedure for "cutting out" a smaller line segment from any given one: for example, Euclid's construction for bisecting a line segment of Proposition I.10.

Thus, whereas we can represent infinite divisibility by $\ulcorner \forall x\ \exists y$ (y is a proper part of x)$\urcorner$, Kant would formulate this proposition by $\ulcorner f_B(x)$ is a proper part of $x\urcorner$, where $f_B(x)$ is the operation of bisection, say.[20] And, in this representation, the idea of infinity is conveyed not by logical features of the relational concept $\ulcorner y$ is a proper part of $x\urcorner$ but by the well-definedness and iterability of the function $f_B(x)$: our ability, for any given line segment x, to construct (distinct) $f_B(x)$, $f_B(f_B(x))$, *ad infinitum.*[21] This, I suggest, is why Kant gives priority to the singular intuition *space,* from which all parts or spaces must be "cut out" by intuitive construction ("limitation"). Only the unbounded iterability of such constructive procedures makes the idea of infinity, and therefore all "general concepts of space," possible. And, of course, it is this very same constructive iterability that underlies the proof-procedure of Euclid's geometry.

· III ·

Even if we are on the right track, however, we have still gone only part of the way towards understanding construction in pure intuition. We can bring out what is missing by three related observations. First, as we noted above, the notions of denseness, infinite divisibility, and (even) constructibility with straight-edge and compass do not amount to full continuity.

20. Kant mentions the Euclidean construction of the bisection operation in the *Groundwork of the Metaphysics of Morals* at 4, 417.18–21; in the *Physical Monadology,* however, he uses a different proof of infinite divisibility due to John Keill. (We understand the output of $f_B(x)$ here to be not the midpoint of x but, say, the left-most half segment of x.)

21. Kant certainly recognizes relations such as the part-whole relation. For Kant, however, our theory of this relation will still be "essentially monadic" in exhibiting no quantifier-dependence: see notes 8 and 14 above. In particular, then, the inference from $\ulcorner f_B(x)$ is a proper part of $x\urcorner$ to $\ulcorner f_B(f_B(x))$ is a proper part of $f_B(x)\urcorner$ is synthetic for Kant (= not "essentially monadic") because (in Hintikka's terminology) "new individuals are introduced." Moreover, it is instructive to contrast this inference with the following example from Leibniz's *New Essays:* If Jesus Christ is God, then the mother of Jesus Christ is the mother of God ([67], vol. 5, p. 461; [71], p. 479—compare the citation in Tait [107], §XIII). If we represent motherhood as a function (and hence as presupposing existence) then Leibniz's argument also "introduces new individuals" and thus should count as synthetic; if we represent motherhood as a relation (and hence as not presupposing existence) then the argument counts as "essentially monadic" or analytic. Since there is no question of *iteration* in Leibniz's argument, however, the functional representation is certainly not required, and it is just this, it seems to me, that distinguishes the argument from $\ulcorner f_B(x)$ is a proper part of $x\urcorner$ to the infinite divisibility of space.

These notions all involve denumerable sets of points which are but small fragments of the set **R** of real numbers. Hence, to understand how full continuity comes in we have to go beyond Euclidean geometry. Second, these notions do not (on a modern construal) exploit very much of polyadic logic: just the logical form $\forall \ldots \forall\exists$; if this were all that was required modern logic would hardly need to have been invented. Third, the procedure of construction with Euclidean tools—with straight-edge and compass—does not really exploit the kinematic element that is essential to Kant's conception of pure intuition: no appeal is made to the idea that lines, circles, and so on are generated by the *motion* of points. So why does Kant think that *motion* is so important?

These three observations are in fact intimately related. For there exists a branch of mathematics which was just being developed in the seventeenth and eighteenth centuries; which does require genuine continuity—"all" or "most" real numbers; whose modern, "rigorous" formulation requires full polyadic logic—much more intricate forms of quantifier-dependence than $\forall. .\forall\exists$; and, finally, whose earlier, "non-rigorous" formulation made an essential appeal (in at least one tradition) to temporal or kinematic ideas—to the intuitive idea of motion. This branch of mathematics is of course the calculus, or what we now call real analysis. It goes far beyond Euclidean geometry in considering "arbitrary" curves or figures—not merely those constructible with Euclidean tools—and in making extensive use of *limit operations.*

From a modern point of view the basic limit operation underlying the calculus is explained in terms of the Cauchy-Bolzano-Weierstrass notion of *convergence.* Moreover, we also appeal to this notion in explaining the distinction between denseness and genuine continuity, in precisely expressing the idea that there are "gaps" in a merely dense set such as the rational numbers. Thus, let $s_1, s_2, \ldots$ be a sequence of rational numbers that converges to π, say—that approaches π as its limit (for example, let $s_1 = 3.1$, $s_2 = 3.14$, and in general $s_n =$ the decimal expansion of π carried out to n places). This sequence of rationals *converges* (to "something," as it were), but in the set Q of rational numbers (and even in the expanded set Q* of Euclidean-constructible numbers) there is no limit point it *converges to.* Such limit points are "missing" from a merely dense set such as the rationals. A truly continuous set contains "all" such limit points.

More precisely, using Cauchy's criterion of 1829, we say that a sequence $s_1, s_2, \ldots$ *converges* if

$$\forall\epsilon\ \exists N\ \forall m\ \forall n\ [m,n > N \rightarrow |s_m - s_n| < \epsilon],$$

where ϵ is a positive rational number and N, m, n are natural numbers.

A sequence $s_1, s_2, \ldots$ *converges to a limit r* if

$$\forall \epsilon \, \exists N \, \forall m \, [m > N \rightarrow |s_m - r| < \epsilon].$$

The problem with a merely dense order is that the first can be true even when the second is not, whereas a continuous order satisfies the additional axiom of *Cauchy completeness*—whenever a sequence converges, it converges to a limit r—which clearly has the logical form $\forall\exists\forall\forall \rightarrow \exists\forall\exists\forall$. Note the additional logical complexity of this axiom: in particular, the use of the strong form of quantifier-dependence, $\forall\exists\forall$.

The increase in logical strength which I find so striking here can be best brought out if we compare the way points are generated by a completeness axiom with the way they are generated by Euclidean constructions (or, from a modern point of view, by the weaker form of quantifier-dependence, $\forall. .\forall\exists$). In the latter case, although the total number of points generated is of course infinite, each particular point is generated by a finite number of iterations: each point is determined by a finite number of previously constructed points. In generating or constructing points by a limit operation, on the other hand, we require an infinite sequence of previously given points: no finite number of iterations will suffice. So limit operations involve a much stronger and more problematic use of the notion of infinity than that involved in a simple process of iterated construction.

Let us now return to Kant and the late eighteenth century. We cannot of course represent the ideas of convergence and transition to the limit by complex quantificational forms such as $\forall\exists\forall$. But the idea of *continuous motion* appears to present us with a natural alternative. Thus, for example, we can easily "construct" a line of length π by imagining a continuous process that takes one unit of time and is such that at $t = 1/2$ a line of length 3.1 is constructed, at $t = 2/3$ a line of length 3.14 is constructed, and in general at $t = n/(n + 1)$ a line of length s_n is constructed, where s_n, again, equals the decimal expansion of π carried out to n places. Assuming this process in fact *has* a terminal outcome, at $t = 1$ we have constructed a line of length π. In this sense, then, we can thereby "construct" any real number.[22]

Here the notion of convergence or approach to the limit is expressed by a temporal process: by the idea of one point moving or becoming closer and closer to a second. This intuitive process of becoming does the work of our logical form $\forall\exists\forall$, as it were. That the limit of a convergent se-

22. Here, and throughout this section especially, I am indebted to clarificatory suggestions from Roberto Torretti.

quence exists is expressed by the idea that any such process of temporal generation has a terminal outcome. This idea does the work of our logical form $\exists\forall\exists\forall$. In particular, then, what we now call the continuity or completeness of the points on a line is expressed by the idea that any finite motion of a point beginning at a definite point on our line also stops at a definite point on our line.[23] What the modern definition of convergence does, in effect, is replace this intuitive conception based on motion and becoming with a formal, algebraic, or "static" counterpart based on quantifier-dependence and order relations.

Now a temporal conception of the limit operation is explicit in the basic lemma Newton uses to justify the mathematical reasoning of *Principia:*

> Quantities, and the ratios of quantities, which in any finite time converge continually to equality, and before the end of that time approach nearer to each other than by any given difference, become ultimately equal. (Book I, §1, Lemma I: [82], p. 73; [83], p. 29)

This lemma is perhaps best understood as a definition of what Newton means by "quantity": namely, an entity generated by a continuous temporal process (it clearly fails for discontinuous "quantities") .

Newton's conception of "quantities" as temporally generated is even more explicit, of course, in his method of *fluxions,* where all mathematical entities are thought of as *fluents* or "flowing quantities." For example:

> I don't here consider Mathematical Quantities as composed of Parts *extremely small,* but as *generated by a continual motion.* Lines are described, and by describing are generated, not by any apposition of Parts, but by a continual motion of Points. Surfaces are generated by the motion of Lines, Solids by the motion of Surfaces, Angles by the Rotation of their Legs, Time by a continual flux, and so in the rest. ([85], p. 141)

Moreover, Kant appears to be echoing these ideas in an important passage about continuity in the Anticipations of Perception:

> Space and time are *quanta continua,* because no part of them can be given without being enclosed between limits (points and instants), and therefore only in such fashion that this part is itself again a space or a time. Space consists only of spaces, time consists only of times. Points and instants are only limits, that is, mere places [Stellen] of their limitation. But places always presuppose the intuitions which they limit or determine; and out of mere places, viewed as constituents capable of being given prior to space or time, neither space nor time can be composed [zusammengesetzt]. Such quantities

23. Compare Heath's interpretation of the intuitive content of Dedekind continuity: "[it] may be said to correspond to the intuitive notion which we have that, if in a segment of a straight line two points start from the ends and describe the segment in opposite senses, they meet in a point" ([46], p. 236).

> [Größen] may also be called *flowing* [*fließende*], since the synthesis (of the productive imagination) in their generation [Erzeugung] is a progression in time, whose continuity is most properly designated by the expression of flowing (flowing away). (A169–170/B211–212)

For Kant, like Newton, spatial quantities are not composed of points, but rather generated by the motion of points.

I take Kant's choice of language to be especially significant here, for his "fließende Größen" is the standard German equivalent of Newton's "fluents."[24] This expression is used, for example, by the mathematician Abraham Kästner in his influential textbooks on analysis and mathematical physics.[25] Kästner's analysis text attempts to develop the calculus from a "rigorous" standpoint that makes no appeal to infinitely small quantities. In this connection he develops a version of Newton's method of fluxions, and, what is more remarkable for a German author of this period, he argues that Newton's fluxions are in some respects clearer and more perspicuous than Leibniz's differentials. Further, he explicitly applauds Collin Maclaurin's attempt, in his monumental *Treatise of Fluxions* (1742), to develop the calculus on the basis of a kinematic conception of the limit operation.[26]

Without going into detail, the most basic ideas of the fluxional calculus are as follows.[27] We start with fluents or "flowing quantities" x, y, conceived as continuous functions of time. We can then form the fluxions or time-derivatives $\dot{x}$, $\dot{y}$, because continuously changing quantities obviously have well-defined instantaneous velocities or rates of change.[28] If we are then given a curve or figure $y = f(x)$ generated by independent motions in rectangular coordinates of the fluents x, y, the *derivative* (slope of the tangent line) will be $dy/dx = \dot{y}/\dot{x}$ (by the "parallelogram of velocities") (see Figure 4). Finally, we can recover the integral from the derivative via the Fundamental Theorem (which is also understood temporally: see

24. Kitcher, in [59], p. 41, has drawn attention to Kant's use of Newtonian terminology and the connection between B211–212 and the fluxional calculus. See also note 29 below.

25. See *Anfangsgründe der Analysis des Unendlichen* (Göttingen, 1761) and *Anfangsgründe der höhern Mechanik* (Göttingen, 1766). Kant was well acquainted with Kästner's works, and an admirer of them. See the reference to Kästner's 1766 work in §14.4 of the *Inaugural Dissertation,* for example: 2, 400.3–4.

26. Philip Kitcher has emphasized to me that Maclaurin's book contains a number of different approaches to the foundations of the calculus. In Maclaurin—and in the Newtonian tradition generally—it is certainly not obvious that the kinematic approach is predominant. My point is simply that this strand of the Newtonian tradition is central to Kant's thinking.

27. See, for example, the very sympathetic account in Whiteside [120], especially §§V, X, XI, making extensive use of the concept of "limit-motions."

28. See the Scholium following Lemma XI of *Principia,* Book I, §1, for example, where Newton appeals to the intuitive idea of instantaneous velocity to justify the existence of the required limits: [82], pp. 87–88; [83], pp. 38–39.

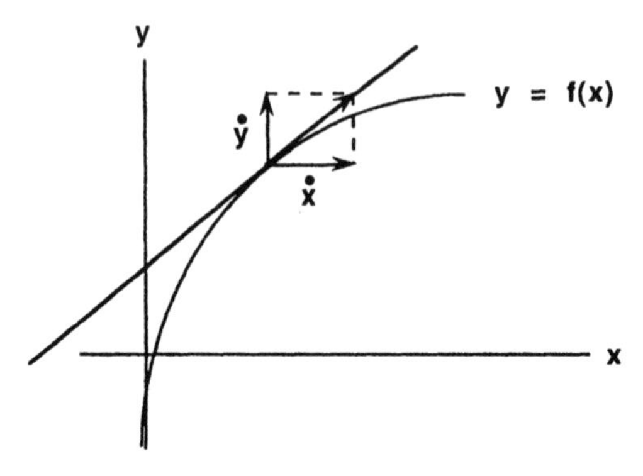

Figure 4

Whiteside [120], pp. 374–376). So the basic notions of the calculus are explained without ever appealing to differentials or infinitely small quantities.

In any case, it is extremely likely that some such understanding of the calculus (the method of fluxions) underlies Kant's insistence on the kinematic character of construction in pure intuition.[29] When he speaks of the "productive synthesis" involved in the "mathematics of extension" Kant is referring to what we would now call calculus in a Euclidean space; he is not simply thinking of Euclidean geometry proper. And, if this is correct, we can better understand why, given the way concepts such as continuity and passage to the limit are understood, there is no possibility of a distinction between pure and applied—uninterpreted versus interpreted—mathematics in the modern sense. The only way one can represent continuity, for example, is to provide what we would now call an intuitive interpretation of the continuity or completeness axiom, an interpretation that necessarily makes that axiom true. In particular, since convergence

29. As far as I know, Kant comes closest to making this explicit in the course of an exchange with August Rehberg in 1790 concerning the nature of irrational numbers and their relation to intuition. In a draft (Reflexion 13) of his reply to Rehberg Kant says: "If we did not have concepts of space then the quantity $\sqrt{2}$ would have no meaning [Bedeutung] for us, for one could then represent every number as an aggregate [Menge] of indivisible units. But if one represents a line as generated through fluxion [durch fluxion], and thus generated in time, in which we represent nothing simple, then we can think 1/10, 1/100, etc., etc. of the given unit" (14, 53.2–7; see also Adickes's note to this passage on pp. 53–54). The entire exchange (Rehberg's letter: 12, 375–377; Kant's drafts (Reflexionen 13–14): 14, 53–59; Kant's reply: 11, 195–199) sheds much light on Kant's view of the relationship between spatio-temporal intuition and arithmetical-algebraic concepts, and calls for a detailed investigation: for a beginning, see Chapter 2 below. Some aspects of the exchange are discussed by Parsons [94]. (I am indebted to Parsons for first calling my attention to the passage at 14, 53.2–7.)

is represented by a *continuous* process of temporal generation, the relevant limit point is automatically generated as well.

Moreover, although the kinematic interpretation of the calculus certainly does not meet modern standards of rigor, it is also not afflicted with the obvious problems about consistency and coherence facing an interpretation based on differentials, infinitesimals, and infinitely small quantities. Indeed, when the kinematic interpretation was explicitly criticized by mathematicians such as D'Alembert and l'Huilier in the late eighteenth century this was not on grounds of coherence and consistency, but rather because it was thought to import a "foreign" or "physical" element into pure mathematics. Pure mathematics should be independent of and prior to mathematical physics; therefore, it should be developed in complete independence of the idea of motion.[30] For Kant, on the other hand, this "mixing" of physical and mathematical ideas is not a defect but a virtue. Since part of the "general doctrine of motion"—namely, pure kinematics or "phoronomy"—is, in effect, also a branch of pure mathematics, it is possible to hold that this part of mathematical physics is a priori as well.[31] So an explicit "mixing" of physical and mathematical ideas is essential to the unity of Kant's system.[32]

30. See the excellent account in Grabiner [39] and, in particular, the quotation from Bolzano's "Rein analytischer Beweis . . . " (1817), §II: "the concept of Time and even more that of Motion are . . . foreign to general mathematics," on p. 53.

31. See §§X–XV of Kästner's analysis text, entitled "Bewegung gehört in die Geometrie." Kästner replies to l'Huilier's criticism of the idea of motion by drawing a sharp distinction between kinematics ("phoronomy")—where one considers the motion of mere mathematical points independently of their physical properties (such as mass); and dynamics—where one explicitly considers both the physical constitution of such points and the forces that produce the motion. The former is a branch of pure mathematics and is therefore a priori; the latter is a branch of physics and is therefore a posteriori. In the third edition of his text (1799), Kästner even refers to Kant's *Metaphysical Foundations of Natural Science* (1786) for a justification of this distinction.

32. In this connection, see the important footnote at B155: "Motion of an *object* [*Objekt*] in space does not belong to a pure science, and consequently not to geometry; for, that something is movable cannot be cognized a priori, but only through experience. But motion, as the *describing* [*Beschreibung*] of a space, is a pure act of successive synthesis of the manifold in an outer intuition in general, and belongs not only to geometry, but even to transcendental philosophy." (That the *describing* of a space = the motion of a mathematical point is confirmed by the Observation to Definition 5 in the first chapter or Phoronomy of the *Metaphysical Foundations of Natural Science:* 4, 489.6–11.) Thus Kant does have a distinction between pure and applied geometry—although it is certainly not *our* distinction. In pure geometry we consider figures generated in "empty" space by the motion of mere mathematical points; in applied geometry we consider the actual sensible objects contained "in" this space. That what holds for mere mathematical points in "empty" space holds also for actual sensible objects found "in" this space (that pure mathematics can be applied) can only be established in transcendental philosophy. See also A165–166/B206–207. (Joshua Cohen, Ralf Meerbote, and Manley Thompson have all emphasized the importance of B155n to me.)

At the same time, however, the difference between the iterative infinity involved in Euclidean constructions and the stronger use of infinity involved in limit operations helps to elucidate the sense in which the kinematic interpretation fails to meet modern standards of rigor. In Euclidean geometry we specify the objects of our investigation—circles, straight lines, and any figures constructible from them—by a well-defined iterative or "inductive" procedure. This specification then underlies our iterative, step-by-step method of proof: the substitution of a previously constructed object—a given finite straight line segment, say—as argument in a further constructive operation—the construction of a circle based on this line segment as radius via Postulate 3, for example. By contrast, in the fluxional calculus we have no such specification: no step-by-step procedure (nor any other precisely defined method) for constructing all fluents or "fließende Größen" has been given.[33] Similarly, our temporal representation of the limit operation does not proceed by repeated application of previously given functions: each new limit has to be constructed "on the spot," as it were. This, in the end, is perhaps the most fundamental advantage of the Cauchy-Bolzano-Weierstrass definition of convergence. For our use of the logical form $\forall\exists\forall$ in an appropriate formal system of quantificational logic permits us to reestablish iterative methods of proof: we can "handle" the points generated via limit operations by rigorous—finitary—deductions.[34]

Be this as it may, the class of curves generated as fluents or "fließende Größen" proves to be inadequate to the needs of mathematics and mathematical physics. The main problem is that, since continuity is explained by continuous motion, *continuity* automatically implies *differentiability* as well (see Whiteside [120], p. 349). A curve generated by continuous motion (drawn by the continuous motion of a pencil, as it were) automatically has a tangent or direction of motion at each point. So the class of continuous curves is assimilated to the class of what we now call *smooth* (differentiable) curves.[35] Actually, this is not quite right, for those who employed the method of fluxions of course knew that there are continuous

33. *We* might conceive "fluents" as smooth (or at least piece-wise smooth) maps from the real numbers ("time") into some smooth manifold (Euclidean three-space, for example). The point, of course, is that precisely these concepts are unavailable to Newton and Kant.

34. This is perhaps part of what Russell had in mind when he praised the "infinitary" power of quantification in 1903: "An infinitely complex concept, although there may be such, can certainly not be manipulated by the human intelligence; but infinite collections . . . can be manipulated without introducing any concepts of infinite complexity" ([102], §72).

35. Thus Kant replies to a question due to Kästner in the *Inaugural Dissertation*, §14.4, by proving that "the continuous motion of a point over all the sides of a triangle is impossible" (4, 400.4–5). Kant's proof consists in the observation that no tangent or direction of motion exists at any vertex of the triangle. From a modern point of view, then, he "assumes" that a continuous map from **R** ("time") into space is also a smooth map.

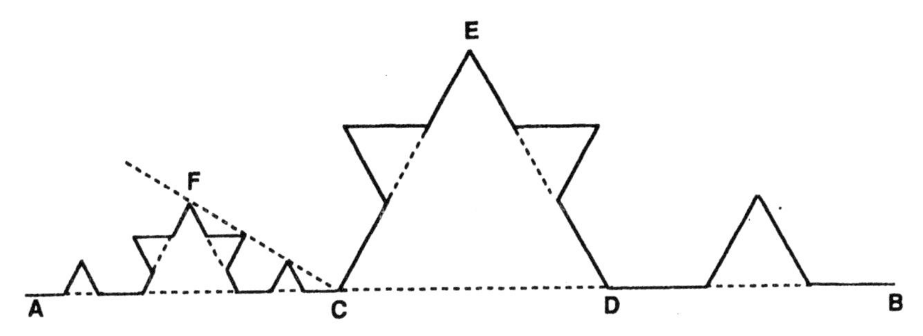

Figure 5

curves that lack tangents at certain points: curves with "cusps" or "corners." However, such curves can be easily comprehended within a kinematic understanding of continuity so long as one can think of them as "pieced together" by a finite number of smooth curves, so long as they have a finite number of "isolated" singular points.

But what happens if we allow such a process of "piecing together" smooth curves to be itself iterated indefinitely: if we apply limit operations to infinite collections of given smooth or "well-behaved" curves? What we get, of course, includes continuous but *nowhere* differentiable curves. The most famous examples of such curves, given by Weierstrass in 1872, are constructed via trigonometric series.[36] Fortunately, however, we can also give simpler, more intuitive examples. Perhaps the simplest is the Koch curve: we start with a horizontal line segment *AB* which we divide into three equal parts by points *C* and *D;* on the middle segment *CD* we construct an equilateral triangle *CED* and erase the open segment *CD;* we repeat the same construction on each of the segments *AC, CE, ED, DB;* finally, we continue this process indefinitely on each remaining segment (see Figure 5). The resulting curve is continuous, but at no point is there a well-defined tangent.[37] (In *every* neighborhood of *C*, for example, both lines *AC* and *FC* intersect infinitely many other points of the curve.) Thus, no finite segment of the Koch curve can be drawn by the continuous motion of a pencil: we must think of each point as laid down independently, as it were, yet nevertheless in a continuous order.

36. See, for example, Boyer [9], pp. 284–285. (As Boyer points out, the first example of a continuous but nowhere differentiable function was given by Bolzano in 1834: [9], pp. 269–270). We should also note that such "pathological" functions arise naturally out of Fourier's work in the 1820s on partial differential equations, work that of course is directly inspired by, and has extremely important applications to, problems in mathematical physics. So the difficulty is in no way confined to pure mathematics.

37. This example was published by H. von Koch in papers of 1903 and 1906. See Eves [29], §13.4. As Eves points out, although the Koch curve is not single-valued, we can easily construct similar single-valued examples.

Such continuous but nowhere differentiable curves clearly exceed the scope of the kinematic interpretation: we cannot understand their continuity via the intuitive idea of continuous motion.[38] To get a mathematical grip on this wider class of curves we need a clear distinction between continuity and differentiability; and this, of course, is one of the main achievements of our modern approach to convergence. We define the *continuity* of a function $f(x)$ at a given point x_0 by an expression of the form $\forall\ \exists\ \forall$ [Conv(s, x_0) & Φ], where s' is a sequence defined algebraically from $f(x)$ and Φ says that s converges *to* $f(x_0)$. We define the *differentiability* of $f(x)$ at a given point x_0 by an expression of the form $\forall\ \exists\ \forall$ [Conv(s', x_0], where s' is a second sequence defined algebraically from $f(x)$. By understanding these notions formally rather than intuitively, we can, for the first time, both clearly and precisely distinguish them and clearly and precisely explore their logical relations: differentiability logically implies continuity but not vice versa, for example.[39]

· IV ·

The present approach to Kant's theory of geometry follows Russell in assuming that construction in pure intuition is primarily intended to explain mathematical proof or reasoning, a type of reasoning which is therefore distinct from logical or analytic reasoning. Again following Russell, we have sought an explanation for this idea in the difference between the essentially monadic logic available to Kant and the polyadic logic of modern quantification theory. Further, we have tried to link this conception of mathematical reasoning with the very possibility of thinking or representing mathematical concepts and propositions. Thus, for example, "I cannot think a line except by *drawing* it in thought" (B154), because only

38. Kitcher [59] stresses the importance of this problem for Kant's conception of pure intuition, a problem that is perhaps even more fundamental than the discovery of non-Euclidean geometries. As Kitcher remarks: "The death blow was not struck by Bolyai, Lobachevski, and Klein but by the men in the tradition which led to Weierstrass's function, continuous everywhere but differentiable nowhere" ([59], p. 41).

39. It is also worth noting that, although the distinction between continuity and differentiability obviously makes an essential (and rather strong) use of polyadic logic, it is not itself a purely logical distinction: the two formulas have the same logical form, they differ only algebraically. However, one finds precisely such a logical distinction in the contrast between *point-wise* and *uniform* properties. For example, the distinction between point-wise and uniform convergence is purely logical: it is a distinction in quantifier order alone. And this distinction, which is at least obscured in the work of Cauchy, is developed with great subtlety and precision by Weierstrass. In Weierstrass's work, we might say, polyadic quantification theory comes fully into its own. It is perhaps no accident, then, that Frege, who was of course intimately acquainted with Weierstrass's foundational contributions, invented the first accurate and complete formulation of quantificational logic in 1879. (In this connection, see the description of Frege's "advanced course on *Begriffsschrift*" in Carnap [17], p. 6.)

this representation permits me to use the concept of line in mathematical reasoning (such as Euclid's or Newton's) where properties like denseness and continuity play an essential role.

Yet Russell's assumption has been vigorously debated. It has been maintained that Kant did not deny, and indeed may have even affirmed, that mathematical inference is logical or analytic; his primary concern, rather, is with the status of the premises or axioms of such inferences. Geometry is synthetic precisely because its underlying axioms are synthetic; the (synthetic) theorems of geometry then follow purely logically or analytically. This anti-Russellian view is clearly and forcefully stated by Beck:

> The real dispute between Kant and his critics is not whether the theorems are analytic in the sense of being strictly [logically] deducible, and not whether they should be called analytic now when it is admitted that they are deducible from definitions, but whether there are any primitive propositions which are synthetic and intuitive. Kant is arguing that the axioms cannot be analytic . . . because they must establish a connection that can be exhibited in intuition.[40]

As Beck indicates, this view is attractive because Kant will not be refuted, as Russell thought, by the mere invention of polyadic logic. For even modern formulations of Euclidean geometry such as Hilbert's will contain primitive propositions or axioms, and pure intuition can be called in to secure their truth (to provide a model, as it were).[41]

Indeed, from this point of view the discovery of logically consistent systems of non-Euclidean geometry should be seen as a vindication of Kant's conception. The existence of such geometries shows conclusively that Euclid's axioms are not analytic and, therefore, that no analysis of the basic concepts of geometry could possibly explain their truth (as Leibniz apparently thought). Assuming that Euclid's axioms are true, then, there is no alternative but to appeal to a synthetic source: hence pure intuition.[42]

40. Beck, "Can Kant's Synthetic Judgements Be Made Analytic?" (1955), reprinted in [4], pp. 89–90. Martin takes this point of view to extremes, viewing Kant as a forerunner of "modern axiomatics": see [73], for example. Needless to say, such a conception is completely antithetical to the present interpretation.

41. Thus Hilbert, in his brief Introduction to [48], refers to Kant and equates the task of axiomatizing Euclidean geometry with "the logical analysis of our spatial intuition."

42. In the context of contemporary discussion this view has been articulated most clearly and explicitly by Brittan [10]. Indeed, after referring to A220–221/B268 ("there is no contradiction in the concept of a figure which is enclosed within two straight lines, since the concepts of two straight lines and of their coming together contain no negation of figure"), Brittan says: "It was Kant's appreciation of the fact that non-Euclidean geometries are consistent (possibly something of which his correspondent, the mathematician J. H. Lambert, made him aware) that, among several different considerations, led him to say that Euclidean geometry is synthetic. The further development of non-Euclidean geometries only confirms his view" ([10], p. 70, n. 4; Brittan follows Martin [73], §2, in this estimate of the Lambert-

On the Russell-inspired interpretation developed here, by contrast, there can be no question of non-Euclidean geometries for Kant. Non-Euclidean straight lines, if such were possible, would have to possess at least the order properties—denseness and continuity—common to all lines, straight or curved. And, on the present interpretation, the only way to represent (the order properties of) a line—straight or curved—is by drawing or generating it in the space (and time) of pure intuition. But this space, for Kant, is necessarily Euclidean (on both interpretations). It follows that there is no way to draw, and thus no way to represent, a non-Euclidean straight line, and the very idea of a non-Euclidean geometry is quite impossible.[43] (Another way to see the point is to note that the anti-Russellian interpretation would reinstate precisely the modern distinction between pure and applied geometry argued above to be unavailable to Kant.)

The anti-Russellian interpretation draws its primary support from B14:

> For as it was found that all mathematical inferences proceed in accordance with the principle of contradiction [nach dem Satze des Widerspruchs fortgehen] (which the nature of all apodictic certainty requires), it was supposed that the fundamental propositions [Grundsätze] could also be recognized from that principle [aus dem Satze des Widerspruchs erkannt würden]. This is erroneous. For a synthetic proposition can indeed be comprehended [eingesehen] in accordance with the principle of contradiction, but only if another synthetic proposition is presupposed from which it can be derived [gefolgert], and never in itself.

Kant seems to be saying that because inference from axioms to theorems was (correctly) seen as analytic, the axioms themselves were (incorrectly) thought to be analytic. But these axioms are really synthetic; for this reason (and only for this reason), so are the theorems. Kant therefore agrees with Russell that the conditional, Axioms → Theorems, is a logical

Kant connection). The idea, apparently, is that two-sided plane figures exist in *elliptic* (positive curvature) space—more precisely, in the subcase of *spherical* space, and Kant is supposed to have learned of this type of non-Euclidean space from Lambert. This idea is most implausible, however, for Lambert of course proved that elliptic space is *impossible;* in that elliptic space—unlike hyperbolic (negative curvature, Bolyai-Lobachevsky) space—does contradict the remainder of Euclid's axioms, in particular, the assumed infinite extendability of straight lines (Postulate 2): see, for example, Bonola [8], §§18–22. (Brittan develops an alternative reading of A220–221/B268 and the Lambert-Kant connection in [11], pp. 74–79.)

43. Compare Beth [6], p. 364: "If one assumes this view of geometrical demonstration [that intuition plays an essential role], then *absolutely nothing follows* from the formal possibility of a non-Euclidean geometry, that is, from the formal independence of the Parallel Postulate relative to the remaining axioms of Euclidean geometry. For if we attempt to answer any geometrical question on the basis of the remaining axioms, we must (according to Kant) first construct the corresponding figure. This construction will proceed according to the antecedent laws of pure intuition, and therefore the Euclidean answer will come out at the end. The distinction between axioms and theorems will therefore obviously collapse."

or analytic truth;[44] his point is simply that the antecedent of the conditional is synthetic.

I do not think this reading of the passage is forced on us. First of all, Kant does not actually say that mathematical inference is analytic, nor that the theorems can be analytically derived. Thus the first sentence may mean only that mathematical proofs necessarily involve logical or analytic steps—and, of course, no logical fallacies.[45] Second, it is assumed that by *fundamental propositions* (Grundsätze) Kant means *axioms,* and this is doubtful. Kant's own technical term for axioms is *Axiomen* (see A163–165/B204–206, A732–733/B760–762), and at A25 he calls the proposition that two sides of a triangle together exceed the third a fundamental proposition (Grundsatz). This latter is of course not an axiom in Euclid, but a basic (and therefore fundamental) theorem (Prop. I.20). So the error Kant is diagnosing here may not be the (really rather ridiculous) mistake of transferring analyticity from inference to premise (axiom), but the more subtle supposition that because logic plays a central role in the proof of basic theorems it is sufficient for securing their truth.

A more fundamental problem for the anti-Russellian reading of B14 is posed by Kant's conception of arithmetic. Kant is supposed to have a more-or-less modern picture of mathematical theories as strict deductive systems. The synthetic character of mathematics depends solely on the synthetic character of the underlying axioms. But this is certainly not Kant's picture of arithmetic. According to Kant, arithmetic differs from geometry precisely in having no axioms, for there are no propositions that are both general and synthetic serving as premises in arithmetical arguments (A163–165/B204–206). Thus our conception of arithmetic as based on the Peano axioms, say, is completely foreign to Kant, and one cannot use the model of an axiomatic system to explain why arithmetic is synthetic: one cannot suppose that arithmetical reasoning proceeds purely logically or analytically from synthetic axioms as premises.[46]

44. But we should remember that the Russell of 1903 still believed that *logic* is synthetic. See [102], §434: "Kant never doubted for a moment that the propositions of logic are analytic, whereas he rightly perceived that those of mathematics are synthetic. It has since appeared that logic is just as synthetic as all other kinds of truth."

45. Compare A59/B84, where the principle of contradiction is said to be a necessary, but insufficient, criterion for *all* truth, and A151/B190, where it is asserted that the truth of *analytic judgements* "can be sufficiently recognized according to the principle of contradiction [nach dem Satze des Widerspruchs hinreichend können erkannt werden]."

46. Of course this difference between arithmetic and geometry is explicitly recognized by Beck: he suggests that Kant's discussion of arithmetic is simply inconsistent with the general account of mathematics at B14 (see [4], p 89). Compare also Brittan [10], pp. 50–51. Martin, on the other hand, goes so far as to attribute an axiomatic conception of arithmetic to Kant: see [73] and especially [74]. Yet Martin relies almost exclusively on the writings of Kant's contemporaries and students—Johann Schultz, in particular—and has no account of Kant's

Yet arithmetic is the very first example Kant uses (at B15–16) to illustrate, and presumably illuminate, the general ideas of B14. Arithmetical propositions such as 7 + 5 = 12 are synthetic, not because they are established by analytic derivation from synthetic axioms (as we would derive them from the Peano axioms, say), but because they are established by the successive addition of unit to unit. This procedure is synthetic, according to Kant, because it is necessarily temporal, involving "the successive progression from one moment to another" (A163/B203).[47] Thus, for example, only the general features of succession and iteration in time can guarantee the existence and uniqueness of the sum of 7 and 5, which, as far as logic and conceptual analysis are concerned, is so far merely possible (non-contradictory). Similarly, only the unboundedness of temporal succession can guarantee the infinity of the number series, and so on.

For Kant, then, arithmetical propositions are established by calculation, a procedure that is sharply distinguished from discursive argument in being essentially temporal. This is why Kant says that the synthetic character of arithmetical propositions "becomes even more evident if we consider larger numbers, for it is then obvious that, however we might turn and twist our concepts, we could never by mere analysis unaided by intuition be able to find the sum"(B16). The reference to larger numbers makes it clear that intuition is not being called in to secure the truth of basic propositions—such as 2 + 2 = 4, perhaps—by "setting them before our eyes." Rather, intuition underlies the *step by step* process of calculation which, in its entirety, may very well not be surveyable "at a glance."[48]

We have now reached the heart of the matter, I think, for it is the idea of a sharp distinction between calculation and discursive argument that is perhaps most basic to Kant's conception of the role of intuition in mathematics. Thus, at A734–736/B762–764 Kant contrasts mathematical and philosophical reasoning. Only mathematical proofs are properly called *demonstrations,* while philosophy is restricted to conceptual or discursive ("acroamatic") proofs. The latter "must always consider the uni-

---

explicit assertion—repeated in the letter to Schultz of November 25, 1788—that arithmetic has no axioms. For this reason he is justly criticized by Parsons, "Kant's Philosophy of Arithmetic," §III, who rightly points out that what is most striking here is the *difference* between Kant and Schultz.

47. Compare A162/B202–203, where a "synthesis of the manifold whereby the representation of a determinate space or time is generated, that is through the combination of the homogeneous [Zusammensetzung des Gleichartigen]" is said to underly all concepts of magnitude, and A143/B182: "Number is therefore simply the unity of the synthesis of the manifold of a homogeneous intuition in general, a unity due to my generating time itself in the apprehension of the intuition." See Parsons, "Philosophy of Arithmetic," §§VI, VII, for a rich and penetrating discussion of such passages.

48. See Young [124] for a very interesting and helpful discussion of the role of calculation in Kant's conception of arithmetic.

versal *in abstracto* (by means of concepts)," the former "can consider the universal *in concreto* (in the single intuition), and yet still through pure a priori representation whereby all errors are at once made visible [sichtbar]" (A735/B763).

That Kant has calculation centrally in mind here is indicated by his reference to the methods employed in solving algebraic equations (A734/B762), a reference which recalls the even more explicit conception of calculation found in the *Enquiry Concerning the Clarity of the Principles of Natural Theology and Ethics* (1764). See, for example, the First Reflection, §2, entitled "Mathematics in its methods of solution [Ausflösungen], proofs, and deductions [Folgerungen] examines the universal under symbols *in concreto;* philosophy examines the universal through symbols *in abstracto*":

> I appeal first of all to arithmetic, both the general arithmetic of indeterminate magnitudes [algebra], as well as that of numbers, where the relation of magnitude to unity is determinate. In both symbols are first of all supposed, instead of the things themselves, together with special notations [Bezeichnungen] for their increase and decrease, their ratios, etc. Afterwards, one proceeds with these signs, according to easy and secure rules, by means of substitution, combination or subtraction, and many kinds of transformations, so that the things symbolized are here completely ignored, until, at the end, the meaning of the symbolic deduction is finally deciphered [entziffert]. (2, 278.12–26)

As the Third Reflection, §1, explains, this "symbolic concreteness" of mathematical proof accounts for the difference between philosophical and mathematical certainty. Since philosophical argument is discursive or conceptual, ambiguities and equivocations in the meanings of general concepts are always possible. Mathematics, on the other hand, works with concrete or singular representations that allow us to be assured of the correctness of its substitutions and transformations "with the same confidence with which one is assured of what one sees before one's eyes" (291.29–30). As Kant puts it in the first *Critique,* the step by step application of the easy and secure rules of calculation "secures all inferences against error by setting each one before our eyes" (A734/B762).[49]

From the present point of view, the point could perhaps be recon-

49. These passages, and the closely related passage at A715–718/B743–746, are illuminatingly discussed by Parsons, "Philosophy of Arithmetic," and especially by Thompson [108], who distinguishes between "diagrammatic" and discursive proofs. Beth [6] was perhaps the first, in the context of contemporary discussion, to emphasize the importance of these passages for understanding Kant's philosophy of mathematics. The sharp distinction Kant draws between discursive ("acroamatic") proof and mathematical proof (demonstration) seems to me to establish *part* of the Russellian assumption beyond the shadow of a doubt: mathematical reasoning *cannot* be purely logical for Kant. What has still to be established is that the inferential use of pure intuition is *primary.*

structed as follows. Mathematical proof, unlike discursive proof, operates not only with *predicates* such as ⌜x is even⌝ and ⌜x is a triangle⌝, but first and foremost with *function-signs* such as ⌜$x + y$⌝ and ⌜the bisector of z⌝. In calculation we form functional terms by inserting particular arguments into the function-signs, we set up equalities (and inequalities) between such functional terms, and we substitute one functional term for another in accordance with these equalities. Since both the arguments and the values of our function-signs are individuals,[50] the procedure of *substitution* is to be sharply distinguished from the *subsumption* of individuals under general concepts characteristic of discursive reasoning. In particular, the essence of the former procedure lies in its iterability: $f(a)$ can be substituted in $f(x)$ to form a distinct functional term $f(f(a))$, while it of course makes no sense at all to subsume the predication $F(a)$ under the predicate $F(x)$.[51] Thus, the essentially "extra-logical" form of inference required is that which takes us from one object a satisfying a condition . . . a . . . to a second object $f(a)$ satisfying another condition ___$f(a)$___, and from there to a third object $f(f(a))$ satisfying ---$f(f(a))$---, and so on.

Now this conception of the role of calculation and substitution in mathematical proof also applies, *mutatis mutandis,* to the case of geometry. In Euclidean geometry we start with an initial set of basic constructive functions: the operation $f_L(x, y)$ taking two points x, y to the line segment between them, the operation $f_E(x, y)$ taking line segments x, y to the extended line segment of length $x + y$, and the operation $f_C(x, y)$ taking point x and line segment y to the circle with center x and radius equal to y. We also have a specifically geometrical equality relation (congruence) and, of course, definitions of the basic geometrical figures (circle, triangle, and so on). Euclidean proof then proceeds somewhat as follows. Given a figure a satisfying a condition . . . a . . . , we construct, by iteration of the basic operations, a new constructive function g yielding an expanded figure $g(a)$ satisfying a condition ---$g(a)$---. From this last proposition we are then able to derive a new condition ___a___ on our original figure a.

50. See also Hintikka, "Kant on the Mathematical Method," §6, for an illuminating discussion of the role of function-signs in Kant's conception of algebraic construction.

51. Compare the conception in the *Tractatus* [123] of mathematics as based on "calculation" and "operations" at 6.2–6.241, along with the distinction between "operations" and "[propositional] functions" at 5.251: "A [propositional] function cannot be its own argument, whereas an operation can take one of its own results as its base." Just as in the *Tractatus,* however, it is hard to see how such a "calculational" conception can yield more than primitive recursive arithmetic: see Thompson [108], n. 21 on p. 341. The essential difference between Kant and the *Tractatus* here is that Wittgenstein also applies the notion of "iterative operations" to *logic* in "the general form of a proposition" (6), whereas Kant uses this notion *to distinguish* logic and mathematics. This, of course, is because Wittgenstein is operating in the context of Frege's much stronger logic, where iterative construction of *propositions* via truth-functions and quantifiers plays a central role.

Whereas the inference from ---*g*(*a*)--- to . . . *a* . . . can be viewed as "essentially monadic," and is therefore analytic or logical for Kant, the inference from . . . *a* . . . to ---*g*(*a*)--- is not: it proceeds synthetically, by expanding the figure *a* as far as need be into the space around it, as it were. Since this procedure is grounded in the indefinite iterability of our basic constructive operations, geometry is synthetic for much the same reasons as is arithmetic; and, therefore, the case of arithmetic is primary.[52]

Confirmation is apparently provided by the discussion at B15–17. For Kant illustrates B14 at great length with the example of arithmetic and only then touches on geometry, almost as a corollary:

> Just as little is any fundamental proposition [Grundsatz] of geometry analytic. That the straight line between two points is the shortest is a synthetic proposition. For my concept of *straight* [*Geraden*] contains nothing of quantity [Größe], but only a quality. The concept of shortest is entirely an addition, and cannot be derived by any analysis of the concept of straight line. The aid of intuition must therefore be brought in, by means of which alone the synthesis is possible. (B16–17)[53]

As the discussion of arithmetic has shown, the general concept of magnitude [Größe] requires an intuitive synthesis (the successive addition of unit to unit). But geometry requires this concept as well (for example, in connecting the notion of straight line with the notion of shortest line). Therefore, geometry, just as much as arithmetic, is a synthetic discipline.[54]

Nevertheless, there is of course an important difference between the two cases. As already noted above, geometry has axioms whereas arithme-

52. Thus I cannot follow Parsons when he draws a sharp distinction between the cases of arithmetic and geometry, and even endorses an interpretation of the geometrical case of the Beck-Brittan variety (see "Philosophy of Arithmetic," §§II, IV, and p. 128 of [92]). On the contrary, I think Kant's views can only be understood if we apply the ideas Parsons has developed for the case of arithmetic to the case of geometry also.

53. See also the *Enquiry,* I, §2, where the discussion follows the same order, and the letter to Johann Schultz of November 25, 1788, where the priority of arithmetic is stated rather explicitly: "General arithmetic (algebra) is an *ampliative* [sich *erweiternde*] science to such an extent that one cannot name another rational science equal to it in this respect. Indeed, the other parts of pure mathematics [reine Mathesis] await their own growth [Wachstum] largely from the amplification [*Erweiterung*] of this general theory of magnitude" (10, 555.10–14).

54. It is interesting to speculate on what exactly Kant has in mind in his example of the geodesicity of straight lines. This proposition appears as neither an axiom nor a theorem in Euclid, but it was stated as an *assumption* by Archimedes (Heath [46], pp. 166–169). Kant does not appear to endorse this idea, and it is perhaps most plausible to suppose that he is referring to the variational methods developed by Euler in 1728 for *proving* geodesicity. That is, we consider the result of integrating arc-length over all possible (neighboring) curves joining two given points, and we look for the curve that minimizes the integral. Here, of course, the idea of "synthesis" (in the guise of integration) is especially prominent. But this is so far just speculation.

tic does not; moreover, geometry uses "ostensive construction (of the objects [Gegenstände] themselves)" (A717/B745) in addition to the "symbolic" or "characteristic" (A734/B762) construction common to algebra and arithmetic. Thus Kant's discussion of algebraic construction has a decidedly "formalistic" tone: we "abstract completely from the nature of the object" (A717/B745) and, as the above passage from the *Enquiry* puts it, "symbols are first of all supposed, instead of the things themselves" and "the things symbolized are here completely ignored." In geometry, on the other hand, such "formalism" is quite inappropriate: geometrical construction operates with "the objects themselves" (lines, circles, and so on).

This difference between arithmetical-algebraic construction and geometrical construction is perhaps most responsible for the confusion that has surrounded Kant's theory. For it begins to look as if geometrical intuition has not merely an inferential or calculational role, but also the more substantive role of providing a model, as it were, for one particular axiom system as opposed to others (Euclidean as opposed to non-Euclidean geometry). Intuition does this, presumably, by placing the objects themselves before our eyes, whereby their specific (Euclidean) structure can be somehow discerned.

From the present point of view, of course, there can be no question of picking out Euclidean geometry from a wider class of possible geometries. Rather, the difference between geometrical and arithmetical-algebraic construction is understood as follows. Geometry, unlike arithmetic and algebra, operates with an initial set of specifically geometrical functions (the operations f_L, f_E, and f_C) and a specifically geometrical equality relation (congruence). To do geometry, therefore, we require not only the general capacity to operate with functional terms via substitution and iteration (composition), we also need to be "given" certain initial operations: that is, intuition assures us of the existence and uniqueness of the values of these operations for any given arguments. Thus the axioms of Euclidean geometry tell us, for example, "that between two points there is only one straight line, that from a given point on a plane surface a circle can be described with a given straight line" (*Inaugural Dissertation,* §15.C: 2, 402.33–34), and they also link the specifically geometrical notion of equality (congruence) with the intuitive notion of superposition (*Prolegomena,* §12: 4, 284.22–26).[55]

55. Serious complications stand in the way of the full realization of this attractive picture. First, of course, Euclid's Postulate 5, the Parallel Postulate, does not have the same status as the other Postulates: it does not simply "present" us with an elementary constructive function which can then be iterated (thus, given two straight lines falling on a third with interior angles on one side together less than two right angles, Postulate 5 not only tells us that we can extend these lines *ad infinitum* on this side, but also says what will happen in

Now one might at first suppose that the case of arithmetic is precisely the same. After all, we need intuitive assurance that the successor function, say, is uniquely defined for all arguments. But the point, I think, is that the successor function is not a specific function at all for Kant; rather, it expresses the general form of succession or iteration common to all functional operations whatsoever. So it is not necessary *to postulate* any specific initial functions in arithmetic: whatever initial functions there may be, the existence and well-definedness of the successor function is guaranteed by the mere form of iteration in general (that is, time). Thus, in explaining why geometry has axioms while arithmetic does not at A164–165/B205–206, Kant refers to the need in geometry for general "functions of the productive imagination" such as our ability to construct a triangle from any three line segments such that two together exceed the third (this functional operation is of course definable, in Euclidean geometry, from the operations f_L, f_E, and f_C: Prop. I.22). The point, presumably, is that no such specific functional operations need be postulated in arithmetic.[56]

In any case, the idea that pure intuition plays the more substantive role of providing a model for one particular axiom system as opposed to others—as the anti-Russellian interpretation requires—is rather obviously untenable and definitely unKantian. The untenability of such a view has been clearly brought out in an instructive article by Kitcher [59].[57] Kitcher supposes that the primary role of pure intuition is to discern the metric and projective properties (the Euclidean structure) of space. We construct geometrical figures like triangles and somehow "see" that they are Euclidean: "[Kant's] picture presents the mind bringing forth its own creations

---

the limit: the two lines must meet, and not simply approach one another asymptotically). Second, as argued in §III, geometry for Kant includes the new calculus: the method of fluxions. And this calculus, unlike Euclidean geometry proper, has no basis at all in a finite set of initial constructive functions. So intuition has the even more substantive role of creating each new object "on the spot," as it were. These two complications reflect deep mathematical problems that are only fully solved in the next century through the discovery of non-Euclidean geometries and the independence of the Parallel Postulate, on the one hand, and through the "rigorization" and eventual "arithmetization" of analysis, on the other. In the end, therefore, the relation between arithmetic and geometry remains a source of fundamental, and unresolved, tensions in Kant's philosophy. Yet it is surely remarkable that these tensions arise precisely in connection with some of the deepest mathematical questions of the time.

56. These ideas can once again profitably be compared with *Tractatus* 6.01–6.031: "Number is the exponent of an operation." For a fuller discussion of arithmetical-algebraic construction see Chapter 2 below.

57. Kitcher himself remains officially neutral on the issue between Beck and Russell. He suggests ([59], §IV) that pure intuition may play a role in proofs, and even makes some interesting remarks about the use of pure intuition in (Kant's conception of) Newton's fluxional reasoning ([59], p. 41). Nevertheless, Kitcher's setting of the problem only makes sense in the context of an interpretation of the Beck-Brittan variety.

and the naive eye of the mind scanning these creations and detecting their properties with absolute accuracy" ([59], p. 50). It is then easy to show that pure intuition, conceived on this quasi-perceptual model, could not possibly perform such a role. Our capacity for visualizing figures has neither the generality nor the precision to make the required distinctions.

Thus, for example, Kant's appeal to the proposition that two sides of a triangle together exceed the third at A25/B39 is considered, and "[w]e now imagine ourselves coming to know [it] in the way Kant suggests. We draw a scalene triangle and see that this triangle has the side-sum property" ([59], p. 44). But this idea quickly founders on Berkeley's generality problem: how are we supposed to conclude that all triangles have the side-sum property and not, say, that all triangles are scalene? (Actually, in this connection a more relevant dimension of generality is *size*. In elliptic—positive curvature—space, triangles that are "small"—relative to the dimensions of the space itself—have the side-sum property while arbitrarily large triangles do not. So one cannot argue from the properties of small, visualizable triangles to the properties of arbitrary triangles.)[58]

It is extremely unlikely, however, that in appealing to intuition at A25/B39 Kant is imagining any such process of "visual inspection." It is much more plausible that, in precise parallel to his discussion of the angle-sum property at A715–717/B743–745, he is referring to the Euclidean *proof* of this proposition (Prop. I.20). We consider a triangle ABC and prolong BA to point D such that DA is equal to CA (see Figure 6). We then draw DC, and it follows that $\measuredangle ADC = \measuredangle ACD$ and therefore $\measuredangle BCD > \measuredangle ADC$. Since the greater angle is subtended by the greater side (Prop. I.19), $DB > BC$. But $DB = BA + AC$; therefore $BA + AC > BC$. Q.E.D. Intuition is required, then, not to enable us to "read off" the side-sum property from the particular figure ABC, but to guarantee that we can in fact prolong BA to D by Postulate 2.[59] (It is precisely this Postulate that fails

58. Closely related considerations are presented by Hopkins in his penetrating criticism of Strawson's attempted reconstruction [54]. The basic point is that "in the small," wherein alone "visual inspection" is possible, Euclidean and non-Euclidean geometries are quite indistinguishable.

59. In Kant's technical terminology, Berkeley's generality problem is solved via the distinction between *images* and *schemata*—only the latter figure essentially in geometrical proof: "No image [Bild] could ever be adequate to the general concept of triangle. For it would never attain the generality of the concept which makes it valid for all triangles—whether right-angled, obtuse-angled, etc.—but is always limited to a part of this sphere. The schema of a triangle can exist nowhere but in thought and signifies [bedeutet] a rule of synthesis of the imagination in respect to pure figures [Gestalten] in space (A141/B180). (Ralf Meerbote emphasized the importance of this passage to me.) As A164–165/B205 makes clear, this "rule of synthesis" is nothing but the Euclidean construction of a triangle from any three line segments such that two together exceed the third of Proposition I.22.

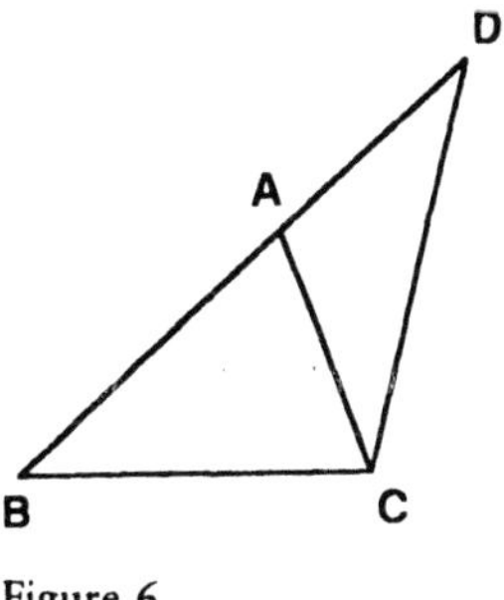

Figure 6

in elliptic space: straight lines are finite—but unbounded—and cannot be indefinitely prolonged at will.)

Now, as Heath observes in his commentary on Proposition I.20, "It was the habit of the Epicureans . . . to ridicule this theorem as being evident even to an ass and requiring no proof" ([46], p. 287), and one might be tempted to suppose that Kant holds a similar view (so that the "visual inspection" metaphor is appropriate after all). This suggestion is immediately quashed by the important discussion of mathematical method at Bxi–xii, however. Kant applauds Diogenes Laertius for naming "the reputed author of even the least important elements of geometrical demonstrations, even of those which, according to ordinary judgement, require absolutely no proof," and concludes:

> A new light dawned on the first man (whether he may be Thales or whoever) who demonstrated the *isosceles triangle;* for he found that he must not inspect what he saw in the figure [Figur], or even in the mere concept of it, and as it were learn its properties therefrom, but he must rather bring forth what he himself has injected in thought [hineindachte] and presented (through construction) according to concepts, and that, in order to know something a priori, he must attribute nothing to the thing except that which follows necessarily from what he himself has placed in it in accordance with his concept.

Kant's example here, of course, is the discovery (sometimes attributed to Thales) of the Euclidean proof that the angles at the base of an isosceles triangle are equal (Prop. I.5),[60] a proof which also proceeds by means of

We might represent it, then, by a constructive function $f_T(x,y,z)$ which, as Proposition I.22 shows, is definable from the basic constructive functions f_L, f_E, and f_C. Kant's point is simply that to do geometry we need such (general!) constructive functions (to represent our "existence assumptions"). So we do not establish geometrical propositions by "inspection" of the resulting images, but by rigorous proof from axioms and definitions.

60. The reference to Proposition I.5 is made explicit in a letter to Christian Schütz of June

an ingenious expansion of our original triangle into several additional triangles via "auxiliary construction." I do not see how there can be any doubt, therefore, that Kant's "new method" of geometry is precisely Euclid's procedure of construction with straight-edge and compass.[61]

Once again, we are forced to conclude that the primary role of pure intuition is to underwrite the constructive procedures used in mathematical proofs. Moreover, when Kant himself uses "visual inspection" and "eye of the mind" metaphors, it is almost always in connection with inference and proof. Thus, in the passage from the *Enquiry*, Third Reflection, §1, quoted above, Kant says that we can check the correctness of algebraic substitutions and transformations "with the same confidence with which one is assured of what one sees before one's eyes." At A734/B762 we are told that the procedure of algebra "secures all inferences against error by setting each one before our eyes." The intuition involved here is not a quasi-perceptual faculty by which we "read off" the properties of triangles from particular figures, but that involved in checking proofs step by step to see that each rule has been correctly applied: in short, the intuition involved in "operating a calculus." The only apparent exception of which I am aware is *Inaugural Dissertation,* §15.C, where Kant says that some Euclidean axioms (Postulates 1 and 3) are "seen, as it were, in space *in concreto.*" Yet the fact that Kant does not use such language in the first *Critique* suggests that he himself became sensitive to its possible misuse.

Indeed, there is no room in the critical philosophy for the picture underlying the anti-Russellian conception of pure intuition. That conception views Euclidean geometry as a body of truths that selects one structure for space from the much wider class of all possible such structures. Since both Euclidean and non-Euclidean axiom systems are consistent, we need

---

25, 1787, where Kant changes "gleichseitiger" in the printed text to "gleichschenkligter": 10, 489.30–32.

61. Hintikka has emphasized the importance of the passage at Bxi–xii (for example, in "Kant's 'New Method of Thought' and His Theory of Mathematics," §2) and the fact that Euclid's proof-procedure provides a model for Kant's notion of construction (especially in "Kant and the Mathematical Method"). Hintikka also rightly emphasizes that Kant's conception of mathematical method is therefore to be sharply *distinguished* from a naive "visual inspection" view. Yet in his zeal to refute Russell's contention that Kant's view of geometry requires an "extra-logical element," Hintikka overlooks the fact that his own reconstruction of the analytic/synthetic distinction (see note 14 above) allows us to do justice to both Russell and Kant: Euclidean constructive proofs do indeed require an "extra-logical element"—if logic, as Kant thought, is syllogistic or (essentially) monadic logic (and this, of course, is precisely Russell's point). See [52], chap. IX, §§4–7, especially p. 218, n. 45, where Hintikka is driven to equate Kant's conception of geometrical reasoning with that of Leibniz and Wolff.

to call on pure intuition to provide a model, as it were, for one system rather than another. As it happens, intuition picks out the Euclidean system.[62] The problem is that Kant has no notion of possibility on which both Euclidean and non-Euclidean geometries are possible. His official notion is "that which agrees with the formal conditions of experience (according to intuition and concepts)" (A218/B265). Mere absence of contradiction is quite insufficient to establish a possibility (A220–221/B267–268); and "To determine [a geometrical figure's] possibility, something more is required, namely, that such a figure be thought under pure [lauter] conditions on which all objects of experience rest" (A224/B271). Accordingly, Kant complains that "the poverty of our customary arguments by which we throw open a great realm of possibility, of which all that is actual (the objects of experience) is only a small part, is patently obvious" and concludes "this alleged process of adding to the possible I refuse to allow. For that which has still to be added would be impossible" (A231/B284).

What produces confusion here is the circumstance that Kant is operating with two notions of possibility: "logical possibility," given by the conditions of thought alone; and "real possibility," given by the conditions of thought plus intuition (compare, for example, Bxxvi,n). One then supposes that the former is a wider genus (containing both Euclidean and non-Euclidean spaces) of which the latter is a species (containing only Euclidean space). But this line of thought employs a notion of *logical possibility* that is completely foreign to Kant. Kant's conception of logic is not that of modern quantification theory, and he can have no notion like ours of all logically possible structures—all models of consistent first-order (or second-order) theories, say. Thus, for example, while there may be no (monadic!) contradiction in the concept of a non-Euclidean figure such as "the concept of a figure which is enclosed within two straight lines" (A220/B268), this does not mean that there is a possible non-Euclidean structure containing such a figure. For a non-Euclidean structure would have to possess the topological properties (denseness and continuity) common to Euclidean and non-Euclidean spaces, and this, for Kant, is impossible. There is only one way even to think such properties: in the space and time of *our* (Euclidean) intuition. Considered independently of *our* sensible intuition, then, the concept of a non-Euclidean

62. This picture is explicit in Kitcher [59], §§I, II, and Brittan [10], chaps. 1–3—both of which make heavy use of contemporary "possible worlds" jargon. The same idea is found in Parsons's more circumspect discussion in "Philosophy of Arithmetic," pp. 117, 128. As Brittan points out ([10], p. 70, n. 4), this picture appears to correspond to Frege's conception of Euclidean geometry: for example, in [33], pp. 20–21. Yet Frege's conception is surely not Kant's, for it is only possible in the context of Frege's much stronger logic.

figure remains "empty" and lacks both "sense and meaning [Sinn und Bedeutung]" (B149).[63]

A closely related point is that pure mathematics is not a body of truths with its own peculiar subject matter for Kant.[64] There are no "mathematical objects" to constitute this subject matter, for the sensible and perceptible objects of the empirical world (that is, "appearances") are the only "objects" there are. For this reason, pure mathematics is not, properly speaking, a body of knowledge (cognition):

> Through the determination of pure intuition we can acquire a priori cognition of objects (in mathematics), but only with respect to their form, as appearances; whether there can be things that must be intuited in this form, is still left undecided. Therefore, mathematical concepts are not in themselves cognitions, except in so far as one presupposes that there are things that can be presented to us only in accordance with the form of this pure sensible intuition. (B147)

Hence, only *applied* mathematics has a subject matter (the sensible and empirical world), and only *applied* mathematics yields a body of substantive truths.[65] Pure mathematics is a mere form of representation (on the present interpretation, a form of reasoning), whose applicability to the chaotic sensible world must be proved by transcendental deduction. In this sense, pure intuition cannot be said to provide a model for Euclidean geometry at all; rather, it provides the one possibility for a rigorous and rational *idea* of space. That there is a model or realization of this idea is not established by pure intuition, but by Kant's own transcendental philosophy.

In the end, therefore, Euclidean geometry, on Kant's conception, is not to be compared with Hilbert's axiomatization, say, but rather with Frege's *Begriffsschrift*.[66] It is not a substantive doctrine, but a form of rational

63. Kant's conception of possibility can then perhaps be explained as follows. Whereas Kant does distinguish between the conditions of *thought* alone and the conditions of *cognition* (thought plus intuition), the former do not correspond to our notion of logical possibility but, rather, to the "empty" idea of the "thing-in-itself." Thus what best approximates our notion of logical possibility is given by the conditions of thought plus *pure* intuition: namely, pure mathematics. "Real possibility," then, is given by the conditions of thought plus *empirical* intuition: namely, (the pure part of) mathematical physics. So "real possibility" most closely corresponds to our notion of "physical possibility."

64. Here I follow Thompson [108], pp. 338–339. See also Parsons [92], pp. 147–149 (Postscript to "Kant's Philosophy of Arithmetic").

65. Of course Kant is not using our *modern* distinction between pure and applied mathematics here: see note 32 above.

66. What I say here actually corresponds more closely to Wittgenstein's conception (in the *Tractatus*) of Frege's *Begriffsschrift* than to Frege's own. Frege's own conception is far less "formalistic." In particular, the laws of logic are in a sense scientific laws like those of any other discipline: it is just that they are maximally general laws (containing variables of

representation: a form of rational argument and inference. Accordingly, its propositions are established, not by quasi-perceptual acquaintance with some particular subject matter, but, as far as possible, by the most rigorous methods of proof—by the proof-procedures of Euclid, Book I, for example. There remains a serious question about Euclid's axioms, of course; when pressed, Kant would most likely claim that they represent the most general conditions under which alone a concept of extended magnitude—and therefore a rigorous conception of an external world—is possible (see A163/B204). And, of course, we now know that Kant is fundamentally mistaken here. In 1854 Riemann developed the general concept of *n-fold extended manifold*—containing three-dimensional Euclidean space as one very special case alongside of more additional possibilities than Kant (or anyone else in the eighteenth century) ever imagined. In 1879 Frege developed a logical framework which makes possible the even more general concept of *relational structure*—under which are subsumed all models for Hilbert's geometry and even, as we now say, all "logically possible worlds." Yet Kant is surely not to be reproached for failing to anticipate the leading logical and mathematical discoveries of a later age; he is rather to be applauded for the depth and tenacity of his insight into the logical and mathematical practice of his own.

---

different levels, but no non-logical constants). See the remarkable series of papers by Ricketts, [99], [100], and [101], which depict both Frege's conception of logic and the internal pressures pushing that conception towards the *Tractatus*.

· CHAPTER 2 ·

Concepts and Intuitions in the Mathematical Sciences

The distinction between concepts and intuitions is basic to the Kantian philosophy. Our representations are of two essentially different kinds for Kant, and, accordingly, we have two essentially different cognitive faculties: understanding and sensibility. Yet both types of representations, concepts and intuitions, must be united in any instance of knowledge: only through the union of the two faculties, sensibility and understanding, can human cognition arise (A51–52/B75–76). Moreover, not only is such a union necessary for knowledge, it is also necessary for our representations to have objective meaning or content:

> We demand in every concept, first, the logical form of a concept (of thought) in general, and second, the possibility of giving it an object to which it may be related [darauf er sich beziehe]. In the absence of such object, it has no sense [Sinn] and is completely empty in respect to content, though it may still contain the logical function which is required for making a concept out of any possible data. Now the object cannot be given to a concept otherwise than in intuition. (A239/B298).

Thus it is intuition, and intuition alone, which provides the field of possible objects for the application of our concepts (see A19/B33); and, without such a field of possible objects to which they can be applied, our concepts are entirely empty of objective meaning.

This conception of a twofold division of representations and cognitive faculties is embodied in the closely related distinction Kant draws between "general" and "transcendental" logic. The former kind of logic "abstracts from all content of knowledge, that is, from all relation of knowledge to the object" (A55/B79); accordingly, it concerns only concepts and understanding alone, considered wholly independently of intuitions and sensibility (see A53/B77). In transcendental logic, on the other hand, "we do not abstract from the entire content of knowledge" (A55/B80); indeed:

> The employment of this pure knowledge [provided by transcendental logic] depends on the following, as its condition: that objects to which it can be applied may be given to us in intuition. For in the absence of intuition all our knowledge is without objects, and therefore remains entirely empty. (B87)

Transcendental logic, then, is concerned with precisely the relation between concepts and intuitions, between understanding and sensibility—although only with the pure or non-empirical (see A55/B79–80) aspects of this relation: "it concerns itself with the laws of understanding and of reason solely in so far as they relate a priori to objects" (A57/B82).

Kant locates his break with the dogmatic metaphysics of the Leibnizean-Wolffian philosophy at just this point. Dogmatism is "the presumption that it is possible to make progress with only a pure knowledge from concepts (philosophical knowledge), according to principles" (Bxxxv). The chief error of dogmatism is its attempt to proceed with general logic alone, independently of transcendental logic:

> There is, however, something so tempting in the possession of an art so specious, through which we give to all our knowledge, however poor and empty we yet may be with regard to its content, the form of understanding, that general logic, which is merely a *canon* of judgement, has been employed as if it were an *organon* for the actual production of at least the semblance of objective assertions, and has thus in fact been misapplied. General logic, when thus treated as an organon, is called *dialectic*. (A60–61/B85)[1]

When we move from general logic to transcendental logic, on the other hand, and thus add the idea of relation to an object to the mere form of thought, we obtain transcendental analytic: the logic of truth (A62/B87).

Now the idea of a transcendental logic is itself closely connected with Kant's notion of pure intuition; for it is the latter notion that first opens up the possibility of a transcendental logic, that is, the possibility of a pure or non-empirical study of the relation between thought and objects: "since, as the Transcendental Aesthetic has shown, there are pure as well as empirical intuitions, a distinction might likewise be drawn between pure and empirical thought of objects" (A55/B79).[2] And, since there is of

1. Comparing this passage with Bxxxvi, where Wolff is named as "the greatest of all dogmatic philosophers," it appears likely that those who are said to misapply general logic are precisely the Leibnizean-Wolffians.

2. Compare also A76–77/B102: "General logic abstracts, as has been repeatedly said already, from all content of knowledge, and expects that representations are to be given to it from somewhere else—wherever it may be—so that it can first transform them into concepts, which proceeds analytically. Transcendental logic, on the other hand, has a manifold of a priori sensibility lying before it, offered to it by Transcendental Aesthetic in order to give material [eine Stoff] to the pure concepts of the understanding, without which it would be without all content and therefore completely empty."

course an intimate connection between Kant's notion of pure intuition and his conception of the mathematical sciences, this latter conception is centrally involved in motivating the distinction between general and transcendental logic. This is not surprising, for it is Kant's conception of the mathematical sciences that provides him with his starting point, both logically and historically, in his polemic against the dogmatic metaphysics of the Leibnizean-Wolffian philosophy. The logical priority may be seen in §V of the Introduction to the first *Critique,* where the mathematical sciences are appealed to in order to show the existence of *synthetic* a priori knowledge, in direct opposition to the conjectures of those who "supposed that the fundamental propositions of the science [of mathematics] can themselves be known to be true through that principle [of contradiction]" (B14). The historical priority may be seen in §1 of the First Reflection of the *Enquiry Concerning the Clarity of the Principles of Natural Theology and Ethics* (1764), where Wolff is criticized for attempting to use an analytic method in mathematics.[3]

The present chapter attempts to clarify the connections among Kant's conception of the mathematical sciences, his doctrine of pure intuition, and his general distinction between concepts and intuitions (between general and transcendental logic). Specifically, I am interested in how exactly the mathematical sciences relate to the pure intuitions of space and time, in the role of pure intuition in providing objects for the mathematical sciences, and in the relationship between concepts and intuitions in these sciences. Lying in the background is the question of the precise force of Kant's appeal to the mathematical sciences in his polemic against the Leibnizean-Wolffian philosophy.

· I ·

The above questions can be illuminated by considering a certain line of thought which aims sympathetically to explicate Kant's doctrine of the synthetic character of all mathematical truth—a line of thought which is, I believe, implicit or explicit in much of the best recent work on the subject.[4] The basic idea is to view Kant's philosophy of mathematics as

3. See 2, 277.21–35. Of course neither the doctrine of pure intuition nor the explicit division of the faculties into sensibility and understanding occurs in the *Enquiry.* These are first officially introduced in the *Inaugural Dissertation* (1770); see also §7 of the latter, where the unfortunate Wolff is once again rebuked for misunderstanding the nature of geometry: 2, 395.9–14.

4. The writers I have in mind here, although I do not intend directly to impute all aspects of the view to be described below to any of them, are as follows: Beck, "Can Kant's Synthetic Judgements Be Made Analytic?" reprinted in [4]; Brittan [10]; Buchdahl [12]; Kitcher [59]; Martin [73] and [74]; and Parsons, "Kant's Philosophy of Arithmetic" (1969),

inspired by what, from a modern point of view, we might call "anti-formalism." Kant, like Frege, is deeply dissatisfied with a picture of mathematics as dealing with mere uninterpreted formalism, and, accordingly, he rejects mere logical consistency as a sufficient condition for mathematical acceptability. Like Frege, he is interested above all in the objective meaning of mathematical signs and the objective content or truth of mathematical assertions. Frege, for example, does not consider the Peano axioms themselves to be an adequate elucidation of the meanings of the basic concepts of arithmetic, nor does such an axiomatization, by itself, have anything to do with the truth of arithmetic. Most fundamentally, such an axiomatization alone is certainly not sufficient, for Frege, to establish the analyticity of arithmetic. On the contrary, it is only Frege's definitions of the concepts of arithmetic within the logical framework of the *Begriffsschrift* that first provides them with objects or objective meaning; it is only his derivation of the Peano axioms within the same framework that establishes their analyticity.

Kant, on the other hand, is of course quite unacquainted with Frege's logicist program. Like Frege, however, he nonetheless requires that mathematics be provided with objective meaning and content. According to the general conception of objective meaning sketched above, then, Kant naturally looks to intuition to provide such meaning, and, since the truth of mathematics is a priori, not empirical, he naturally looks to pure intuition.[5] According to Kant's general conception of objective meaning, in other words, it is pure intuition alone that can provide the concepts of mathematics with *objective reality*. Similarly, since the objective truth of mathematics is not a matter of mere logical consistency, it can only be classified as synthetic truth—and, indeed, as synthetic a priori truth.

By the same token, since mathematical truth is not reducible to mere logical consistency, it is necessary to distinguish between *logical possibility* and *real possibility*. The former involves the constraints provided by mere thought alone, the constraints provided by mere general logic: that is, agreement with the principle of contradiction. The latter adds to these constraints also the conditions of a possible intuition.[6] In particular, only

---

reprinted (with a Postscript) in [92] (Parsons endorses aspects of the view to be described here for the case of geometry, but draws a sharp distinction between this case and the case of arithmetic; my own view owes much to Parsons's treatment of the latter).

5. Thus Frege himself, apparently rejecting a logicist account of geometry parallel to his own account of arithmetic, also turns to spatial intuition to provide geometry with objective meaning and truth: see, e.g., [33], §14. This circumstance is not lost on those who follow the present line of thought: see, e.g., Brittan [10], p. 70, n. 4.

6. See, e.g., Bxxvi,n; see also the first postulate of modality at A218/B265–266: "that which agrees with the formal conditions of experience (according to intuition and concepts) is *possible*."

the conditions of pure intuition can provide criteria for real possibility and objective reality in the case of mathematics:

> It is, indeed, a necessary logical condition that a concept of the possible must not contain any contradiction; but this is by no means sufficient to determine the objective reality of the concept, that is, the possibility of such an object as is thought through the concept. Thus there is no contradiction in the concept of a figure that is enclosed within two straight lines, since the concepts of two straight lines and of their coming together contain no negation of figure. The impossibility arises not from the concept itself, but in connection with its construction in space, that is, from the conditions of space and of its determination. (A220–221/B268)

Hence it appears that what is decisive for the synthetic character of mathematics is that there are logical possibilities, such as the two-sided plane figure, that are nonetheless mathematically impossible: their impossibility consists precisely in their failure to conform to the conditions of pure intuition.

The example of the two-sided plane figure naturally leads to a further, and very striking, idea: the existence of consistent systems of non-Euclidean geometry not only fails decisively to refute Kant's conception of mathematics, it actually conforms to that conception perfectly and in fact provides rigorous confirmation for Kant's view. For the existence of such systems shows precisely that Euclidean geometry is not logically necessary and that there are indeed logically possible spaces that do not satisfy Euclid's axioms. Yet Kant, as we know, is not interested in mere formal systems; he is interested in objective meaning and objective truth. A synthetic source must therefore be found that provides objective reality for the concepts of geometry and which, as it were, constitutes a model for one logically possible system of geometry in preference to the others. Such a model is provided by the space of pure intuition, which, for Kant, is necessarily Euclidean.[7]

This line of thought thus represents a very attractive attempt at making sense of Kant's conception of mathematics. It is particularly sensitive to the important developments in the foundations of mathematics that have taken place since Kant's day, and it harmonizes with Kant's general conception of objective meaning as involving a necessary interplay between concepts and intuitions, understanding and sensibility. Nevertheless, it cannot be correct, I think, and consideration of precisely where it breaks

7. Aspects of this last idea are found in Brittan [10], chap. 3; Kitcher [59], §II; Parsons, "Philosophy of Arithmetic," §II. Martin [73], §3, goes so far as to see an anticipation of non-Euclidean geometry in Kant's own example of the two-sided plane figure—since such figures do of course exist in elliptic (positive curvature) space (more precisely, in *doubly elliptic* or spherical space).

down can lead us to a better appreciation of the intricacies of Kant's position.

The main problem is that Kant himself does not employ pure intuition to provide objects and objective reality for the concepts of mathematics—or indeed for any concepts. On the contrary, objects for any concept whatsoever can only be found in *empirical* intuition. Thus the passage from A239/B298 quoted above continues as follows:

> Now the object cannot be given to a concept otherwise than in intuition, and, if a pure intuition is still a priori possible prior to the object, even this intuition itself can acquire its object, and therefore objective validity, only through the empirical intuition of which it is the mere form. Therefore all concepts, and with them all principles, even such as are possible a priori, relate to empirical intuitions, that is, to data for a possible experience. Apart from this relation they have no objective validity, and are a mere play of representations—of either the imagination or the understanding respectively. Take for instance the concepts of mathematics, considering them first of all in their pure intuition. Space has three dimensions; between two points there can be only one straight line; etc. Although all these principles, and the representation of the object with which this science occupies itself, are generated in the mind completely a priori, they would still mean [bedeuten] absolutely nothing, were we not yet able to exhibit their meaning [Bedeutung] in appearances (empirical objects). (A239–A240/B298–299)

Therefore it is not pure intuition, but only empirical intuition that is capable of providing a model for the truths of mathematics.

It follows that pure intuition, at least by itself, can in no way demonstrate or exhibit the real possibility of mathematics. For Kant, such real possibility means that mathematics is applicable to empirical objects (objects of experience), and this cannot be shown except on the basis of transcendental philosophy itself: officially, at A165–166/B206–207. Indeed, this point plays a crucial role in the transcendental deduction (§22), and essentially the same point is made in the discussion of the first postulate of modality, several pages after the two-sided plane figure example:

> It does, indeed, seem as if the possibility of a triangle could be known from its concept in itself (since it is certainly independent of experience); for we can in fact give it an object completely a priori, that is, can construct it. But since this is only the form of an object, it would remain a mere product of the imagination, and the possibility of its object would still be doubtful. To determine its possibility, something more is required, namely, that such a figure be thought under pure conditions on which all objects of experience rest. That space is a formal a priori condition of outer experiences, that the image-forming [bildende] synthesis through which we construct a triangle in imagination is precisely the same as that which we exercise in the apprehension of an appearance, in making for ourselves an empirical concept of it—

> these are the considerations that alone enable us to connect the representation of the possibility of such a thing with the concept of it. (A223–224/B271)

The real possibility of a triangle is not shown merely by our ability to construct that concept in pure intuition, but only by Kant's own transcendental proof that the objects of empirical intuition are necessarily subject to the conditions of pure intuition. In other words, real possibility here depends entirely on *applied* mathematics. Yet Kant clearly holds that pure intuition is a necessary condition for *pure* mathematics, quite independently of all questions concerning the application of the latter to objects of experience.[8] Considerations of objective reality and real possibility can therefore not themselves explain Kant's doctrine of pure intuition.[9]

These points show, I think, that the line of thought we have been considering is inadequate as an interpretation of Kant's texts. Even if this were not so, however, it would still be unsatisfactory as a reconstruction of Kant's position. We would, to be sure, have an explanation of the *synthetic* character of mathematical truth, but we would then be also quite unable to explain its *a priori* character. Indeed, on the conception of real possibility under consideration, it is very hard to see how we could know mathematical truths at all. The underlying idea is that the real possibilities are a subclass of the much wider class of logical possibilities: there are logically possible states of affairs that are nonetheless not really possible. Pure intuition then selects a narrower range of possibilities from the logical possibilities—it declares that certain logically possible states of affairs are not really possible—and thus generates the synthetic truths of mathematics. But how, on the present conception, is pure intuition supposed to perform this selection; and how are we supposed to have cognitive access to it?

Consider the case of geometry, for example. It is supposed to be logically possible that space have either a Euclidean or a non-Euclidean structure. Only the Euclidean structure is really possible, however, and pure intuition is supposed to tell us this. Yet it is extremely difficult to conceive, I think, how pure intuition could perform such a discrimination. Thus, it

8. See A25/B39 or A716–717/B744–745, for example,where pure intuition is invoked to explain the mere possibility of geometrical demonstrations. Again, it is only at A165/B206 that we learn that "pure mathematics, in its complete precision, [is] applicable to the objects of experience" and "what geometry asserts of pure intuition is therefore undeniably valid of empirical intuition." Pure geometry and pure intuition have presumably been already sufficiently expounded: namely, at B40–41.

9. The point that it is empirical intuition alone which is capable of providing objects for mathematics and, consequently, that real possibility is established only in transcendental philosophy—not by mathematical construction itself—is due to Thompson [108]: see pp. 338–339 in particular. It is instructively discussed in the Postscript to Parsons, "Philosophy of Arithmetic." Brittan acknowledges the point at pp. 64–65 of [10].

is supposed to be logically possible that any particular triangle I construct or encounter should be either Euclidean or non-Euclidean: the sum of its angles could (logically) be equal to 180° or could (logically) be equal to 180° ± .000001°, say. How is pure intuition possibly able to distinguish these two cases? Of course, within Euclid's axiomatization of geometry, we can *prove* that the angles of any triangle we construct must sum to exactly 180° (*Elements*, Book I, Prop. 32).[10] On the present conception, however, other axiomatizations, essentially different from Euclid's, are supposed to represent logical possibilities as well. The problem is then to see how pure intuition can select one such possibility from the class of all logically possible axiomatizations, and this, I think, is quite impossible.[11] Moreover, the point is actually independent of the question of Euclidean as opposed to non-Euclidean geometry, for it arises also for neutral properties of space such as infinite divisibility. Again, it is supposed to be logically possible that any given line segment be divisible to infinity or only divisible to a length of .000001 cm, say. How is our spatial intuition possibly able to discriminate between these two cases?[12]

The difficulty is a general one. Since logic or the understanding alone is supposed to present us with a wider range of states of affairs than does mathematics, we are able to think or represent possibilities that conflict with those admitted by mathematics. We then invoke a second faculty, pure intuition or pure imagination, to rule out these deviant states of affairs. Although we can *think* such deviant states of affairs, we cannot intuit or imagine them. But how are we supposed to know, and know a priori, that our intuition or imagination has exactly the required structure? The problem is that if we can think deviant possibilities, we can also think a deviant imagination within which such possibilities can be intuited or constructed, and it then becomes absolutely unintelligible how we are supposed to know that *our* imagination, which is now conceived as just one among many alternatives, is not itself a deviant one.[13] The assumed

10. Kant discusses this proof at A716–717/B743–745.

11. Essentially this problem is raised by Kitcher [59], §V, as a refutation of Kant's position. Hopkins [54] puts it forward (more properly, I think) as a criticism of reconstructions that invoke "visual" or "phenomenal" geometry to explicate Kant's preference for Euclid.

12. This question is raised by Parsons in "Infinity and Kant's Conception of the 'Possibility of Experience' " (1964), reprinted in [92]; it is also discussed in Kitcher [59], §5.

13. Of course Kant in some sense leaves room for the possibility of different forms of intuition, and therefore different forms of imagination, from our own (see, for example, A27/B43, B72, A286/B342–343). At the same time, however, he insists that we can in no way conceive what the structure of such deviant imaginations may be. See especially B139: "we cannot form the least concept of any other possible understanding, either of such as is itself intuitive or of any that may possess an underlying mode of sensible intuition which is different in kind from that in space and time" (compare also A230/B283). By contrast, on the conception of pure intuition I wish to reject, we can form a definite concept of, for

a priori character of mathematics thereby itself becomes absolutely unintelligible.[14]

Finally, recall that one of the most interesting aspects of Kant's conception of the synthetic character of mathematics, in the present connection, is its role in motivating his rejection of the dogmatic metaphysics of the Leibnizean-Wolffian philosophy. In particular, the synthetic character of mathematics apparently provides Kant with a starting point—with a basis, as it were—for moving from general to transcendental logic. On the line of thought under consideration, however, it is hard to see how this can work; for we explain the synthetic character of mathematics simply as an application of Kant's general conception of objective meaning or objective reality. This conception, in turn, simply assumes the distinction between general and transcendental logic. Certainly, if one is already convinced of the inadequacy of general logic for supplying us with a useful notion of objective meaning, one might very well look to intuition and thus arrive at the synthetic character of mathematics (and at transcendental logic). But this is precisely what the Leibnizean-Wolffian philosophy is not convinced of, and so the problem is to see how the synthetic character of mathematics can move us beyond this philosophy without simply begging the question.

· II ·

Thus far our considerations have been largely negative: we have argued that a certain initially attractive conception of the relationship between pure intuition and the mathematical sciences is not in fact tenable. We have also confined our attention almost exclusively to the science of geometry. By now considering what Kant has to say about the science of arithmetic, we can, I hope, both deepen our understanding of the inadequacies of the line of thought examined above and also point the way to a better interpretation.

Geometry, for Kant, is the science of space, and it is precisely this that lends plausibility to the above line of thought: the role of our pure intu-

---

example, a non-Euclidean space, and intuition has merely the role of showing that this definite *concept* lacks an *object.* On this conception, therefore, we can conceive determinate (e.g., non-Euclidean) deviant possibilities and thereby conceive a determinate (e.g., non-Euclidean) deviant imagination. Kant's view, as I understand it, is that *this* is certainly not possible. Rather, we can provide room for different forms of sensibility only from the point of view of transcendental philosophy, where we discover that the faculty of sensibility is both distinct from and less general than the faculty of understanding.

14. It is thus noteworthy that Brittan, for example, explicitly distances his interpretation of Kant's conception from what he calls the "epistemological" use of aprioricity: [10], p. 24, n. 43.

ition of space is to provide an object—a model, as it were—for the science of geometry. Now arithmetic, for Kant, is of course also a synthetic a priori science. So, following the above line of thought, one naturally looks to pure intuition to provide an object—a model—for the science of arithmetic as well. And where is such a model to be found? The obvious suggestion is time: arithmetic is the science of our pure intuition of time in precisely the same way as geometry is the science of our pure intuition of space. This suggestion, however, faces immediate, and insurmountable, difficulties.

As all careful writers on the subject have observed, Kant does not in fact say that arithmetic stands to time as geometry does to space.[15] In the Transcendental Aesthetic, §5 (The Transcendental Exposition of the Concept of Time) corresponds to §3 (The Transcendental Exposition of the Concept of Space), where the synthetic a priori knowledge of geometry is explained in terms of the pure intuition of space. In §5, however, arithmetic is not mentioned; instead, the synthetic a priori science whose possibility is explained by the pure intuition of time is identified as "the general doctrine of motion" (B49). The same idea is found in the *Inaugural Dissertation,* §12: "PURE MATHEMATICS considers *space* in GEOMETRY, *time* in pure MECHANICS" (2, 397.28–29); and in the *Prolegomena,* §10: "above all, however, pure mechanics can attain its concepts of motion only by means of the representation of time" (4, 283.17–21).[16] The science of time, for Kant, is therefore not arithmetic, but rather pure mechanics or the pure doctrine of motion.

Indeed, it is well that Kant does not consider arithmetic to be the science of time, for it is obvious that the pure intuition of time does not constitute a model for arithmetic. A model for arithmetic must contain the successor function, and thus a distinguished unit, but it is clear that there is no distinguished unit in time itself: on the contrary, the temporal unit depends entirely on arbitrary choice. Moreover, the same is also true of space: neither space nor any line in space comes equipped with a distinguished unit. Hence, while the pure intuition of space can indeed be conceived as a model for Euclidean geometry (which itself posits no distin-

15. See, e.g., Kitcher [59], §III; Parsons, "Philosophy of Arithmetic," p. 133—Parsons in fact uses this asymmetry between geometry and arithmetic as the starting point for his seminal study: cf. §§III–V.

16. To be sure, this last passage from the *Prolegomena* also mentions arithmetic. The full passage reads: "Geometry is based on the pure intuition of space. Even arithmetic attains its concepts of number through successive addition of units in time; above all, however, pure mechanics can attain its concepts of motion only by means of the representation of time." Thus there is no doubt that arithmetic *involves* time. Even this passage suggests, however, that it is pure mechanics which most properly *considers* time (as an object, as it were)—this will be discussed more fully below.

guished unit, of course), neither the pure intuition of time nor the pure intuition of space can be conceived as a model for arithmetic.[17]

Moreover, there are a number of important passages where Kant explicitly asserts that arithmetic does not concern temporal objects of intuition at all.[18] The first is found in a letter to Johann Schultz of November 25, 1788:

> Time, as you rightly remark, has no influence on the properties of numbers (as pure determinations of magnitude), as it does on the properties of any alteration (as a *quantum*), which is itself only possible relative to a specific constitution of inner sense and its form (time). And the science of number, notwithstanding the succession that any construction of magnitude requires, is a pure intellectual synthesis that we represent to ourselves in thought. (10, 556.36–557.6)

Thus numbers, unlike alterations, are not temporal objects; and the reference to alteration here fits very well with Kant's insistence, noted above, that it is not arithmetic but pure mechanics or the pure doctrine of motion that deals with time: the intuition of motion, for Kant, is what alone makes possible the concept of alteration (see B48–49, B291–292). Moreover, the idea that the science of number is a "pure intellectual synthesis" echoes the *Inaugural Dissertation,* §12:

> In addition to these concepts, there is a certain concept which in itself is indeed intellectual, but whose actualization in the concrete requires the assisting concepts of time and space (successively adding the elements of a plurality and setting them next to one another simultaneously); this is the concept of *Number,* which is treated by ARITHMETIC. (2, 397.29–33)

This passage appears to suggest that the science of number is itself entirely independent of intuition, and that only its application will concern intuitive objects—objects which are to be counted, say.[19] One might then be

17. Kitcher, appealing to *Inaugural Dissertation,* §12, suggests that arithmetic deals with both space *and* time, with what he calls "'combinatorial' features of space-time": [59], pp. 34–35. In light of the above, however, it is difficult to see what this might mean (space-time fails to provide a model of arithmetic for just the same reason). *Inaugural Dissertation,* §12, will be considered further below.

18. The importance of such passages has been emphasized by Parsons: see "Philosophy of Arithmetic," §VI; [94], pp. 109–121, §§II, III.

19. The same suggestion appears to be made immediately following the above cited passage from the letter to Schultz: "But in so far as magnitudes *(quanta)* are to be determined thereby, they must be given to us in such a way that we are able to apprehend their intuition successively—and thus this apprehension is subject to the condition of time. Therefore, we are then still able to subject no object, except those of a possible *sensible* intuition, to our estimation of magnitude by means of numbers; and it thus remains a principle without exception that mathematics extends only to *sensibilia*" (557.6–12). Kant appears to be

tempted to think that *pure* arithmetic has nothing at all to do with time: time only enters the picture in the first place because the empirical objects of *applied* arithmetic are necessarily temporal.

Perhaps the most interesting passage, however, occurs several paragraphs earlier in the same letter to Schultz:

> Arithmetic certainly has no *axioms*, because it properly speaking has no *quantum*—i.e., no object [Gegenstand] of intuition as magnitude—for its object [Object], but merely *quantity* [*Quantität*]—i.e., a concept of a thing in general through the determination of magnitude. (555.34–37)

Kant is here clearly thinking of pure arithmetic, for applied arithmetic certainly does have objects of intuition as its subject matter. Pure arithmetic, on the other hand, does not have such objects *(quanta)*; in their place it has merely a concept, the concept of *quantity*. But how exactly are we to understand the distinction between *quanta* and *quantity?*

The distinction is made in the *Critique* at A163–165/B204–206, in the course of an explanation of the difference between geometry and arithmetic. After referring explicitly to the axioms of geometry Kant says:

> These are the axioms that properly speaking concern only magnitudes *(quanta)* as such.
>
> In what concerns magnitude *(quantitas)*, however—that is, the answer to the question: "What is the magnitude of something?"—there are no axioms in the proper sense, although there are a number of propositions that are synthetic and immediately certain *(indemonstrabilia)*. (A163–164/B204)

As examples of the latter type of propositions Kant then cites "numerical formulas [Zahlformeln]" such as $7 + 5 = 12$. Such synthetic propositions are not *axioms* because they, unlike the axioms of geometry, are not general (A164/B205). Thus Kant is here making essentially the same point as in the letter to Schultz: geometry concerns *quanta* and has axioms; arithmetic, which has no axioms, accordingly does not concern *quanta* but merely *quantity*. Here we learn in addition that it is geometry that does have *quanta*—objects of intuition as magnitudes—for its object.

The distinction between *quanta* and *quantity* occurs again in an important passage in the Discipline of Pure Reason:

> Mathematics, however, does not only construct magnitudes *(quanta)*, as in geometry, but also mere magnitude *(quantitatem)*, as in algebra, wherein it completely abstracts from the constitution of the object that is to be thought in accordance with such a concept of magnitude. Thereupon it chooses a certain designation [Bezeichnung] for all constructions of magnitudes in gen-

reiterating the point, emphasized in §I above, that mathematics applies only to objects of *empirical* intuition.

> eral (numbers, as for addition, subtraction, etc.), extraction of roots; and, after it has also specified [bezeichnet] the general concept of magnitudes according to their various ratios, it then presents in intuition all operatations through which the magnitude is generated and altered, in accordance with certain general rules. For example, where one magnitude is to be divided by another, it sets both of their characters together in accordance with the form that designates division [der bezeichnenden Form der Division]; etc. And thus it arrives by means of a symbolic construction—just as well as does geometry by an ostensive or geometrical construction (of the objects themselves)—at what discursive knowledge by means of mere concepts could never attain. (A717/B745)

Here the distinction between *quanta* and *quantity* is associated with two different types of construction: "ostensive or geometrical" and "symbolic" (several pages later this latter is also referred to as "characteristic construction": A734/B762). It is clear, moreover, that the latter type of construction is much more abstract than the former: symbolic construction "abstracts completely from the constitution of the object," whereas ostensive construction deals with "the objects themselves."

Although Kant mentions algebra, not arithmetic, in the above passage, it is likely that both arithmetic and algebra are to be included under symbolic or characteristic construction. For the initial distinction between *quanta* and *quantity* is, as we have seen, also specifically linked to arithmetic by Kant. Further, in the above passage itself Kant explicitly mentions numbers (Zahlen). Finally, there are a number of passages where algebra and arithmetic are closely associated. Thus, in our letter to Schultz, Kant's first example of a mathematical science is "general arithmetic (algebra)" (555.10). And, in §2 of the First Reflection of the *Enquiry,* Kant appeals to "arithmetic, both the general arithmetic of indeterminate magnitudes and that of numbers—where the ratio of the magnitude to unity is determinate" (2, 278.18–19).[20] So algebra appears to be a kind of arithmetic: "the general arithmetic of indeterminate magnitudes."

What then is the difference between algebra and arithmetic—where the latter is construed more narrowly as the science of numbers? The characterization of algebra as "general arithmetic" might tempt one to suppose that, whereas arithmetic proper deals with particular numbers by means of numerical formulas such as $7 + 5 = 12$, algebra generalizes

20. The passage continues: "In both, symbols are first supposed, instead of the things themselves, together with particular designations [Bezeichnungen] of their increase and decrease, their ratios, etc. Then, one proceeds with these signs, according to easy and secure rules, by means of substitution, combination or subtraction, and many kinds of transformations, so that the things designated [bezeichnen] are themselves here completely ignored, until, at the end, the meaning of the symbolic deduction is finally deciphered." Thus both arithmetic and algebra are clearly subsumed under symbolic construction here.

over all numbers. Algebra is concerned with generalizations such as the commutative law of addition, for example, and its distinguishing feature is thus the use of variables (as opposed to particular numerals, say).[21] It would then follow that algebra is an essentially more abstract science than arithmetic,[22] and the sense in which algebra "abstracts completely from the constitution of the object" would be correspondingly clear.

This suggestion is problematic, however. Kant, as we have seen, appears to put algebra and arithmetic on the same level of abstraction: both appear to "abstract completely from the constitution of the object." Indeed, arithmetic has no axioms precisely because it too "has no object of intuition as magnitude for its object." Moreover, as A163–165/B204–206 explains, the reason arithmetic has no axioms is that there are no general laws of number analogous to the general laws of geometry (Kant's example here is Proposition I.22 of Euclid).[23] It would then appear that generalizations such as the commutative law of addition have no clear place at all for Kant.[24]

An alternative suggestion is the following: "General arithmetic (algebra)" goes beyond arithmetic in the narrower sense, not by generalizing over it, but by considering a more general class of magnitudes. In particular, in the "arithmetic of numbers" we are limited to cases where "the ratio of the magnitude to unity is determinate," whereas in "general arithmetic" we can also consider "indeterminate magnitudes." And this means, I suggest, simply that the arithmetic of numbers is concerned only with rational magnitudes, whereas general arithmetic or algebra is also con-

21. Aspects of this idea can be found in Kitcher [59], pp. 36–37, and in Young [124], pp. 34–36. It is not clear that either fully endorses the idea, however. As Young points out, the idea is suggested by §6 of Hintikka, "Kant on the Mathematical Method" (1967), reprinted in [53].

22. This conclusion is explicitly endorsed by Young [124], pp. 44–55.

23. This point is entirely in accord with the letter to Schultz, in which, immediately after denying that arithmetic has axioms, Kant says: "It has, on the other hand, *postulates*—i.e., immediately certain practical judgements" (555.37–556.1). For, as examples of such "postulates," Kant again gives particular numerical formulas such as 3 + 4 = 7.

24. This point is illuminatingly discussed by Parsons, "Philosophy of Arithmetic," §III. Note that it will not do to count such generalizations as belonging to algebra but not to arithmetic; for A163–164/B204 claims that there are no axioms "in what concerns magnitude *(quantitas)*," and Kant explicitly includes algebra under *quantitas* at A717/B745. Thus A163–165/B204–206 leaves no room for general laws of algebra corresponding to general laws of geometry either. Kant's point, rather, is that in the case of geometry *(quanta)* there are general constructions—such as the Euclidean construction of the triangle of Proposition I.22—whereas in the case of arithmetic and algebra *(quantitas)* there are no such general constructions, but only particular numerical formulas or "postulates" such as 7 + 5 = 12 or 3 + 4 = 7. Moreover, by means of such general constructions, geometry generates a specific domain of objects in a way in which arithmetic and algebra do not. This last idea will be explored further below.

cerned with irrational or incommensurable magnitudes. Thus general arithmetic or algebra corresponds approximately to the Eudoxean-Euclidean theory of ratios or proportion developed in Book V of the *Elements;* the arithmetic of numbers corresponds approximately to the special theory of numerical magnitudes developed in Books VII–IX.

This suggestion receives confirmation from an exchange between Kant and August Rehberg in 1790 concerning the nature of irrational magnitudes. In his letter to Rehberg Kant argues that the understanding can think "the mere concept of a square root of a positive number = $\sqrt{a}$, as it is represented in algebra" (11, 208.32–33) and, indeed, can think such a root "in the *number series,* between two of its terms" (208.19).[25] Specifically, the square root of a can always be thought as "the mean geometrical proportional between 1 and a" via "the equation $1:x = x:a$" (207.31–32). This equation, however, does not give the root a determinate ratio to unity, for we know only that the two ratios $1:x$ and $x:a$ are equal:

> Therefore, from the fact that every number must be able to be represented as the square of some other as root, it can *not be concluded,* according to the principle of identity, from the concept of the problem—namely, to conceive the two equal (but indeterminate) factors of a given product—that the root must be rational, i.e., have a countable [auszählbar] ratio to unity. For in this [concept] absolutely no determinate ratio to unity is given, but only their ratios [of the factors] to one another. (208.11–18)

In algebra, then, we represent magnitudes by their ratios to one another, without (as in arithmetic) thereby necessarily representing a determinate ratio of such magnitudes to unity. As Kant says in an earlier draft of his letter to Rehberg:

> Algebra is properly speaking the art of bringing the generation of an unknown magnitude through enumeration [Zählen] under a rule, merely by means of the given ratios of magnitudes independently of any actual number. (14, 54.9–13)

This fits very well with the passage about algebra from A717/B745, where Kant first mentions numbers, then the extraction of roots, and finally says that algebra "has specified the general concept of magnitudes according to their various ratios."[26]

25. By contrast, this is not true of the concept of the square root of a negative number, which is simply self-contradictory: 209.35–36.

26. I thus adopt the reading of this passage in the original text in preference to the alternate readings according to which "the extraction of roots" is to appear in the same list of operations as "addition, subtraction, etc." (Kemp Smith opts for the latter reading). The point is that addition and subtraction keep one within the domain of numbers, whereas the extraction of roots does not; the latter, in fact, leads one naturally to the more general theory

My suggestion, then, is that by "determinate ratio to unity" Kant here means rational ratio to unity,[27] for even two incommensurable magnitudes certainly have a (definite) ratio—this, in fact, is the whole point of the theory of ratios.[28] As we have seen, Kant characterizes such rational ratios as "countable [auszählbar] ratios," and he says also that for such ratios we have a "complete number-concept [vollständig Zahlbegriff]":

> . . . the understanding, which arbitrarily makes for itself the concept of $\sqrt{2}$, can not also bring forth the complete number-concept, namely, by means of its rational ratio to unity, but rather, as if guided by another faculty, it must give way in this determination to follow an infinite approximation to a number. (11, 208.23–28)

Such an infinite approximation is given, for example, by what we would now call the decimal expansion of $\sqrt{2}$:

> . . . the parts of the unit, which are to serve as the denominators of a series of fractions decreasing to infinity, are allowed to grow in accordance with a certain proportion—e.g., by decades; and this series, because it can never be completed, although it can be brought as near to completion as one wishes, expresses the root (but only in an irrational way). (209.25–29)

This series or infinite approximation is "itself no number, but only the rule for approximation to the number" (210.13–14); or, as Kant puts it in a draft of his letter: "it is actually no number, but only a determination of magnitude by means of a rule of enumeration [Zählen]" (14, 57.6–7).

---

of ratios. Compare also A734/B762, where Kant describes the procedure of algebra as a "characteristic construction, in which one presents in signs the concepts—above all those of the ratios of magnitudes—in intuition," and §4 of the First Reflection of the *Enquiry*, where Kant discusses "the general theory of magnitude (which is properly speaking general arithmetic)" in terms of "the increase and decrease of magnitudes, their decomposition into equal factors by means of the doctrine of roots" (2, 282.16–19). Earlier, in §1, he speaks of magnitudes in terms of "their increase or decrease, their ratios, etc."—cf. note 20 above.

27. As Howard Stein has emphasized to me, this is certainly not the only possible reading of the passage at 11, 208.11–18; for we could also plausibly interpret "determinate ratio" as *known* or *given* ratio—in which case $\sqrt{2}$ would have a determinate ratio to unity after all, only one that is not given (analytically) in the equation $1:x = x:2$. On the other hand, Kant appears consistently to deny that $\sqrt{2}$ is a *number* (for example, immediately before our passage and in passages cited below), and, as we have seen, he characterizes the arithmetic of numbers as that in which the ratio of magnitude to unity is determinate. This seems to suggest that the ratio of $\sqrt{2}$ to unity is not determinate and to confirm my reading.

28. Thus the heart of the Eudoxean-Euclidean theory of ratios is given in Definitions 4 and 5 of Book V of the *Elements*. Definition 4 states that two magnitudes a and b *have a ratio* if for some number n, na (that is, a added to itself n times) is greater than b—and vice versa with a and b reversed (Archimedean property). Definition 5 states that, given four magnitudes a, b, c, and d, the ratio $a:b$ is the *same ratio* as the ratio $c:d$ if for all pairs of numbers m, n we have either (i) $ma > nb$ and $mc > nd$, or (ii) $ma = nb$ and $mc = nd$, or (iii) $ma < nb$ and $mc < nd$. Here see the lucid exposition in Stein [106].

In the case of an irrational magnitude such as $\sqrt{2}$, Kant concludes, we are therefore "not capable of adequately presenting the concept of such a quantity [Quantität] in number-intuition [Zahlanschauung]," and, since the understanding is therefore not able "to expect that such a *quantum* can be a priori given," the understanding "is indeed not even empowered to assume the possibility of an object = $\sqrt{2}$." On the contrary, the possibility of such magnitudes can be exhibited only in geometrical construction (in this case, of the diagonal of a square), which alone can show that such magnitudes are "not merely thinkable, but are also to be given adequately in intuition" (11, 210.16–23).

We are back, then, to the distinction between *quanta* and *quantity*, which, as before, is closely linked to the distinction between algebra and arithmetic (symbolic construction), on the one hand, and geometry (ostensive construction), on the other. We are now in a position better to understand this distinction, however, as well as the intimately related question concerning the abstractness of algebra and arithmetic.

Just as in Euclid, Kant sees the objects of the theory of magnitude, both the general theory of (rational or irrational) magnitudes and the special theory of numerical magnitudes, as given independently, from outside the theory itself, as it were.[29] These objects are conceived first and foremost as spatial magnitudes, such as lengths, areas, and volumes, which are independently given by the science of geometry. The theory of magnitude takes such an independently given object—a given finite line segment, for example—as input and yields a definite answer to the question "What is the magnitude of this object?" as output.[30] This is accomplished by arbitrarily choosing a unit for the magnitude in question (thus if the magnitude is the diagonal of a square, say, we might choose the side of the square as unit) and then, by calculation, attempting to express it as a sum of such units. If the magnitude and the unit are commensurable, arithmetic yields a determinate whole number or fraction; if not, algebra (the theory of ratios) nonetheless allows us to find a definite rule of approximation by numbers (including fractions), a rule of approximation which can be made as accurate as one wishes.

29. See again Stein [106]. As Stein points out, perhaps the chief difference between the Eudoxean-Euclidean theory of ratios and the modern theory of the real numbers due to Cantor and Dedekind lies precisely here. The Cantor-Dedekind theory explicitly postulates that every "cut" determines a real number, whereas the theory of ratios leaves all questions concerning the *existence* of ratios corresponding to such "cuts" entirely open: these questions are decided from outside the theory itself.

30. Compare *Prolegomena*, §20, for example, where the determination of the length of a line segment is given as the example of the application of the concept of magnitude (4, 301.14–302.2).

On this conception, therefore, algebra and arithmetic are not conceived as we would understand them today: that is, as bodies of general truths concerning specific domains of objects—the domains of natural, integral, rational, or real numbers, for example. The only mathematical science that concerns a special domain of objects is geometry, and neither arithmetic nor algebra has a special domain of its own.[31] Instead, algebra and arithmetic are conceived as techniques of calculation for solving particular problems, for finding the magnitudes of any objects there happen to be—where the latter are not given by the sciences of arithmetic and algebra themselves. To be sure, general laws—what Kant himself calls "general rules" at A717/B745—will be involved in such calculations, and it is precisely these which are now understood as quantifications over a special domain. The present conception, however, is still at a lower level of abstraction: the "general rules" of calculation are not to be thought of as quantified propositions in the modern sense.[32]

It follows that the abstractness of algebra and arithmetic does not consist in their generality, but rather in the fact that they, as techniques of calculation, are in turn independent of the specific nature of the objects whose magnitudes are to be calculated. In the theory of magnitude itself we assume absolutely nothing about the nature and existence of the magnitudes to be thereby determined: we merely provide operations (such as addition, subtraction, and also the extraction of roots) and concepts (above all the concept of ratio) for manipulating any magnitudes there

31. From a modern point of view, we could perhaps reconstruct Kant's conception of arithmetic as involving a sub-system of primitive recursive arithmetic (such as Robinson arithmetic) where generality is expressed by means of free variables and there are no true quantifiers. In such a system we can prove all numerical formulas, such as $7 + 5 = 12$, and all particular instances of the commutative, associative, etc. laws—but not these laws themselves. To prove such general laws we need the principle of *mathematical induction,* which expresses quantification over the numbers. Algebra would then be understood as also involving mere free variable generality, and as differing from arithmetic only in its use of the additional operation of root-extraction and the additional concept of ratio. (Note, however, that the definitions constituting the heart of the Eudoxean-Euclidean theory of ratios themselves require explicit quantification over the numbers—compare note 28 above.)

32. Kant explicitly appeals to one general law in this connection, namely to "Theaetetus's Theorem" (Euclid, Prop. X.9): if a (whole) number does not have a square root in the whole numbers, the square root is irrational (see 11, 209.31–34; 14, 57.20–58.7). For us such a general proposition depends ultimately on mathematical induction (in the form of the least-number principle) and thus stands in conflict Kant's non-axiomatic conception of arithmetic (compare notes 24 and 31 above). Kant is most likely again following the example of Euclid, where no new axioms—beyond the axioms of geometry—are explicitly introduced for either algebra (the theory of ratios) or arithmetic. Instead, Euclid (as we would put it) tacitly introduces the least-number principle in Proposition VII.31, in the course of proving the decomposition of any number into powers of primes.

may be. What magnitudes there are is settled outside the theory itself—for Kant, by the specific character of our intuition. Thus, while our primary examples of magnitudes are spatial magnitudes, such as lengths, areas, and volumes, other types of magnitudes are perfectly conceivable. Indeed, we might say that one important aspect of the progress of science (mathematization) is precisely the extension of the concept of magnitude beyond its original, purely spatial domain.[33]

We can now understand Kant's terminology. *Quanta,* objects of intuition as magnitudes, are just the particular magnitudes there happen to be. These are given, in the first instance, by the axioms of Euclid's geometry, which postulate the construction (from a modern point of view, the existence) of all the relevant spatial magnitudes. *Quantity,* the concept of a thing in general through the determination of magnitude, comprises the operations and concepts invoked by arithmetic and algebra for manipulating, and thereby calculating the specific magnitude of any magnitudes which happen to exist. Yet, since we do not here postulate the construction (existence) of any particular magnitudes, there are no axioms for *quantity.* The application of the science of *quantity,* unlike that of the science of geometry, is therefore not limited to the specific—that is, spatial—character of our intuition; in this sense it provides us with the concept of a thing in general.[34]

· III ·

We have argued above that the abstractness of algebra and arithmetic consists in the fact that they, unlike geometry, do not assume anything specific about the nature and existence of the objects of our intuition. Algebra and arithmetic comprise the rules and operations for calculating any magnitudes there may happen to be, that is, for answering the question "What is the magnitude of something?" In particular, algebra and arithmetic do not assume that the magnitudes to be manipulated are necessarily spatial magnitudes, that they are necessarily subject to the science

33. Thus, in modern physics we subsume *motions* and hence also *times* under the concept of magnitude—a point which is centrally important to Kant and which will be further discussed below.

34. More properly, perhaps, we are provided with the concept of *an object of intuition in general.* See A143/B182: "Number is nothing other than the unity of the synthesis of the manifold of a homogeneous intuition in general"; and B162: "Precisely the same synthetic unity, however, if I abstract from the form of space, has its seat in the understanding, and is the category of the synthesis of the homogeneous in an intuition in general, i.e., the category of *magnitude.*" See also Parsons, "Philosophy of Arithmetic," pp. 134–135, for a discussion of this difficult matter.

of geometry. Similarly, algebra and arithmetic do not assume that the objects of intuition to which they are to be applied are necessarily temporal magnitudes either, if only because spatial magnitudes, unlike alterations, say, are not themselves temporal objects.[35] Does it follow, then, that arithmetic and algebra are wholly independent of pure intuition—independent of the pure intuition of time, in particular? Is the view briefly mentioned above (§II), according to which only *applied* arithmetic involves the intuition of time, therefore correct?

This idea is also not consistent with Kant's texts. Thus, in a draft of his letter to Rehberg, Kant asserts that:

> The objects of arithmetic and algebra are not under time-conditions according to their possibility; nevertheless, the construction of the concept of magnitude in its representation through the synthesis of the imagination—namely composition—without which no object of mathematics can be given, still is. (14, 54.5–9)

In other words, although the objects to which algebra and arithmetic apply *(quanta)* are not necessarily temporal, there is still a temporal element in the construction of the concept of magnitude *(quantity)* itself. This temporal element is involved in what Kant, in the first *Critique,* calls the "pure schema of magnitude":

> But the pure *schema* of *magnitude (quantitas)* as a concept of the understanding is *number,* which is a representation that unites [zusammenbefasst] the successive addition of unit to (homogeneous) unit. Therefore number is nothing other than the unity of the synthesis of the manifold of a homogeneous intuition in general, in such a way that I generate time itself in the apprehension of the intuition. (A142–143/B182)

Note that the temporal element here, successive addition of unit to unit, does not refer to the objects of intuition *(quanta)* and is therefore not simply a matter of applied arithmetic. On the contrary, what pertains to such objects of intuition is the "pure image of magnitude":

> The pure image of all magnitudes *(quantorum)* of the outer senses is space, of objects of the senses in general, however, is time. (A142/B182)

And, as A140–142/B179–181 explains, images are to be sharply distinguished from schemata: "the *image* is a product of the empirical capacity of the productive imagination, the *schema* of sensible concepts (as of figures in space) is a product and as it were a monogram of the pure a priori imagination" (A141–142/B181). In any case, the point here is not

35. See again the draft of Kant's letter to Rehberg: "In spatial representation certainly nothing of time is thought" (14, 54.1).

that the objects of the theory of magnitude *(quanta)* are temporal, but that the concept of magnitude *(quantity)* is itself in some way temporal.[36] Indeed, in his letter to Rehberg Kant goes further:

> However, that the understanding, which arbitrarily makes for itself the concept of $\sqrt{2}$, can not also bring forth the complete number-concept, namely, by means of its rational ratio to unity, but rather, as if guided by another faculty, it must give way in this determination to follow an infinite approximation to a number—this is in fact based on successive progression as the form of all enumeration [Zählen] and number-magnitudes [Zählgrössen], as well as the condition lying at the basis of this generation of magnitude: time. (11, 208.23–31)

Thus, the fact of the irrationality of $\sqrt{2}$, which is presumably a fact of *pure* arithmetic, is itself based on successive enumeration and hence on time. Even apart from all questions concerning the objective reality of $\sqrt{a}$ (as shown by geometrical construction, for example), the pure intuition of time is therefore necessarily required for determining whether $\sqrt{a}$ is rational or irrational.[37]

I propose to interpret these passages as follows. Lying at the basis of all operations with the concept of magnitude—all calculations undertaken in the sciences of arithmetic and algebra—is the number series: the series of what we now call the natural numbers. And this series, for Kant, can in turn itself only be represented by means of a progression in time: the successive addition of unit to unit. In particular, it is only the necessarily temporal activity of progressive enumeration that allows us to find or determine the result of any calculation. Such an activity is required for even the simplest possible numerical formulas, since, as Kant puts it in our letter to Schultz:

> . . . if I view 3 + 4 as the expression of a *problem*—namely, to find for the numbers 3 and 4 a third = 7 that can be considered as the *complementum ad totum* of the others—then the solution is effected through the simplest action, which requires no particular formula of resolution, namely through

36. See also A242/B300: "No one can explain the concept of magnitude in general except in such a way as this: that it is the determination of a thing so that how many times One [a unit] is posited in it can thereby be thought. But this how-many-times is based on successive repetition, and therefore on time and the synthesis (of the homogeneous) in time."

37. See again the letter to Rehberg: "But as soon as, instead of *a*, the number of which it is the sign is given, in order not merely *to designate* [*bezeichnen*] its root, as in algebra, but also *to find* it, as in arithmetic, then the condition of all number-generation [Zahlerzeugung], time, is hereby unavoidably the basis—and in fact time as pure intuition, in which we can be instructed, not only concerning the given number-magnitude but also concerning the root, whether it can be found as a whole number, or, if this is not possible, only by means of an infinitely decreasing series of fractions, and thus as irrational number" (209.2–11).

> the successive addition that brings forth the number 4, only set into operation as a continuation of the enumeration of the number 3. (10, 556.2–8)

This, in fact, is why 3 + 4 = 7 is synthetic: why, in Kant's words, it is a "*postulate,* i.e., an immediately certain practical judgement" (556.1). For I need assurance that a number 3 + 4 can be constructed (for us, that it exists); and this assurance can only be grounded on the possibility of *successive repetition* of any given operation. In the present case, I am given the two numbers 3 and 4, each in terms of the possibility of repeating the operation of addition of one (for us, the successor function) a given finite number of times; to represent 3 + 4, then, I have only to perform the operation corresponding to 4 succeeding (as a continuation of) the operation corresponding to 3.

In such simple cases the activity of progressive enumeration (successive iteration of operations) terminates in a finite number of steps. However, there are also cases where the activity of progressive enumeration does not and cannot so terminate. The determination of an irrational magnitude is just such a case, but here the operation to be iterated takes a more complicated form: express the unit as a sum of n equal parts and attempt to exhibit the given magnitude as a sum of m such parts for some finite m (in which case the magnitude corresponds to the rational ratio $m:n$, of course), if this does not succeed (which can itself be determined in a finite number of steps if the unit and the given magnitude *have* a ratio) try again with a subdivision of the unit into $n + 1$ equal parts, and so on. That the given magnitude is irrational consists in the fact that this iterative procedure does *not* terminate in a finite number of steps; and this, I suggest, is why Kant claims that the irrationality of $\sqrt{2}$, for example, is "based on successive progression as the form of all enumeration."[38]

Successive progression is also the basis for *finding* the actual magnitude of a given irrational magnitude, for answering the question "What is its magnitude?" Only here the answer is not given by a definite number or fraction—not by a "complete number-concept"—but rather by an infinite series of fractions. In the case of $\sqrt{2}$, for example, we have the familiar decimal expansion:

$$s_1 = 1.4, \quad s_2 = 1.41, \quad s_3 = 1.414, \ldots$$

In this case I obtain each element of the series by successively inserting finite decimals into the "formula of resolution" $x^2 = 2$. Such a series is certainly not itself a number, but it is still "a determination of magnitude

38. See also the draft of the letter to Rehberg where Kant expresses the notion of irrationality as that of "a finite magnitude . . . whose concept still falls between all given divisions of the unit in number-series" (58.24–59.1).

by means of a rule of enumeration." In particular, although the entire series can of course not be given in a finite number of steps, I can still approximate the magnitude as closely as I wish in a finite number of steps.[39] I thereby obtain a definite concept of the magnitude of an irrational magnitude (although this concept is not, of course, a "complete number-concept"). Thus Kant claims that "without arithmetic (still prior to algebra) we would be able to have no concept of the magnitude of the diagonal of the square" (14, 54.21–55.2).

As we have seen above, Kant also claims that only geometry can show the real possibility of the concept of $\sqrt{2}$. He repeats this claim in the very passage last cited: "that such a magnitude is possible would not be known by us without geometry" (54.20–21). We are now in a position, I think, to appreciate the full force of this claim. To be really possible in this context, I suggest, is to be constructible by a procedure of successive iteration in a finite number of steps. We can therefore know the real possibility of a *quantum* with an integral or fractional magnitude independently of the science of geometry. For, no matter what the particular character of our intuition may be, we can (provided successive iteration of a unit is possible at all) construct such a magnitude in a finite number of steps; this is what it means to be "adequately presented in number-intuition [Zahlanschauung]" (11, 210.20–21). This much, in other words, is guaranteed by the mere concept of *quantity,* no matter what *quanta* there are. The possibility of a *quantum* with irrational magnitude, however, cannot be known in this way; for here an appeal to mere *quantity* or "number-intuition" is necessarily non-terminating. In this case, therefore, we must appeal to geometry in order to construct such a magnitude in a finite number of steps: for example, via Euclid's construction of the square in Proposition I.46, which easily yields a construction of the diagonal if we then apply Postulate 1.

This last point is of considerable importance, I think, for it shows that geometry, like algebra and arithmetic, also centrally involves successive progression or repeated iteration. In particular, Euclid's axiomatization is based on three initial operations, given by Postulates 1, 2, and 3: (i) drawing a line segment connecting any two given points, (ii) extending a

39. Kant holds, as we have seen, that successive enumeration underlies the procedure of determining whether a given magnitude is rational or irrational as well. This may appear strange, since if the magnitude is not rational we will *not* in general find this out in a finite number of steps, either by generating the decimal expansion or by successively dividing the unit as above. Here Kant has in mind appealing to "Theaetetus's Theorem": if a root of a whole number is not found in the whole numbers, then it is irrational. His idea is to examine successively all numbers smaller than n so as to determine by the multiplication algorithm whether $\sqrt{n}$ is rational (11, 209.15–34). "Theaetetus's Theorem" itself leads to problems for Kant's general view, however: see note 32 above.

line segment by any given line segment, (iii) drawing a circle with any given point as center and any given line segment as radius. We are then allowed to iterate or successively repeat operations (i), (ii), and (iii) any finite number of times (and in any order), and this procedure generates all the objects required for Euclidean geometry: that is, to be an object of Euclidean geometry is just to be constructible by means of the operations (i), (ii), and (iii) in a finite number of steps.[40] Hence "ostensive or geometrical" construction is based on successive progression just as much as is the "symbolic" or "characteristic" construction common to algebra and arithmetic; and it follows that, although the objects of geometry are not themselves necessarily temporal, geometrical construction is nonetheless a temporal activity:

> In spatial representation certainly nothing of time is thought, but it is in the construction of the concept of a certain space, e.g., a line. All magnitude is generation in time by means of repeated position in time. (14, 54.1–4)

The necessarily temporal concept of magnitude in this way applies to geometry as well, and this is in fact the ground to which Kant appeals at B16 in first arguing that geometry—just as much as arithmetic—is synthetic.[41]

The difference between geometrical construction and the symbolic construction common to algebra and arithmetic is then simply that geometrical construction necessarily starts from the given operations (i), (ii), and (iii): they serve as given fixed inputs, as it were, for our general capacity successively to iterate any operation (to insure that anything is actually constructed we also need a given finite line segment or two given distinct points). In algebra and arithmetic, by contrast, we specify no such fixed inputs and concern ourselves only with the form of successive progression common to any and all iterative procedures. In other words, symbolic construction differs from ostensive construction solely in virtue of the greater abstractness of algebra and arithmetic as articulated in §II above.[42]

40. See Chapter 1 above for a more detailed attempt to flesh out the significance of this for Kant's conception of geometry.

41. "Just as little is any fundamental proposition of pure geometry analytic. That the straight line between two points is the shortest is a synthetic proposition. For my concept of *straight* contains nothing of magnitude, but only a quality. The concept of shortest is entirely an addition, and cannot be derived by any analysis of the concept of a straight line. The aid of intuition must therefore be brought in, by means of which alone the synthesis is possible." Compare also *Prolegomena,* §20.

42. There are actually two distinguishable, although closely related, aspects to symbolic construction. On the one hand, in *finding* the magnitude of anything we will employ the successive progression underlying the number series: either by generating a whole number or fraction in a finite number of steps or by generating an infinite approximation to an irrational number. On the other hand, however, successive iteration is also employed in the

Thus, for example, we might construct a definite model for arithmetic by starting with a given finite line segment as unit, and then extending the segment by iteration any finite number of times according to operation (ii). We might also proceed, however, by starting with the construction of a unit square, and then iterating this construction any finite number of times. Other models—even non-spatial ones—are of course also conceivable. Arithmetic, we might say, is not concerned with any particular such construction, but only with the general iterative form common to all such constructions.[43]

We have now reached the question lying behind these considerations, a question we have been postponing. Arithmetic, to be sure, essentially

---

mere manipulation of signs in algebraic formulas: such "operation of a calculus" is also an iterative, step by step procedure. Kant does not clearly distinguish these two aspects: for example, he appeals to the latter at A717/B745 in his remarks about "setting characters together." As we now understand the matter there is a good reason for such conflation, for any formal calculus is itself a model for arithmetic. This point has been well emphasized in the present context by Parsons: see "Philosophy of Arithmetic," pp. 138–139; [94], p. 118. See also Thompson [108], pp. 336–338.

43. For the importance of the notion of progressive iteration in Kant's conception of arithmetic, I am indebted, above all, to Parsons, "Philosophy of Arithmetic," especially §VII. I am also indebted to the insightful discussion of calculation and symbolic construction in Young [124]. I cannot follow these writers, however, when they suggest that *ostensive* construction also has a place in arithmetic (see Parsons [94], p. 111; Young, [124], §III). Here they have in mind such simple arithmetical truths as $7 + 5 = 12$; and the idea, apparently, is that in such cases intuition is not only involved in the step by step procedure of calculation but also plays a role in "seeing," as it were, that the proposition is true. This suggestion seems to me to conflict with the evidence reviewed above. At A717/B745, for example, Kant explicitly contrasts the symbolic construction involved in *quantity* with "ostensive or geometrical construction." In the letter to Schultz, Kant asserts that arithmetic, unlike geometry, deals with *quantity* not *quanta,* and he illustrates this with the extremely simple successive generation involved in $3 + 4 = 7$. The role of intuition here is not to "see" that the formula is true, but to guarantee the construction or generation (the existence) of the required sum. The contrary idea seems to me to arise from concentrating on possible outcomes for the construction rather than on the construction itself and to rest, in the end, on the supposition that arithmetic, like geometry, has *quanta* or special objects of its own: namely, finite collections (see Parsons [94] and §§II–III; Young [124], §§III–IV—this, in fact, is why Young thinks that algebra is more abstract than arithmetic: "algebra is a more abstract study than either arithmetic or geometry. It does not deal with particular species of quantity, such as the length exhibited in line segments or the discrete quantity exhibited in collections of things," [124], p. 44). Yet it is a mistake, I think, to attribute anything like our concept of cardinal number to Kant and, in particular, to view finite collections as *quanta* for arithmetic. At A170–171/B212, for example, Kant explicitly denies that an aggregate of coins may be considered a *quantum* and concludes that "appearance . . . as a *quantum* is always a continuum." Once again, therefore, the only *quanta* are the continuous magnitudes supplied in the first instance by geometry. (There are, however, passages from the Reflexionen and from the metaphysics lectures that apparently support the idea that *quanta* can be *discrete:* see Parsons [94], §II.)

involves the notion of progressive iteration; but why should this idea, in turn, essentially involve time? Is not progressive iteration in fact a much more abstract concept than any temporal concept? The answer to this question has been well expressed by Parsons: "finite iteration is an abstract counterpart of the notion of successive repetition. But to describe it in abstract terms was quite beyond the logical and mathematical resources of Kant and his contemporaries; the task was first accomplished in the 1880s by Frege and Dedekind" ([94], p. 116).[44] Thus we can now articulate abstract theories of progressive iteration either by formulating a set of axioms for arithmetic, or by attempting explicitly to define the natural numbers—on the basis of the theory of the ancestral, say. On either approach we will make an essential appeal to the modern logic first adequately formulated by Frege in 1879: to at least polyadic quantification theory on the first approach, to set theory or some form of higher-order quantification on the second. However, since general logic, for Kant, is given basically by Aristotelian syllogistic, there is no room, on his conception, for a purely conceptual representation of progressive iteration in mere general logic: pure intuition (and therefore transcendental logic) must be called in.

In particular, it is impossible within mere syllogistic logic adequately to represent the essential idea of the infinite or indefinite extendibility of the number series: such an idea requires polyadic quantificational dependence of the form $\forall . \forall\exists$. Since such quantificational forms are of course not available in syllogistic logic, Kant naturally holds that the idea of indefinite iteration cannot be captured in mere general logic. What allows us to think or represent such indefinite iteration is thus taken to be the pure intuition of time: the form of inner sense in which *all* of our representations must necessarily be found (A33–34/B49–51). In other words, for whatever operation we may think, there is always time for iterating or repeating that operation as many times as we wish. It is this fact about the pure intuition of time as the form of inner sense, and this fact alone, that first allows us to represent the idea of progressive iteration:

> . . . that one can require a line to be drawn to infinity *(in indefinitum)*, or that a series of alterations (e.g., spaces traversed by motion) is to be infinitely continued, still presupposes a representation of space and of time that can only belong to intuition, namely, in so far as it is in itself bounded by nothing; for it could never be inferred from concepts. (*Prolegomena*, §12: 4, 285.1–7)

44. Parsons is not of course to be held responsible for the way in which I elaborate this idea below. In particular, it may well be that Parsons has in mind the *higher-order* character of Frege's and Dedekind's theories, while I concentrate below on polyadic *first-order* quantification. On this issue see Tait [107], §XVII.

The pure intuition of time is therefore presupposed in any representation whatever of progressive iteration or the number series; in this sense, time is "the formal condition of all series" (A411/B438).[45]

· IV ·

A very different conception of the relationship between pure intuition and the mathematical sciences from that sketched in §I above emerges from our discussion. The science of arithmetic, as we have seen, is quite independent of the specific character of the objects of our intuition: in an important sense, arithmetic does not concern objects of intuition at all. Nevertheless, the pure intuition of time is still presupposed by the science of arithmetic, not to provide that science with an object or model, as it were, but rather to make the science itself possible in the first place. Without the pure intuition of time, according to Kant, the basic idea underlying the science of arithmetic, the idea of progressive iteration, could not even be thought or represented—whether or not this idea has a corresponding object or model. Thus, for example, whereas geometry is required in order to provide objective meaning for the concept of $\sqrt{2}$ (by constructing the diagonal of a square), arithmetic is required for us to have a "concept of the magnitude of the diagonal of the square" (14, 54.21–55.2). The successive iteration made possible by the pure intuition of time, in other words, is a necessary condition for our possession of the *concept* of magnitude *(quantity)* itself: without such iteration we would be quite unable even *to think* the magnitude of any given thing.[46]

As a matter of fact, the situation with respect to the science of geometry and the pure intuition of space is closely analogous. The Euclidean constructions underlying pure geometry do not serve, in the first instance, to guarantee the objective reality of (as it were, to provide a model for) the science of geometry. As we saw in §II, such an object (or model) can only be provided by empirical intuition, and the constructions of pure geometry secure objective reality only in conjunction with transcendental philoso-

45. These ideas about the representation of infinity are more extensively developed in Chapter 1 above, concentrating primarily on the case of geometry (in the representation of infinite divisibility, for example).

46. See also, once again, A241/B300. Parsons conceives time, not as itself a model for arithmetic (as we saw above, this is absurd), but rather as providing "a universal source of models for the numbers" such that "models for the numbers can be constructed in it if any can be constructed at all": see "Philosophy of Arithmetic," p. 140; [94], p. 118. I would prefer to conceive time rather as a "universal source of representations for the number series," in that any *representation* of the numbers must be constructed in time. From a modern point of view, however, any representation of the numbers—via canonical numerals, say—must itself provide a model for arithmetic: compare note 42 above.

phy itself. In this sense, geometrical construction provides us with only a necessary but insufficient condition for objective meaning: we are provided, strictly speaking, not with objects, but with "mere forms of object" (A239/B298, A223–224/B271). Nevertheless, the Euclidean constructions of pure geometry are still necessary even to think or represent geometrical ideas in the first place—whether or not such ideas have objective reality or corresponding (empirical) objects. Thus, for example, the only way to represent the infinite extendibility of a line segment is by means of Euclid's Postulate 2 (see *Prolegomena,* §12); the only way to represent the infinite divisibility of a line segment is by means of a construction such as Euclid's construction of the bisector in Proposition I.10 (see B40); and so on. In other words, just as in the case of arithmetic and the pure intuition of time, the pure intuition of space is necessary to make geometrical *concepts* first possible.[47]

Kant's language consistently accords with this conception. Thus, in §24 of the second edition transcendental deduction, Kant says:

> We can think no line without *drawing* it in thought; we can think no circle without *describing* it; we can absolutely not represent the three dimensions of space without *setting* three lines perpendicular to one another at the same point. (B154)

The activity of geometrical construction is therefore necessary for even the *thought* of such ideas. Essentially the same point is made in the Axioms of Intuition:

> I can represent no line to myself, no matter how small, without drawing it in thought, that is gradually generating all its parts from a point, and thereby first recording this intuition. . . . The mathematics of extension (geometry), together with its axioms, is based on this successive synthesis of the productive imagination in the generation of figures, which express the conditions of a priori sensible intuition under which alone the schema of a pure concept of outer appearance can arise. (A162–163/B203–204)

Schemata of geometrical concepts, in other words, not only serve to contribute towards the objective reality of such concepts, but are also essential to our rigorous representation of the concepts themselves. This is asserted quite explicitly in the case of the concept of a circle and Euclid's Postulate 3 in the Postulates of Empirical Thought:

> Now a postulate in mathematics is a practical proposition that contains nothing but the synthesis whereby we first give ourselves an object and generate its concept, e.g., to describe a circle with a given line from a given point, and such a proposition can therefore not be proved, because the procedure

47. Compare A25/B39: "an a priori intuition (that is not empirical) underlies all concepts of space."

> it requires is precisely that by which we first generate the concept of such a figure. (B287)

The Euclidean construction of the circle is necessary for giving ourselves an object (or rather the "mere form of an object") corresponding to this concept; it is also necessary for "generating" the concept itself—and it is precisely for this reason that this postulate cannot be proved.

I think that there can be little doubt, moreover, that Euclidean constructions of figures are in fact what Kant means by the schemata of the corresponding geometrical concepts. Any particular figure produced by such a construction counts as an *image* of the corresponding concept, but it is the general procedure for producing any and all such figures that is the *schema* of the concept:

> In fact our pure sensible concepts are not based on images of objects, but on schemata. Absolutely no image of a triangle could ever be adequate to the concept of a triangle in general. For it would never attain the generality of the concept, which makes it hold for all cases—right-angled, oblique-angled, etc.—but would always be limited to a part of this sphere. The schema of the triangle can never exist anywhere but in thought, and it signifies a rule of synthesis of the imagination with respect to pure figures in space. (A140–141/B180)

And what exactly is this "rule of synthesis of the imagination"? Kant makes it clear in the Axioms of Intuition, in the course of explaining the generality of geometry (as contrasted with the singularity of arithmetical formulas):

> If I say: by means of three lines such that two taken together are greater than the third a triangle can be drawn; then I have here the mere function of the productive imagination, whereby the lines can be drawn greater and smaller, and so can be made to meet in accordance with all possible angles. (A164–165/B205)

Thus, the schema of the concept of a triangle is just the Euclidean construction of the triangle of Proposition I.22.

Arithmetical concepts have schemata too, of course, but they are correspondingly more abstract. In particular, they, unlike geometrical constructions, in no way presuppose the representation of space. Just as in the geometrical case, however, such schemata must be sharply distinguished from images:

> Thus if I set five points one after another, · · · · · , this is an image of the number five. On the other hand, if I only think a number as such—which can now be five or one hundred—then this thought is more the representation of a method for representing an aggregate (e.g., one thousand) in an image

according to a certain concept than the image itself, which in our last case could scarcely be surveyed and compared with the concept. (A140/B179)

The schema corresponding to the concept of a particular number *n*, such as 1,000, for example, does not consist of an aggregate of *n* objects (points, say); rather, it consists in the procedure by which any such aggregate can be enumerated. And this representation, I suggest, is just the idea of *n* iterations or repetitions of an arbitrary operation.[48] As we have seen above, there is also a schema for the general concept of magnitude itself, namely, the representation of number in general, "which is a representation that unites the successive addition of unit to unit . . . [in] a homogeneous intuition in general" (A142–143/B182). The schema of the general concept of magnitude is therefore just the idea of successive iteration as such. This representation, as we have seen, presupposes the pure intuition of time for Kant, but it is entirely independent of the pure intuition of space.

It is precisely in schemata, both geometrical and arithmetical, that the peculiar connection between mathematical concepts and pure intuition is to be found. Schemata are not themselves objects for mathematical concepts. Such objects are only to be found in empirical intuition, and images alone, not schemata, belong to empirical intuition (A141–142/B181). Whereas the images that result from the application of mathematical schemata *in concreto* do in fact play a crucial role in providing our mathematical concepts with objective meaning,[49] schemata themselves play quite a

48. In the case of large numbers such as 1,000, we can make this surveyable by means of decimal notation: thus, 1,000 iterations can be conceived as ten iterations of the operation of performing ten iterations of the operation of performing ten iterations. Compare A78/B104: "thus our enumerating [Zählen] (as is seen above all in larger numbers) is a *synthesis according to concepts,* because it takes place in accordance with a common ground of unity (e.g., the decade)." (The example of the five points does not have to be read as suggesting that Kant's numbers are cardinal numbers of discrete collections after all—contrary to note 43 above. For it is not the collection of five points itself, I think, but rather the particular *act* of setting these points one after another that constitutes the image of the number five. This particular act then embodies or exemplifies the general schema of fivefold iteration.)

49. Thus the passage from A239–A240/B298–299 cited above (§I) continues as follows: "One therefore demands that an isolated concept *be made sensible,* i.e., that an object corresponding to it be exhibited in intuition, because, without this, the concept (as one says) would remain without *sense* [*Sinn*], i.e., without meaning [Bedeutung]. The mathematician fulfills this demand by means of the construction of the figure, which is an appearance present to the senses (although it arises a priori). The concept of magnitude in the same science seeks its support [Haltung] and sense [Sinn] in number, and this in turn in the fingers, in the beads of the abacus, or in strokes and points, which are placed before the eyes. The concept is still generated a priori, together with the synthetic principles or formulas derived from such concepts; but their use and their relation to ostensible objects can in the end be

different role. Their role is to provide something essential to the mathematical concepts themselves: namely, the possibility of a kind of rigorous representation of—more precisely, the possibility of a kind of rigorous reasoning with—these concepts that goes far beyond the resources of mere general logic as Kant understands it.

To illustrate, let us return to the example of the Euclidean construction of the circle from A234/B287. Kant says that this construction "first generates the concept of such a figure." Why exactly does he say this? The underlying idea, I think, is that the existential proposition corresponding to the construction—that for any point and any line there is a circle with the given point as center and the given line as radius—cannot be conceptually expressed for Kant. In mere syllogistic logic this existential proposition cannot, strictly speaking, even be stated (as we would now put it, it involves the form of quantificational dependence $\forall\forall\exists$). The only way even to think or represent this proposition—so as, in particular, to engage in rigorous geometrical reasoning thereby—is by means of the construction itself. In other words, Euclid's Postulate 3 does not simply assert something about circles as a mere fact, as it were, rather it alone makes it possible even to have the thought in question in the first place. The only way I can represent the existence of the required circles is by actually being in possession of the construction, and this circumstance automatically makes the thought in question true.

An analogous point holds for symbolic or arithmetical construction. Consider the existential proposition corresponding to the successor function: namely, for every n there is a number $n + 1$. Again, this existential proposition cannot, strictly speaking, even be expressed in mere general logic as Kant understands it (it involves the logical form $\forall\exists$, of course). The only way even to think or represent this proposition—so as, in particular, to engage in rigorous arithmetical reasoning thereby—is by means of our possession of the successor function itself: in Kant's terms, by our capacity successively to iterate any given operation. This, for Kant, presupposes the pure intuition of time, and it again follows that the only way I can represent the existence of the required numbers (which amounts to a representation of the infinity of the number series) automatically makes that very representation true.[50]

---

sought nowhere but in experience, whose possibility (according to its form) they contain a priori."

50. There is also a crucial disanalogy between this case and the case of geometry. Geometry starts from certain specific initial operations or functions and thereby iteratively generates a specific domain of objects (more precisely, of "mere forms of objects"). The successor function, however, is not, for Kant, a specific function at all—but instead expresses the general capacity to iterate as such. Thus in arithmetic no specific domain of objects is

On this conception of the relationship between mathematical concepts and pure intuition, Kant's position turns out to be much stronger than it appeared in §I above. In particular, it is now understandable why and how Kant thinks that mathematics is not only synthetic but also a priori. The rigorous representation of mathematical concepts and propositions requires schemata: constructions in pure intuition. In the case of arithmetic we are given the general capacity for successively iterating any operation in time; in the case of geometry we are given in addition certain fixed, specifically spatial operations as input: the constructions underlying Euclid's geometry. We can only think the propositions of mathematics by, as we might now put it, *presupposing* the required constructions, and it follows, as we have seen, that the true propositions of mathematics are then necessarily true.[51] If we can even think the infinite divisibility of a line segment, for example, it follows, via Euclid's Proposition I.10, that the segment is in fact infinitely divisible. The truth of the true propositions of mathematics follows from the mere possibility of thinking or representing them, and this is the precise sense in which we know them a priori. In other words, the a priori status of mathematics rests, in the end, on a kind of transcendental argument or transcendental deduction, and it in no way depends on the kind of quasi-perceptual "a priori visualization" which we argued in §I above to be entirely unintelligible.[52]

---

iteratively generated. The numbers whose "existence" is in question here are not in any sense objects, but mere "places" in an iterative series—whatever objects may or may not occur as elements of such a series.

51. What about the false propositions of mathematics? They too can only be adequately thought or represented by means of the constructions that guarantee the truth of the true propositions of mathematics; the conditions for adequately thinking and reasoning with any mathematical proposition thereby imply their impossibility. The situation is precisely analogous, in fact, to the status (for us) of a logical contradiction. (Here I am indebted to Howard Stein, who pointed out that an earlier formulation of this point implied that the negation of a thinkable proposition might not itself be thinkable.)

52. See the Preface to the *Metaphysical Foundations of Natural Science* (1786), for example: "to know something a priori is to know it from its mere possibility" (4, 470.18–19); and compare also A88/B121. In the case of geometry, however, the independence of the Parallel Postulate creates insurmountable problems for this conception: the mere possibility of representing a triangle via Proposition I.22, for example, does not automatically imply that the sum of the angles must be 180° (Proposition I.32: the Parallel Postulate is first used in Proposition I.29). It therefore seems to me that the mere existence of consistent systems of non-Euclidean geometry does defeat Kant's view after all: once such systems are on the scene it is impossible to see how we know the truth of one particular system a priori. For Kant, however, Euclidean geometry is (quite reasonably, I think) the *only* geometry; and so the only possibility for rigorous geometrical reasoning is given by Euclid's axiomatization. For Kant, then, it makes perfect sense to view all the theorems of Euclidean geometry as true a priori.

Logical possibility and real possibility are therefore connected in a more subtle fashion than appeared in §I. Although concepts that are mathematically impossible can indeed be logically consistent, it does not follow that there are logically possible states of affairs corresponding to such concepts. For the only way adequately to represent any mathematical state of affairs is by means of the schemata of the concepts in question found in pure intuition, and so the conditions for the possibility of even thinking or representing (purported) mathematically impossible states of affairs themselves rule out such states of affairs. Thus, for example, the concept of a line segment that is not infinitely divisible is logically consistent: no contradiction can be derived from this concept within mere general logic as Kant understands it. Nevertheless, the conditions for the possibility of thinking or representing any line segment whatsoever automatically provide me with Euclid's bisection construction as well, and so a line segment that is not infinitely divisible is entirely impossible. In other words, I do not think "The line may be infinitely divisible and then again it may not," and wait for pure intuition to decide between these alternatives (as we saw above, this idea is quite unintelligible). Rather, in even thinking the proposition that the line is infinitely divisible in the first place, I thereby secure its truth: the contrary supposition is thus not, properly speaking, an alternative possibility at all.[53]

We are now in a position, finally, better to appreciate the force of Kant's conception of the relationship between mathematics and pure intuition in motivating his rejection of the dogmatic metaphysics of the Leibnizean-Wolffian philosophy. In particular, Kant's polemic does not proceed by simply assuming what the Leibnizean-Wolffian philosophy explicitly denies: namely, that general logic is itself inadequate for providing us with a useful notion of objective meaning. Kant's conception of the role of pure intuition in mathematics does not directly invoke his general conception

53. The concept of the two-sided rectilinear figure of A220–221/B268 is handled in the same way. To be sure, no contradiction can be derived from this concept in mere general logic. Nevertheless, pure intuition does not somehow "show" us the impossibility—whatever this might mean. Rather, it is impossible simply because it conflicts with Euclid's axiomatization of geometry. And, in this connection, we should remember that the possibility of such a figure—and so of elliptic space—is incompatible with the axioms of Euclidean geometry quite independently of the Parallel Postulate: it conflicts, in fact, with Postulate 2, which asserts the infinite extendability of finite line segments. (More precisely, *singly elliptic* space contradicts Postulate 2—but also contains no two-sided plane figures, since it satisfies the assumption of the uniqueness of the straight line segment connecting any two points implicit in Postulate 1. *Doubly elliptic* space, on the other hand, which does contain two-sided plane figures, contradicts both Postulate 1 and Postulate 2. Thus, whereas doubly elliptic space can of course be visualized in two dimensions as the surface of a Euclidean sphere, singly elliptic space can be visualized in two dimensions as the surface of a Euclidean hemisphere on which antipodal points of the bounding "equator" are identified.)

of objective meaning and objective reality at all; for this latter concerns empirical intuitions and images, not pure intuitions and schemata. Rather, Kant's theory of the synthetic character of mathematics is an attempt to show the inadequacy of mere general logic directly, quite independently of the relation of our concepts to possible (empirical) objects. Mere general logic is entirely inadequate for even the *representation* of mathematical concepts and propositions, and this circumstance motivates a move to pure intuition and transcendental logic all by itself. To be sure, after we have made this decisive break, we are subsequently able to develop a characteristically Kantian notion of objective meaning and objective reality. For it then turns out that the mathematical schemata of pure intuition have a dual role: they serve to generate the concepts of pure mathematics, and also, when embodied in particular constructive activities *in concreto*, to provide objects (namely, images) for these concepts. That we thereby obtain a coherent notion of objective reality, however, only follows if we are also able to show "that the image-forming synthesis through which we construct a triangle in imagination is precisely the same as that which we exercise in the apprehension of an appearance, in making ourselves an empirical concept of it" (A224/B271). And we are able to show this, in turn, only when we move beyond the theory of pure mathematics to the deeper mysteries of the transcendental deduction.

· V ·

We have concentrated above on the relations between the sciences of arithmetic (including "general arithmetic" or algebra) and geometry to pure intuition: the relationship between arithmetic (and algebra) and the pure intuition of time, the relationship between geometry and the pure intuition of space. We saw in §II, however, that there is a third mathematical science that is also essentially connected to pure intuition for Kant: namely, "pure mechanics" or "the general doctrine of motion." Indeed, it is this science, not arithmetic, that stands to the pure intuition of time as geometry does to the pure intuition of space. By briefly considering the role of this latter science in Kant's philosophy, we can, I hope, further illuminate his general conception of the role of mathematics and pure intuition.

What does it mean to say that pure mechanics stands to time as geometry does to space? As we saw above, geometry supplies *quanta* or "objects of intuition as magnitudes" for the general theory of magnitude (general arithmetic). Geometry enables us to construct specifically spatial magnitudes, such as lengths, areas, and volumes, and, by means of the Euclidean metric of space, geometry makes it possible to consider such spatial objects as magnitudes. For, in order to consider an object as a magnitude, we must view it as a "composition of homogeneous [units] [Zusammensetzung des Gleichartigen]" (B202–203). In the case of our paradigmatic spatial magnitudes, then, we require the notion of *congruence:* that is, the notion of two spatial lengths, areas, volumes, and so on having the same magnitude.[54] In other words, the units that are to be composed or successively added must be congruent or have the same magnitude if their number is to indicate the magnitude of the object in question.

Now arithmetic, as we have also seen, does not consider time itself as a *quantum* or object of intuition as magnitude. The successive iteration on which arithmetic is based necessarily takes place in time, but the objects that are manipulated thereby do not themselves need to be temporal objects: in the case of our intuition, in fact, such *quanta* are primarily spatial objects. Nevertheless, arithmetic itself is entirely independent of the specifically spatial character of our intuition: in this sense, it requires only a manifold of intuition in general (A143/B182, B162). Kant clearly also holds, however, that in order to consider time itself as a *quantum* or object of intuition as magnitude—in order to apply the concept of magnitude to time itself—the mediation of spatial intuition is absolutely necessary:

> . . . the possibility of things as magnitudes, and thus the objective reality of the category of magnitude, is only able to be exhibited in outer intuition; and only by means of the latter is it afterwards also applied to inner sense. (B293)

That this application of the concept of magnitude to inner sense consists precisely in the determination of temporal duration, as a magnitude, is explicitly stated in §24 of the transcendental deduction:

> . . . we must always derive the determination of temporal intervals [Zeitlänge], and also the temporal positions [Zeitstellen] for all inner perceptions, from that which outer alterable things present to us. (B156)

And this means, I suggest, simply that the notion of temporal congruence or sameness of magnitude for temporal intervals can only arise in connection with spatial intuition: time by itself, as it were, cannot support a metric.

How, then, does the mediation of spatial intuition enable us to endow time with a metric? How does it generate a notion of temporal congruence? In the texts in which the passages just cited occur, it is very strongly suggested that this happens precisely through the idea of motion. Thus B292 asserts that:

54. Compare *Metaphysical Foundations,* 4, 493.14–17: "Complete similarity and equality [Gleichheit], in so far as it can be known only in intuition, is *congruence.* All geometrical construction of complete identity rests on congruence." Compare also *Prolegomena,* §12, for the importance of equality of magnitude [Gleichheit] in geometry.

> . . . in order to make inner alterations themselves thinkable, we must make time, as the form of inner sense, conceivable to ourselves figuratively as a line, and the inner alteration by means of the drawing of this line (motion),

and at B156 we have:

> . . . time, although it is certainly no object of outer intuition, cannot be made representable to us, except under the image of a line, in so far as we draw it.

Most striking, perhaps, is B154–155:

> I cannot think time itself, except by attending, in the *drawing* of a straight line (which is to serve as the outer figurative representation of time), merely to the act of synthesis of the manifold by which we successively determine inner sense, and thereby attending to the succession of this determination in inner sense. Motion, as act of the subject (not as determination of an object),* and thus the synthesis of the manifold in space—if we abstract from the latter and attend merely to the act by which we determine *inner* sense according to its form—such motion in fact first produces the concept of succession.

In other words, it is the idea of the rectilinear motion of a mathematical point that first enables us to represent time itself as an object of intuition[55]—and, presumably, first enables us to represent time as a *quantum* or object of intuition as magnitude.

I propose to interpret this last idea as a reference to inertial motion: the privileged state of force-free "natural" motion which is basic to modern physics. For, given the idea of an inertial trajectory or inertially moving mathematical point ("particle"), we can then derive the temporal metric from the spatial metric: equal or congruent temporal intervals are those during which an inertially moving point traverses equal or congruent spatial intervals. In other words, we do not define inertial motion in terms of a pre-existing temporal metric; we do not conceive inertial trajectories as those rectilinear trajectories that traverse equal distances in equal times. Rather, we first pick out a class of privileged inertial trajectories (in modern terminology, we endow space-time with an "affine structure"), and then define the notion of temporal equality in terms of these trajectories. This, in any case, would explain why Kant holds that the law of inertia is itself an a priori law (Kant's second law of mechanics in the *Metaphysical*

55. That we are here dealing with the motion of a mathematical point follows from the footnote to this passage, where the motion in question is characterized as the "*describing* of a space," in conjunction with *Metaphysical Foundations*, 4, 489.6–11: "In phoronomy, since I am acquainted with matter through no other property but its movability, and therefore may consider it only as a point, motion can be considered merely as the *describing of a space*—although in such a way that I do not merely attend to the space that is described, as in geometry, but also to the time in which, and thus to the speed with which, a point describes a space."

Foundations of Natural Science: 4, 543.15–20). Moreover, that Kant has inertial motion in mind here is further supported by the fact that the intuition of "the motion of a point in space" in the drawing of a line is said to be the "intuition corresponding to the concept of *causality*" at B291–292, whereas it is precisely the law of inertia that realizes the category of causality in the *Metaphysical Foundations.*[56]

It is important to note, however, that the pure representation of inertial motion (in modern terminology, of an "affine space-time") can only be applied so as to yield an actual determination of temporal intervals if we have a means for picking out or constructing a privileged frame of reference. (Relative to *what* do inertially moving bodies traverse equal distances in equal times?) Moreover, we also need a theory of *force* or deviation from the natural inertial state, for we are not in fact confronted with any actual cases of force-free motion in nature. Both, for Kant, require the consideration of actually existent forces, and it follows that in order to apply the pure representation of inertial motion we need to move beyond pure mechanics to *empirical* physics. How this move is made, at least in the case of the Newtonian theory of universal gravitation, is depicted in the *Metaphysical Foundations,* but these matters lie beyond the bounds of our present discussion.[57]

56. See 551.9–14. This interpretation is also consistent with Kant's seemingly peculiar procedure in the first chapter or Phoronomy of the *Metaphysical Foundations.* He there characterizes phoronomy as "the pure theory of magnitude *(mathesis)* of motion" (489.11–12); and, since "the determinate concept of a magnitude is the concept of the generation of the representation of an object by means of the composition of homogeneous [units] [Zusammensetzung des Gleichartigen]" (489.12–14), he then proceeds to derive the classical principle of the composition of velocities (490–493). He emphasizes, however, that this principle is not a "mechanical" principle governing "moving forces," but rather a "*mathematical construction* . . . that only serves to make intuitive what the object (as *quantum*) *is*" (495.28–38). In other words, it is only the classical or Galilean velocity addition law, as derived from the manipulation of relative spaces according to the classical or Galilean transformation law, which first enables us to conceive motion (speed) as itself a magnitude. It follows that the traditional definition $v = s/t$ (see 4, 484.37) does not by itself suffice; and this can only mean, I suggest, that we have no independent way of conceiving time as a magnitude: on the contrary, time becomes a magnitude only after motion does.

57. For further discussion of the place of the Newtonian theory of universal gravitation in the *Metaphysical Foundations,* see Chapter 3 below. I believe it is this twofold role of the representation of motion—pure and applied, as it were—that explains Kant's remarks at A41/B58, where he asserts that transcendental aesthetic contains only space and time "because all other concepts belonging to sensibility, even that of motion, which unites both elements, presuppose something empirical." This is at first sight puzzling, for Kant has previously given precisely "the general doctrine of motion" as an example of the synthetic a priori knowledge yielded by time (B48–49). At B155n, however, Kant explains that the concept of motion can be understood in two ways: "motion of an *object* in space" is certainly empirical, yet "motion, as the *describing* of a space, is a pure act of successive synthesis of the manifold in outer intuition in general through the productive imagination, and belongs

Be this as it may, we can meanwhile observe that Kant's conception of the relation of pure mechanics to the pure intuition of time illuminates central features of his general doctrine of pure intuition: specifically, the important distinction he draws between "form of intuition" and "formal intuition" at B160n. In the case of time, the basic point emerges in the course of Kant's correspondence with Rehberg:

> Not the time-magnitude [Zeitgröße] (for this would contain a circle in the explanation) but rather merely the time-form [Zeitform] is taken into account in the estimation of magnitude [Großenschätzung]. However, without space time itself would not be represented as a magnitude; and, in general, this concept would have no object. (14, 55.4–7)

As the context makes clear, by estimation of magnitude Kant means the procedures of enumerating and measuring made possible by arithmetic and algebra. These procedures are thus independent of the metric of time—of the consideration of time itself as a magnitude—and involve only the fact that our representations, whatever objects or content they may have, are temporally ordered.[58] Arithmetic and algebra, therefore, depend only on time as a form of intuition: as the form of inner sense (A33–34/B49–51). By contrast, to represent time as itself a magnitude (as "time-magnitude"), the representation of space must also be considered—presumably, in the pure theory of motion. It is this pure theory of motion that alone enables us to consider time as a formal intuition: that is, as an object of intuition or as itself an intuition (B160).[59]

The science of geometry, however, considers spatial objects as *quanta:* objects of intuition as magnitudes. Geometry presupposes that spatial objects have metrical structure—that they are themselves magnitudes—and thus it deals throughout with objects of intuition or with intuitions themselves. It follows that the science of geometry presupposes that space is

not only to geometry but even to transcendental philosophy." This last kind of motion can therefore be considered by a "pure science." Nevertheless, even this last kind of motion does not belong *wholly* to transcendental aesthetic, for, as "figurative synthesis," it represents an "action of the understanding on the sensibility and the first application of the understanding to objects of our possible intuition (which is at the same time the ground of all other applications)" (B152).

58. This makes sense, because our capacity successively to enumerate does not depend on the metrical properties of time: it requires only that a representative of the number series (considered as a type of linear ordering) be embeddable into time, and thus that time have the structure of a linear ordering (more precisely, that time contain a linear ordering as a substructure).

59. Compare B154: "inner sense contains the mere *form* of intuition, but without combination of the manifold in this form, and thus it still contains absolutely no *determinate* intuition—which is only possible by means of the consciousness of the determination of the manifold through the transcendental action of the imagination (synthetic influence of the understanding on inner sense), which I have called figurative synthesis."

not merely a form of intuition but also an object of intuition in its own right—that is, geometry presupposes space as a formal intuition:

> Space, represented as *object* (as is actually required in geometry), contains more than the mere form of intuition—namely, it contains *uniting* [*Zusammenfassung*] of the manifold according to the given form of sensibility—so that the *form of intuition* yields only a manifold, but the *formal intuition* yields unity of representation. (B160n)

To consider space as merely a form of intuition, by contrast, is simply to conceive any particular representation of outer sense as itself containing a spatial ordering. Starting with a particular representation of a body, for example, I can "abstract out" the contributions of understanding and sensibility so as to be left solely with the pure intuition of space—as a mere form of sensibility (A20–21/B35). Yet this fact about my representations of outer sense does not by itself enable me to "unite their manifold." For this something more is required:

> Thus the mere form of outer sensible intuition, space, is still absolutely no cognition; it yields only the manifold of a priori intuition for a possible cognition. In order to cognize anything in space, e.g., a line, I must *draw* it, and therefore produce a determinate combination of the given manifold, in such a way that the unity of this act is simultaneously the unity of consciousness (in the concept of a line), and thereby first cognize an object (a determinate space). (B137–138)

It is the construction of geometrical concepts, then, that first enables me to conceive space as an object, that is, as a formal intuition, and it is in this sense that pure mechanics stands to time as geometry stands to space.[60]

60. Once again, however, we should remember that geometrical construction by itself, strictly speaking, provides only a necessary but insufficient, condition for conceiving space as an object. For geometrical construction actually provides only the "mere form of an object," and to obtain objects in the full-blooded sense we need to show that the space of geometrical construction is precisely the same space as the form of outer sensible (i.e., empirical) intuition: "The synthesis of spaces and times, as the essential form of all intuition, is that which at the same time makes possible all apprehension of appearance, and therefore every outer experience, and consequently all cognition of objects of outer sense; and what mathematics proves of the former in its pure employment, necessarily holds also of the latter" (A165–166/B206). Thus only the *physical* space of *applied* geometry counts as an object in the full-blooded sense. That geometrical construction nonetheless contributes towards the consideration of space as an object in the full-blooded sense depends, I think, on the circumstance that geometrical construction involves *both* time and space and that the "*drawing* of a line" thus exemplifies "figurative synthesis"—which "belongs not only to geometry but even to transcendental philosophy." This, in turn, means that we are necessarily involved with both *mathematical* and *dynamical* synthesis here (B201n), and it follows that the consideration of space as a formal intuition in the end leads us to the spatio-*temporal* structure underlying physical dynamics. See Chapter 4 below for further discussion.

Finally, we should note that the necessary connection between the pure intuition of time and the pure intuition of space effected by the general doctrine of motion or pure mechanics plays a central role in Kant's refutation of idealism. Kant makes the point himself in a striking passage from his letter to Rehberg, a passage which nicely brings together the considerations with which we have last been occupied:

> The necessity of the connection of both sensible forms, space and time, in the determination of the objects of our intuition—so that time, if the subject makes it itself an object of its representation, must be represented as a line in order to cognize it as *quantum*, just as, conversely, a line can only be thought as a *quantum* by being constructed in time—this insight into the necessary connection of inner sense with outer sense even in the time-determination of our existence appears to me to aid in the proof of the objective reality of the representations of outer sense (contrary to psychological idealism), which, however, I cannot pursue further here. (11, 210.24–34)[61]

The idea, presumably, is that it is only the general doctrine of motion, which unites time and space, that first enables me to consider any temporal interval—including the temporal interval comprising my own existence as a temporally extended thing—as a *quantum* or object of intuition as magnitude. And this in turn is necessary if I am to represent any temporally extended thing, including myself, as a determinate object of experience. It follows that I cannot determine my own existence in time, as an object of experience, independently of objects outside me in space. But we too are unable to pursue this thought further here.

61. This point is remarked upon by Parsons [94], pp. 117–118 and n. 58 on p. 121; compare also "Philosophy of Arithmetic," p. 133.

· CHAPTER 3 ·

Metaphysical Foundations of Newtonian Science

The science for which Kant aims to provide "metaphysical foundations" in the *Metaphysical Foundations of Natural Science* is Newtonian science: in particular, the science of Newton's *Principia* (1687). This is indicated by the many explicit references to Newton and the *Principia* scattered throughout the *Metaphysical Foundations,* and, more important, by its content—which centrally involves both Newton's laws of motion (especially in Chapter 3 or Mechanics) and the theory of universal gravitation (especially in Propositions 5–8 and the General Observation to Chapter 2 or Dynamics). Moreover, it is quite clear that Newton's *Principia* serves as the model for scientific achievement during the whole of Kant's long career: from *Thoughts on the True Estimation of Living Forces* (1747) and *Universal Natural History and Theory of the Heavens* (1755) to the unpublished *Opus postumum* (1796–1803). Thus, whether or not one wishes to extract philosophical morals from the *Metaphysical Foundations* that transcend the specific content of Newton's *Principia,* there can be no doubt at all that this work is paradigmatic for Kant.

But what does it mean to supply Newtonian science with metaphysical foundations, and why is this enterprise so important to Kant? It is helpful, I think, to distinguish two different aspects of Kant's project. On the one hand, Newton's *Principia* represents a realization of the transcendental principles (as the *Metaphysical Foundations* puts it, the "general metaphysics") contained in the first *Critique.* As such, it provides Kant's system with an "example *in concreto*" that confers "sense and meaning" on the exceedingly abstract concepts and principles of transcendental philosophy:

> And so a separate metaphysics of corporeal nature does excellent and indispensable service to *general* metaphysics, in so far as the former furnishes examples (instances *in concreto*) in which to realize the concepts and propositions of the latter (properly transcendental philosophy), that is, to provide a mere form of thought with sense and meaning. (4, 478.15–20)

In this sense, an investigation of the foundations of Newtonian science is indispensable for a full understanding of Kantian metaphysics.[1]

On the other hand, Kant sees Newtonian science as in need of a critical or metaphysical analysis, an analysis that reveals the origin and meaning of its basic concepts and principles. Such science is inextricably entangled with metaphysical issues; it therefore requires the service of transcendental philosophy in making these issues more explicit and placing them in their proper context:

> Thus these mathematical physicists could certainly not avoid metaphysical principles, and among those certainly not such as make the concept of their proper object, namely matter, a priori suitable for application to outer experience: as the concepts of motion, the filling of space, inertia, etc. However, they rightly held that to let merely empirical principles govern these concepts would be absolutely inappropriate to the apodictic certainty they wished their laws of nature to possess; they therefore preferred to postulate such principles, without investigating then in accordance with their a priori sources. (472.27–35)

(Newton is clearly paradigmatic of the "mathematical physicists" in question here; Newton's laws of motion are apparently paradigmatic of those principles which the "mathematical physicist" simply "postulates" and the philosopher "investigates according to their a priori sources.") In this sense, Newton's *Principia* serves as the object of an important application of transcendental philosophy, and this application of metaphysics to physics is necessary for a full understanding of physics itself.[2]

· I ·

Let us begin, then, by asking why Kant views Newtonian science as in need of critical analysis or metaphysical foundations: What is lacking in the *Principia* as Newton wrote it? There is a tendency to locate Kant's disagreement with Newton at the level of matter theory: specifically, in the contrast Kant sets up between a "mathematical-mechanical" conception of matter and a "metaphysical-dynamical" conception of matter in the General Observation to Dynamics of the *Metaphysical Foundations*. Kant wishes to "banish the so-called solid or absolute impenetrability from natural science, as an empty concept" (523.24–25), and to replace

1. This aspect of the *Metaphysical Foundations* is stressed in Plaass's penetrating study [97], especially §0.5. In particular, Plaass notes the connection between the above-cited passage and the General Note to the System of Principles of the first *Critique* ([97], p. 20). For an opposing view, see Hoppe [55], §II.2, p. 41.

2. Again, the importance of understanding this twofold role of the *Metaphysical Foundations*—as both *realization* and *application* of transcendental philosophy—is stressed by Plaass [97], e.g., p. 68.

this concept with a "dynamical" conception of "relative impenetrability" (Definition 4 of the Dynamics: 501–502) based on a "fundamental force" of repulsion. Since Newton himself appears to embrace "absolute impenetrability"—the "solid, massy, hard, impenetrable, moveable Particles" of Query 31 of the *Opticks* ([84], p. 400)—it appears that Kant's central problem with Newtonian science revolves around its "atomistic" conception of matter. Kant is then seen as opposing such Newtonian "atomism" with a "dynamistic" conception of matter growing out of a broadly Leibnizian approach to natural philosophy.[3]

There is no doubt that Kant does oppose an "atomism" that assumes "absolute impenetrability" as an original and essential property of matter. There is also no doubt that Kant's disagreement with Newton involves broadly Leibnizian strands of thought. Yet the idea that Kant's central disagreement with Newton is located here, at the level of matter theory, seems to me to be profoundly misleading.

First of all, it is far from clear that Kant himself has Newton in mind as a representative of the "mathematical-mechanical" conception of matter.[4] Kant names only Democritus and Descartes (533.2–3) as exponents of the "mechanical natural philosophy" (532.36) he opposes; and he describes this philosophy as wishing to view all actions of matter as arising from the sizes, shapes, and motions of elementary particles—as "machines" (533.12–14)—which are thereby deprived of all "proper forces" (525.15). This is certainly not Newton's view. Further, Kant criticizes this "mechanical natural philosophy" in essentially Newtonian terms: it permits too much "freedom for the imagination" (525.17–19) in "feigning [Erdichtung]" hypotheses (532.13–19), because the sizes, shapes, and motions of its elementary particles are inaccessible to experiment. The "dynamical style of explanation [dynamische Erklärungsart]," by contrast, is "far more suitable and favorable to experimental philosophy, in that it leads directly to the discovery of the proper moving forces of matter and their laws" (533.21–24). And, as the context makes clear, the central example of such a discovery is precisely Newton's theory of universal gravitation (534.15–18). Indeed, one of the main goals of the Dynamics as a whole is to defend the Newtonian attraction—as a true "fundamental force" acting immediately at a distance—against "all sophistries [Vernunfteleien] of a metaphysics that misunderstands itself"

3. See Okruhlik [88], § II; Harman [43], §§IV.5–7; McMullin [76], pp. 119–123; and the Introduction to Ellington's (1970) translation of the *Metaphysical Foundations*.

4. In view of Kant's Observation to Proposition 1 of the Dynamics (497.30), Plaass's suggestion that Lambert is Kant's main target here appears extremely plausible ([97], §0.6, p. 22).

(523.26–29—apparently a reference to Leibniz). So it is most implausible to locate Kant's primary disagreement with Newton here.[5]

Second, Kant's one *explicit* criticism of Newton in the *Metaphysical Foundations* is of quite a different character. This criticism is found in Observation 2 to Proposition 7 of the Dynamics, where Kant disputes the "common opinion" that Newton is able to do without the assumption of an immediate attraction at a distance as an essential property of matter—by leaving room for a possible explanation of gravitation in terms of the pressure exerted by an aether, for example. If Newton—"with the most rigorous abstinence of pure mathematics" (514.30–31)—forbears from postulating such an immediate and essential attraction, then, according to Kant, he is left with no way of "grounding" the proposition that gravitational attraction is directly proportional to mass. In this connection, Kant singles out Proposition VI, Corollary II of Book III of *Principia,* along with Newton's attempt to compare the masses of Jupiter and Saturn by means of the accelerations of their satellites in the Corollaries to Proposition VIII, for special criticism. Kant then remarks, in the Note to Definition 7, that the proportionality of gravitational attraction to mass can only be derived with the help of his own Proposition 7: "The attraction essential to all matter is an immediate action of one matter on another across [durch] empty space" (512.17–19). The significance and force of this Kantian criticism are not immediately obvious, of course, but it is clearly not an external criticism based on a "dynamistic" ("Leibnizean") metaphysics. Rather, Newton is criticized for not daring to be "Newtonian" enough: in denying that gravitation is essential to matter Newton is "set at variance with himself [ihn mit sich selbst uneinig machte]" (515.32).

Finally, we should remember that Kant's critical analysis of Newton's *Principia* is an application of transcendental philosophy, an application that is also supposed to serve as a realization of that philosophy which illustrates its fundamental concepts and principles *in concreto.* From this point of view we should hardly expect matter theory to be central. Rather, we should expect Kant's primary object of concern to be the spatio-temporal framework of the *Principia:* specifically, the notions of absolute

5. Indeed, when one compares Kant's description of the "dynamical style of explanation [dynamische Erklärungsgründe]" towards the end of the General Observation to Dynamics (534.20–30) with Newton's description of his own methodology in Query 31 of the *Opticks* ([84], pp. 401–402), and compares both with the Preface to Kant's *Theory of the Heavens* where he explains that his two forces of attraction and repulsion are "both borrowed from the philosophy of Newton" (1, 234–235), it then becomes difficult to resist the conclusion that Kant's "dynamical natural philosophy" (532.39–40) is modeled precisely on that of Newton.

space and absolute time that are fundamental to Newton's presentation of his theory.[6] These notions, as employed by Newton, can of course find no place in the critical philosophy, and Kant is therefore faced with the problem of capturing the content of Newton's theory without relying on such metaphysically suspect notions. For Kant absolute space and absolute time are not possible objects of experience. How then can the *Principia,* which is entirely based on these notions, find such brilliantly successful application to experience? Here is Kant's "Leibnizean" problem. Here is where Kant needs to find a middle ground between Newtonian "absolutism" and Leibnizean "relationalism."

The overall structure of the *Metaphysical Foundations* confirms this diagnosis. Kant begins the very first chapter, the Phoronomy, by distinguishing between absolute and relative space, and by arguing, in Observation 2 to Definition 1, that an absolute space can be no object of experience.[7] Accordingly, he enunciates a thoroughgoing relativity principle:

> Every motion, as object of a possible experience, can be viewed arbitrarily as motion of the body in a space at rest or as the contrary motion of the space in the opposite direction with the same speed. (487.16)

Yet Kant immediately points out, in the Observation to this Principle, that qualification is required in the case of curvilinear motion (for example, the daily rotation of the earth cannot be ascribed to the surrounding space in which the fixed stars are at rest) and, therefore, that thoroughgoing relativity of motion holds only from a purely phoronomical or kinematical point of view (488.26–38). Thus, in the third chapter or Mechanics, where the Newtonian laws of motion are first introduced, Kant observes that thoroughgoing relativity of motion cannot hold where actual moving forces (and actual causal relations) are involved (footnote to the Proof of Proposition 4: 547.7–10). Finally, the fourth chapter or Phenomenology purports to be a systematic discussion of the entire issue. The Definition initiating this chapter characterizes matter as "the movable, in so far as it, as such a thing, can be an object of experience" (554.6–7), and Kant accordingly emphasizes the need, within experience, to draw a distinction between true (or actual) and apparent motion. The General Observation to Phenomenology then outlines a procedure for implementing this distinction (in the course of which, for example, the daily rotation of the earth is determined to be true or actual) by "reducing all motion and rest to

6. In this connection, see Okruhlik [88], pp. 254–255, 265; Palter [89].

7. See 481.28–31: "To assume an absolute space—i.e., one such that, since it is not material, it can also be no object of experience—as *given for itself* is, since it can neither be observed in itself nor in its consequences (motion in absolute space), to assume something for the sake of experience—which latter must still always be erected without it. Absolute space is therefore *in itself* nothing and absolutely no object . . . ."

absolute space" (560.5–6)—where the latter is enigmatically characterized as an "idea of reason" (559.8).

It is not at all clear what precisely Kant has in mind here, of course. It is clear, however, that questions of absolute versus relative space and absolute versus relative motion are the central questions to be clarified in any attempt to understand Kant's critical analysis or metaphysical foundations of Newtonian science. This, in any case, is how I shall proceed in what follows.

· II ·

The idea of absolute motion with respect to absolute space lies at the basis of Newtonian physics: there is an infinite three-dimensional Euclidean space within which moving bodies satisfy the laws of motion. According to the law of inertia (first law of motion), for example, a body acted upon by no external forces moves with constant speed along an infinite Euclidean straight line. Yet we are given neither a Newtonian absolute space nor cases of rectilinear inertial motion in our actual experience of nature. All we actually observe are cases of non-inertial motion given relative to some physically specified frame of reference: for example, the orbits of the heavenly bodies observed relative to the earth and the fixed stars. How, then, is it possible to apply the Newtonian idea of absolute motion with respect to absolute space to our actual experience of nature?

Newton himself is certain that his abstract spatio-temporal framework can in fact be applied to our experience. For, although it is by no means an easy or trivial matter, it is possible *to infer* the true or absolute motions from their observable effects:

> It is indeed a matter of great difficulty to discover, and effectually to distinguish, the true motions of particular bodies from the apparent; because the parts of that immovable space, in which bodies truly move, do not affect our senses. However, the situation is not completely desperate; for we have some arguments to guide us, partly from the apparent motions, which are the differences of the true motions; partly from the forces, which are the causes and effects of the true motions. ([82], pp. 52–53; [83], p. 12)

Indeed, the central aim of the *Principia* is to carry out just such an inference:

> But how we are to obtain the true motions from their causes, effects, and apparent differences, and the converse, shall be explained more at large in the following treatise. For to this end it was that I composed it. ([82], p. 53; [83], p. 12)

The inference in question is carried out explicitly in Book III, where Newton applies his laws of motion to empirically given "phenomena" so as to derive first the inverse-square law and then the law of universal gravitation. The argument then culminates in Proposition XI: "That the common center of gravity of the earth, the sun, and all the planets, is immovable" ([82], p. 586; [83], p. 419).[8]

Newton's procedure for inferring "the true motions from their causes, effects, and apparent differences" is thereby vindicated in our actual experience of nature. We begin with *apparent* motions: the purely relative motions observable in the solar system. From these, however, we are able to determine the true or absolute motions: in particular, that the earth truly rotates around the sun and not vice versa. And it is in this way that we can finally settle the issue of heliocentrism in favor of Copernicus (more precisely, in favor of Kepler). Antecedent to the argument of Book III of *Principia,* on the other hand, there simply is no way to determine whether the earth truly rotates around the sun or vice versa.[9] Moreover, if all we really had available to us were apparent or merely relative motions, there would not even be an issue to be settled here. It is no wonder, then, that Newton himself has no qualms whatsoever about the actual empirical applicability of his abstract spatio-temporal framework.

Now Kant's transcendental philosophy is of course centrally concerned with elucidating the conditions for the application of spatio-temporal notions in experience. The fourth chapter or Phenomenology of the *Metaphysical Foundations* is centrally concerned with deriving true motions from apparent motions:

> If however the movable *as such a thing,* namely, according to its motion, is to be thought of as *determined*—i.e., for use in a possible experience—then it is necessary to indicate the conditions under which the object (matter) must be determined one way or another through the predicate of motion. Here there is no question of transforming illusion [Schein] into truth, but rather of appearance [Erscheinung] into experience [Erfahrung]. (Observation to the Definition: 555.2–7)

And, as Kant indicates, the derivation of true motions from merely relative or apparent motions is itself a matter of constituting *experience* from mere *appearance.* But how, according to Kant, is the notion of true motion to be understood? Since, as we have seen, Kant rejects Newtonian absolute space, it is entirely unclear how he is entitled to appeal to this notion here.

My suggestion is that we view Kant, in the Phenomenology in particular, as attempting to turn Newton's argument of Book III of *Principia* on its head. Newton begins with the ideas of absolute space and absolute

8. See Stein [104] for a lucid and penetrating analysis of Newton's argument from a modern point of view.

9. Kant, for one, is well aware of the fundamental importance of the *Principia* in this regard: see the Preface to the second edition of the first *Critique* at Bxxii,n.

motion, formulates his laws of motion with respect to this pre-existing spatio-temporal framework, and finally uses the laws of motion to determine the true motions in the solar system from the observable, so far merely relative or apparent, motions in the solar system. Kant, on the other hand, conceives this very same Newtonian argument as a *constructive procedure for first defining the concept of true motion.* This procedure does not find, discover, or infer the true motions; rather, it alone makes an objective concept of true motion possible in the first place.

Thus, for example, Newton conceives his laws of motion as asserting facts, as it were, about antecedently well-defined true motions—motions defined relative to absolute space. Kant, since he rejects absolute space, conceives the laws of motion rather as conditions under which alone the concept of true motion has meaning: that is, the true motions are just those that *satisfy* the laws of motion. In particular, Kant exploits the idea—which goes back to his *New System of Motion and Rest* of 1758—that one can use Newton's third law of motion, the equality of action and reaction, to characterize a privileged frame of reference for describing the true motions in a system of interacting bodies. This, of course, will be the center of mass frame of the system, wherein every acceleration a_A of body A with mass m_A is counterbalanced by an acceleration a_B of body (or system of bodies) B with mass m_B such that $m_A a_A = -m_B a_B$ (compare the discussion of Proposition 4 of the Mechanics: 544–551). In such a center of mass frame apparent motions necessarily count as true motions (Proposition 3 of the Phenomenology: 558).

Kant thus views the laws of motion as definitive or constitutive of the spatio-temporal framework of Newtonian theory, and this, in the end, is why they count as a priori for him. Using the laws of motion we do not then find, discover, or infer that the center of mass of the solar system is in a state of absolute rest; rather, the center of mass of the solar system yields that frame of reference wherein the concepts of true or absolute motion and rest are *defined.* For Kant, however, the center of mass of the solar system yields only an approximation to the desired frame of reference. Indeed, Kant's early writings on gravitational astronomy make it quite clear that neither the earth, the sun, the center of mass of the solar system, nor even the center of mass of the Milky Way galaxy defines the desired frame of reference.[10] Only "the common center of gravity of all matter" (563.4–5) defines such a privileged frame, and this point is of course forever beyond our reach. For Kant, then, Newton's theory does

10. See the *New System of Motion and Rest* and especially the *Theory of the Heavens,* Second Part, Chapter VII: as the planets orbit the sun, and the stars in our galaxy orbit the center of the Milky Way, so the Milky Way itself orbits about a common center of the galaxies, and so on *ad infinitum.*

not require the actual existence of such a privileged frame; rather, it specifies a constructive procedure for finding better and better approximations—a procedure which never actually fully attains its goal. Thus, if we think of Kant's "absolute space" as the ideal end-point of this constructive procedure—the privileged frame of reference towards which it "converges," as it were—it becomes clear why "absolute space" in this sense is characterized as an idea of reason.[11]

Kant's explicit discussion in the Phenomenology is at least consistent with the above suggestion. Like Newton, Kant begins with "appearances [Erscheinungen],"[12] which involve purely relative motions. These "appearances" are then to be "transformed into experience [Erfahrung]" (554.9–555.13), yielding true or actual motions that are no longer merely relative. For example, "the circular motion of a matter, as opposed to the contrary motion of the [surrounding] space, is an actual predicate of the former" (Prop. 2: 556.30–32). Further, Kant's tools for effecting this "transformation" are precisely Newton's laws of motion. Finally, Kant explicitly, and approvingly, refers to Newton's Scholium to the Definitions of *Principia,* where we are told how "to discover, and effectually to distinguish, the true motions of particular bodies from the apparent" (compare the Observation to Proposition 2 at 557.32–558.6 and the second footnote to the General Observation at 562.32–40). All this lends plausibility to the idea that Newton's argument in Book III, where the general notions of the Scholium are actually put into practice, is Kant's model.

But how does Kant's procedure for "reducing all motion and rest to absolute space" actually go? We proceed in three distinct steps or stages,

11. From a modern point of view there is of course no *single* privileged frame of reference at all—even conceived as an idealization to which we approximate. Rather, there is a *class* of privileged frames, the so-called *inertial frames,* and each member of this class of frames represents the true motions equally well. Indeed, Newton himself already explictly recognizes this circumstance in Corollary V to the laws of motion: "The motions of bodies included in a given space are the same among themselves, whether that space is at rest, or moves uniformly forwards in a right line without any circular motion" ([82], p. 63; [83], p. 20). The laws of motion are therefore equally valid in all inertial frames; and, from a modern point of view, we can therefore understand these laws as *implicitly defining* the inertial structure (affine structure) of Newtonian space-time (see, e.g., Stein [104]). Kant, on the other hand, does not explicitly acknowledge Corollary V and instead views the constructive procedure sketched above as constitutive (in the limit, as it were) of a *single* privileged inertial frame (the center of mass frame of all matter in the universe). This frame is then a precise correlate of Newton's absolute space. (I am indebted to Robert DiSalle and Howard Stein for correcting my earlier exposition, which too hastily assimilated Kant's privileged frame of reference to the modern idea of an arbitrary inertial frame.)

12. As Robert Butts and Ralf Meerbote have emphasized to me, it would not be appropriate to translate *Erscheinung* as "phenomenon" here (see A249, where Kant explicitly distinguishes the two). Yet Kant's use of *Erscheinung* here corresponds precisely to *Newton*'s "phenomenon."

corresponding to the three modal categories of possibility, actuality, and necessity. We begin by assuming an arbitrary body to be at rest and by referring all other possible motions to the frame of reference thereby specified. This first, "phoronomic" step therefore involves "merely relative motion and rest" (560.14–15), and all such merely relative motions count as *possible*. In the second stage we use Newton's first and second laws of motion (Kant's second law of mechanics: 557.10–16) to determine states of true or *actual* rotation (for example, "the circular motion of two bodies around a common central point": 557.34–558.1). Finally, we apply the equality of action and reaction (Kant's third law of mechanics: 558.12–15) to conclude that for all states of actual motion of one body with respect to another a corresponding motion of the second with respect to the first is *necessary*.

I suggest that Kant's first, "phoronomic" step parallels the "phenomena" with which Newton begins Book III. For here Newton indeed begins with "merely relative motion and rest": namely, with the purely relative motions of satellites with respect to their primary bodies in the solar system. In particular, at this point Newton leaves it entirely open whether the earth truly rotates around the sun or vice versa (Phenomenon IV: [82], p. 561; [83], p. 404). Without yet settling the issue of true motion or rest, however, we can meanwhile observe that the satellites in question all satisfy Kepler's laws of planetary motion *relative to* their primary bodies: the satellites of Jupiter and Saturn follow Keplerian orbits relative to their primary bodies and the fixed stars, the five primary planets and the earth likewise follow Keplerian orbits relative to the sun, the moon follows a Keplerian orbit relative to the earth, and the sun also follows a Keplerian orbit relative to the earth if we take the earth to be at rest (Tychonic system).[13] Only a later application of the laws of motion can then decide which of these so far merely relative motions correspond to true motions.

Kant, in the ensuing discussion, begins by taking the earth to be at rest. In this initial frame of reference the Tychonic system (more precisely, the Tychonic-Keplerian system) therefore accurately depicts the observable relative motions of the heavenly bodies. Moreover, we have another significant group of relative motions in this initial frame of reference: the motions of freely falling bodies, projectiles, pendulums, and so on relative to the earth, which Galileo has shown how to describe accurately by means of an acceleration a_g directed towards the earth's center. If we now *assume* that this acceleration is itself true or actual—in effect, that this initial frame of reference represents a sufficient approximation to our ideal privileged frame of reference for the purpose of describing these Galilean

13. This assertion needs qualification in reference to Kepler's *third* law: see note 17 below.

phenomena—we are then in a position to discern the true or actual rotation of the earth with respect to the fixed stars. To use Kant's own example (561.21–35), we drop a stone into a deep tunnel directed towards the earth's center and observe its trajectory.[14] The stone's motion is governed by the acceleration a_g directed towards the center and a horizontal velocity v_S due to the earth's easterly rotation. But this linear velocity v_S is greater at the surface than the corresponding linear velocity v_I of the walls of the tunnel in the earth's interior. Therefore, the stone does not fall straight down through the tunnel as it would if there were no rotation, but deviates towards the eastern wall.[15] Using this or related tests (for example, a Foucault pendulum) we can determine the true daily rotation of the earth and are thus in a position to take the fixed stars as truly at rest (non-rotating).

Note that we have applied Kant's second law of mechanics as our criterion for the presence of true or actual forces (accelerations)—in this case, the force of gravity or weight—and have accordingly assumed that our initial frame of reference is sufficiently close to being privileged for this purpose. This assumption is of course subject to correction as we proceed. In effect, we have already made one required correction, for our new frame of reference takes the earth to be at rest in space (no linear velocity), but corrects for its daily rotation. That is, our new frame of reference takes the surrounding space of the fixed stars to be at rest (no rotation), but is still fixed at the center of the earth. (The Galilean acceleration a_g is of course unaffected.)

This frame of reference is then *one* of the frames of reference within which Newton describes his "phenomena" of Book III. For, as noted above, Newton leaves the choice between Kepler and Tycho open at this stage. In effect, therefore, we are simultaneously considering both a frame of reference fixed at the center of the sun (in which the fixed stars are at rest) and a frame of reference fixed at the center of the earth (in which

14. The effect Kant is describing here is due to what we now call "Coriolis force." Interestingly enough, the problem of giving an exact description of the actual path of a body falling into the rotating earth is the subject of an exchange between Newton and Hooke in 1679–80 which plays an important role in the development of the theory of gravitation. The correct solution of the original problem is first given by Coriolis in 1835: see Koyré [61], chap. V.

15. Conversely, as Kant also points out, we can imagine the stone rising vertically above the earth's surface, whereupon "it does not remain over the same point of the surface, but distances itself from this point from east to west" (561.21–35) The Akademie edition has "from west to east" here, on the grounds that Kant is really supposed to be imagining that the stone is *dropped* from a tower, say (see 648–649 for Höfler's discussion). It seems to me, however, that the idea of a body rising *vertically* under the constraint of gravity is an important one for Kant: it is introduced previously in the Observation to Definition 3 of the Phoronomy (485.27–30: here the earth's rotation is not mentioned).

the fixed stars are at rest). If we provisionally assume that *both* frames sufficiently approximate to our ideal privileged frame for describing the states of true rotation in the solar system, it now follows purely mathematically that the satellites of Jupiter and Saturn experience inverse-square accelerations towards their respective planets (Prop. I: [82], p. 564; [83], p. 406), the five primary planets experience inverse-square accelerations towards the sun (Prop. II: [82], pp. 564–565; [83], p. 406), the earth experiences an inverse-square acceleration towards the sun (Prop. II under the Keplerian interpretation of Phenomenon IV), the sun experiences an inverse-square acceleration towards the earth (Prop. II under the Tychonic interpretation on Phenomenon IV), and the moon experiences an inverse-square acceleration towards the earth (Prop. III: [82], pp. 565–566; [83], pp. 406–407). Further, observation also enables us to conclude that this last inverse-square acceleration of the moon towards the earth is such that, when the distance in question shrinks to the radius of the earth (when the moon is "brought down" to the earth's surface), it is there precisely equal to the Galilean acceleration a_g of terrestrial gravity (Prop. IV: [82], pp. 566–570; [83], pp. 407–409). For this reason, we are now entitled to conclude that the "centripetal force" (acceleration) responsible for the orbits of the heavenly bodies is precisely the "same force" (acceleration) as terrestrial gravity.[16]

All this holds only under the assumption that our two present frames of reference are sufficiently close to being privileged so that the accelerations in question can be taken to be true or actual accelerations. Yet neither frame of reference is truly privileged, of course, and application of Newton's third law (Kant's Prop. 3) permits us to see this. For the accelerations ("centripetal forces") we have just described are not yet counterbalanced by opposite accelerations according to the equality of action and reaction. Consider the earth-sun system for a moment. Under the Tychonic interpretation of our "phenomenon" the sun experiences an inverse-square acceleration towards the earth. Under the Keplerian interpretation of this same "phenomenon" the earth experiences an inverse-square acceleration directed towards the sun. But which description is correct? Which frame of reference represents a better approximation to our ideal privileged frame? According to the equality of action and reaction *neither* description can be precisely correct, for we must simultaneously ascribe inverse-square accelerations directed towards the common center of mass of the earth-sun system to *both* the sun and the

16. Kant himself, in a short introductory section to the First Part of the *Theory of the Heavens* entitled "Brief sketch of the most necessary basic concepts of the Newtonian Natural Philosophy, which are required for the understanding of what follows," provides a lucid and masterful exposition of these steps of the Newtonian argument (1, 244–245.5).

earth—where the acceleration of the earth is to the acceleration of the sun as the mass of the sun is to the mass of the earth. If the mass of the sun is much greater than the mass of the earth, then, it follows that the Keplerian description is in fact much closer to the truth.[17]

However, we are not yet in a position rigorously to estimate the respective masses of the sun and the earth. In order to do this, it turns out, we need to apply the equality of action and reaction to all the inverse-square accelerations ("centripetal forces") in our system, and, moreover, we also need to assume that *each* body in the solar system experiences an inverse-square acceleration towards *all* other bodies.[18] In this way we obtain the law of universal gravitation and, in particular, the result that the acceleration of body *A* towards body *B* is directly proportional to the mass of body *B*. This last result now enables us rigorously to estimate the masses of the bodies in the solar system (Corollaries to Prop. VIII: [82], pp. 578–584; [83], pp. 415–417) and, accordingly, rigorously to determine the center of mass of the solar system: it falls sometimes within, sometimes without, the surface of the sun, but never very far from the sun's center (Prop. XII: [82], pp. 586–587; [83], pp. 419–420). Finally, from the point of view of this last frame of reference it turns out that the accelerations described above in the first two stages are indeed approximately true or actual, and we are thus in a position to discharge the provisional assumptions with which we began.[19]

17. This choice is confirmed by the fact that, if we consider the earth to be a satellite of the sun, Kepler's third law holds for all of the sun's satellites By contrast, if we consider the sun to be a satellite of the earth, Kepler's third law will not hold for the earth's "satellites" (although the time of the sun's period will indeed by proportional to the 3/2th power of its mean distance from the earth, and similarly for the moon's period, the constant of proportionality will not be the same in both cases: the former represents the gravitational field of the sun whereas the latter represents that of the earth). As Howard Stein has emphasized to me, this is an extremely subtle and important point; for, as we shall see, Kepler's third law itself plays a crucial role in developing our final, and truly rigorous, estimate of the respective masses of the earth and the sun (see, in particular, note 27 below).

18. This is of course the most difficult step in the entire argument. How do we get from mutual inverse-square accelerations between the sun and each planet to mutual inverse-square accelerations between *each* body in the solar system and *every* other body? Kant and Newton have very different ways of bridging this gap, I think: see §III below.

19. A delicate issue arises here, for we can also show that the relative motions of our first, "phoronomic" stage *cannot* be exactly true: Kepler's laws fail due to the planetary perturbations. Yet this in no way compromises the argument. For first, what is derived at our second, "dynamical" stage is the *existence* of an inverse-square force (centripetal acceleration) directed towards the center of each primary body—and this remains true at our final, "mechanical" stage as well; and second, we infer the properties of this force from the statement that satellites approximately obey Kepler's laws *and would exactly obey them if the force in question were the only force acting*—and this statement also remains exactly true at our third stage (where we show that the deviations from Kepler's laws result *entirely* from the perturbing gravitational forces due to the other primary bodies in the system).

Thus, by carefully applying the laws of motion to our initial "phenomena" we have obtained both a frame of reference that can be taken as privileged to a very high degree of approximation (the center of mass frame of the solar system) and an important law of nature: the law of universal gravitation.[20] By "reducing all motion and rest to absolute space" we have obtained true motions from apparent motions and have "transformed appearance [Erscheinung] into experience [Erfahrung]." Yet, as we observed above, this procedure has still not reached its final goal. For Kant, the center of mass of the solar system is not strictly privileged: the solar system itself experiences a slow rotation around the center of mass of the Milky Way galaxy, and the latter experiences a slow rotation around the center of mass of the entire cosmic system of the galaxies. In the end, only the forever unreachable "common center of gravity of all matter" can furnish us with a truly privileged frame of reference, and our procedure for "reducing all motion and rest to absolute space" never terminates: "absolute space" is an idea of reason.

· III ·

There is a serious problem facing the above reconstruction of Kant's procedure, however. For the most interesting and important step in this reconstruction—the step that proceeds from the observable (Keplerian) relative motions in the solar system to the law of universal gravitation and the center of mass frame of the solar system, as in *Principia*, Book III—does not explicitly occur in Kant's text. In fact, although Kant refers to Newton's Scholium to the Definitions, he does not explicitly refer to Book III at all in the Phenomenology. And, although Kant provides a lengthy discussion of the determination of the earth's true rotation, he does not then describe the remaining motions in the solar system. Rather, he immediately jumps to the "universe [Weltganze]" or "cosmic system [Weltgebaude]" as a whole and "the common center of gravity of all matter" (562.26–563.9). Accordingly, he speaks only of "heavenly bodies [Weltkörper]" in general (563.26–564.33), and makes no explicit mention of the planets, the sun, or even the moon. Finally, nowhere in the Phenomenology (or, in fact, in the *Metaphysical Foundations* itself) does Kant explicitly refer to Kepler's laws and Newton's ensuing derivation of the

20. In this procedure the inverse-square law of gravitation is entirely determined by our initial data: the observable Keplerian relative motions ("phenomena"). I therefore agree with Brittan [10], pp. 140–142, that the inverse-square law is certainly not synthetic a priori for Kant. See also Buchdahl [12], p. 647; Kitcher [60], p. 187; Okruhlik [88], p. 252. In my view, the situation is quite otherwise with the laws of motion, however. (For a contrary view of the inverse-square law, see Plaass [97], §6.5, and for further discussion see Chapter 4 below.)

law of universal gravitation from these laws. Since the latter derivation is of course the centerpiece of *Principia,* Book III, is it not completely inappropriate to view this Newtonian text as a model for Kant's procedure?

Now, as we have seen, there is some reason for Kant to wish to de-emphasize the solar system and to proceed rapidly to "the common center of gravity of all matter." It is essential to Kant's view of "absolute space" as an idea of reason that the center of mass frame of the solar system is not strictly privileged and, indeed, that no strictly privileged frame is actually accessible to us. For Kant, the process of obtaining the true motions from the apparent motions necessarily never terminates, and it is essential to Kant's conception, therefore, that the Newtonian procedure for obtaining the true motions is to be pushed far beyond *Principia,* Book III. Yet this circumstance, important as it is, hardly provides a sufficient explanation for Kant's apparent failure even to mention Book III here.

In this connection there is one passage in the General Observation to Phenomenology that can perhaps be of help to us. This passage, which contains an explicit reference to Newton's *Principia* as well as a footnote thereto, occurs at the very end of Kant's long paragraph on the earth's rotation. After remarking that the earth's true rotation "rests on the representation of the mutual continuous distancing [Entfernung] of each part of the earth (outside the axis) from any other part lying opposite on a diameter at the same distance from the center" (561.39–562.2), Kant says:

> For this motion is actual in absolute space, in that thereby the loss of the imagined distance, which gravity acting alone would produce in the body, is continuously compensated—and, in fact, without any dynamical repulsive cause (as one can see from the example chosen by Newton at *Princ.Ph.N.* pag. 10 Edit. 1714*). Therefore, this loss is compensated by actual motion which, however, is determined with respect to the inside of the moved matter (namely its center) and is not referred to the surrounding space. (562.2–8)[21]

The footnote to Newton runs as follows:

> *At that place he says: "It is indeed a matter of great difficulty to discover, and effectually to distinguish, the true motions of particular bodies from the

21. "Denn diese Bewegung ist im absoluten Raume wirklich, indem dadurch der Abgang der gedachten Entfernung, den die Schwere für sich allein dem Körper zuziehen würde, und zwar ohne alle dynamische zurücktreibende Ursache (wie man aus den von Newton Princ.Ph.N. pag. 10 Edit. 1714* gewählten Beispiele ersehen kann), mithin durch wirkliche, aber auf den innerhalb der bewegten Materie (nämlich das Zentrum derselben) beschlossenem, nicht aber auf den äusseren Raum bezogene Bewegung, kontinuierlich ersetz wird." This difficult sentence is completely garbled in the translation of Ellington (1970, pp. 129–130). The year "1714" should of course be "1713."

> apparent; because the parts of that immovable space, in which bodies truly move, do not affect our senses. However, the situation is not completely desperate." [Kant of course quotes the original Latin.] He then lets two globes connected by a cord rotate around their common center of gravity in empty space, and shows how the actuality of their motion together with its direction can nonetheless be discovered through experience. I have sought to show this under somewhat altered circumstances by means of the earth moving around its axis as well. (562.32–40)

Thus Kant is here referring to the final paragraph of Newton's Scholium to the Definitions mentioned above.

In the main passage Kant is clearly envisioning a counterbalancing of gravitational and centrifugal force. Imagine a stone resting on the earth's surface, which, for present purposes, can be considered as a part of the earth. The centrifugal force of the earth's rotation would, in the absence of any counterbalancing forces, produce a "continuous distancing" from the surface: the stone would fly off into space. But the force of gravity is more than enough to compensate for this centrifugal tendency, and hence the stone stays put. Similarly, in Newton's example, the mutual rotation of the two globes would, in the absence of the cord connecting them, also produce a "mutual continuous distancing." The same rotation then produces a detectable tension in the cord, and this allows us to determine the true mutual rotation of the globes. For Kant's purposes, what is essential is that this tension does not arise from the action of any actual "fundamental forces"—there is no "dynamical repulsive cause"—but simply from the motion itself.

The question to ask ourselves here, I think, is this: What in Kant's example corresponds to the tension in the cord in Newton's example? What is the detectable effect of the rotation in question that permits it to be "discovered through experience"? Imagine again our stone resting on the earth's surface: the force of gravity prevents its centrifugal acceleration from hurling it off into space. But what would happen if there were no rotation and therefore no centrifugal acceleration? Nothing, of course, for the force of gravity would still hold the stone firmly in place. So far, then, we have no effect corresponding to the detectable tension in the cord in Newton's example, an effect that would vanish if there were no rotation. Suppose, however, we imagine the stone to be lifted a small distance off the surface (compare note 15 above), and to be given sufficient horizontal velocity so as to orbit the earth as a little satellite. *This* rotational motion would then parallel the case of the two globes, and the centrifugal acceleration in question would now produce an observable effect: the stone would maintain itself above the earth's surface by means of its motion alone ("without any dynamical repulsive cause"). Thus, if *this* motion were to cease, the stone would drop to the surface.

If we compare this last idea to *Principia,* Book III, Proposition IV—especially the Scholium to this Proposition, where Newton has us consider a "little moon" skimming over the surface of the earth ([82], pp. 569–570; [83], p. 409)—it appears possible that what Kant really has in mind here is precisely Newton's "moon test." By imagining the moon "deprived of all motion" and thus brought down to the surface of the earth, we can compare the moon's centrifugal acceleration with terrestrial gravity and thereby identify the latter force with the centripetal force acting on the moon. Moreover, this moon test is the key step, of course, in connecting the earth's gravitational acceleration a_g with the Newtonian inverse-square acceleration governing the orbital motions in the solar system. It appears possible, therefore, that, by bidding us to run the moon test in reverse, as it were, Kant is attempting to suggest just this crucial step here. This possibility is reinforced by the fact that the earth-moon system is a clear analogue of Newton's rotating globes: the two bodies rotate around their common center of mass, and gravity is the "somewhat altered circumstance" which takes the place of the tensed connecting cord (see Kant's following paragraph: 562.12–13). It is further reinforced by Kant's consistent tendency to draw an explicit parallel between two bodies rotating about a common center and a single body rotating about its axis (see the Observation to Proposition 2, for example: 557.34–558.1).[22]

In any case, let us adopt this suggestion as a provisional hypothesis and see if it can generate any further insight into Kant's text. If Kant is in fact alluding to the moon test here, then one naturally wants to know whether he refers to the moon test anywhere else in the *Metaphysical Foundations.* The answer is affirmative, for there is an apparently quite explicit allusion to the moon test in Observation 1 to Proposition 7 of the Dynamics, where Kant has us imagine that "the earth and moon touch one another" (523.25). This allusion is followed by the above-mentioned Observation 2, where Kant refers to Book III, Proposition VI (there Newton appeals to the moon test in extrapolating the properties of celestial attraction from those of terrestrial gravity), and then discusses Newton's attempt to compare the masses of Jupiter and Saturn in the Corollaries to Proposition

22. Of course this parallel strikes us as odd. What does the (monthly) orbital rotation of the moon have to do with the (daily) axial rotation of the earth? It becomes more understandable, however, when we recall that the *Theory of the Heavens,* Second Part, Chapter IV, presents a hypothesis according to which the axial rotation of a central body actually arises from the orbital rotation of its proto-satellites As a central body forms by means of attraction from a "cloud" of matter, this matter is set into rotation as it falls. Matter with insufficient orbital velocity to counterbalance the gravitational attraction falls into the central body, and imparts its (relatively slow) orbital rotation to the central body as axial rotation. Matter with sufficient orbital velocity to counterbalance the gravitational attraction becomes satellites. (See 1, 285–286.)

VIII. In these two Observations it appears that Kant is calling our attention to Propositions IV–VIII of *Principia,* Book III, which are precisely the key steps in the argument for the law of universal gravitation. Kant here aims to elucidate the significance of the argument and, in Observation 2, to make an important criticism of Newton's actual procedure.

Kant's criticism, it will be recalled, is that Newton's refusal to count gravitational attraction as an essential property of matter leaves him with no way of "grounding" the proposition that gravitational attraction is directly proportional to mass (so that the acceleration of body *A* towards body *B* is directly proportional to the mass of body *B*). If Newton denies that gravitation is essential to matter, then, according to Kant:

> He could absolutely not say that the attractive forces of two planets, e.g., that of Jupiter and Saturn, which they manifest on their satellites (whose masses one does not know) at equal distances, relate to one another as the quantity of matter of these heavenly bodies, if he did not assume that they, simply as matter and therefore according to a universal property of matter, would attract other matter. (515.32–37)

In particular, Kant cites Newton's well-known words from the Advertisement to the second edition of the *Opticks:* "And to show that I do not take *Gravity* for an *essential* Property of Bodies, I have added one Question concerning its Cause" (515.28–30)[23]—where the reference is of course to Query 21. Kant's claim, therefore, is that Newton's attempt to leave open the possibility of explaining gravitational attraction by means of an "Aetherial Medium" as in Query 21 necessarily sets him "at variance with himself."

It is crucial to Newton's argument that he be able rigorously to compare the masses of the different bodies in the solar system. He does this, in the Corollaries to Proposition VIII, by comparing the accelerations of various satellites towards their respective primary bodies (these accelerations are in turn calculated from the distances and periodic times of the satellites). Thus we can compare the respective masses of the sun, Jupiter, Saturn, and the earth by means of the respective accelerations of Venus, the satellites of Jupiter, the satellites of Saturn, and the moon. This calculation assumes that the masses in question are directly proportional to the corresponding accelerations at equal distances of the satellites in question; in other words, that the acceleration of any satellite is proportional to m_p/r^2, where m_p is the mass of the primary body and r is the distance of the satellite from the primary body. This supposition, according to Kant, itself involves or presupposes the proposition that gravitational attraction is essential to matter, that is, it presupposes the truth of his own Proposition 7 of the

23. Kant emphasizes *gravitatem* and *essentiales* in his quotation from the Latin.

Dynamics together with the following Proposition 8: "The original force of attraction, on which the possibility of matter as such itself rests, extends immediately from each part of matter to every other part in the universe to infinity" (516.23–26).

To understand Kant's point here we must carefully distinguish two different properties of gravitational acceleration. On the one hand, gravitational acceleration is independent of the mass of the body being attracted (all bodies fall the same in a gravitational field), so that the acceleration of the attracted body is given by k/r^2, where k is a constant depending only on the attracting body (and r is again the distance of the attracted body from the attracting body). That the attraction of the sun for its planets and the attractions of Jupiter and Saturn for their satellites have this property follows from Kepler's third law.[24] That the attraction of the earth for the moon has this property follows from the moon test and is thereby confirmed by the well-known fact that the terrestrial gravitational acceleration is independent of mass (a_g is constant for all bodies). Newton records these results in the second clause of Proposition VI. On the other hand, however, it is also true that gravitational acceleration is directly proportional to the mass of the *attracting* body. If k_A/r^2 represents the gravitational "acceleration-field" of body A and k_B/r^2 represents that of body B,[25] then the mass of body A is to the mass of body B as k_A/k_B—in other words, the constant k characterizes the mass of the attracting body. Newton demonstrates this second property of gravitational acceleration in Proposition VII. What is important to note here is that the second property alone is at issue in Kant's criticism;[26] accordingly, his target is Newton's Proposition VII.

Newton derives Proposition VII from Proposition LXIX of Book I,[27]

24. Kepler's third or harmonic law states that if r, R and t, T are, respectively, the mean distances and periods of two concentric orbits, then $r^3/R^3 = t^2/T^2$. Assuming circular orbits for simplicity, and using the circumstance that here $v = 2\pi rt$, it follows from the fact that $a = v^2/r$ that the two accelerations stand in the ratio $1/r^2:1/R^2$. Hence the variation in acceleration depends *only* on distance.

25. The notion of "acceleration-field" is taken from Stein [104], p. 178. The gravitational *force* is of course given by $F_{grav} = Gm_Am_B/r^2$. Yet, since F_{grav} on m_B is equal to m_Ba_B, we have $a_B = Gm_A/r^2$ *independently* of m_B. This is what allows us to speak of an *acceleration-field* here.

26. Kant never expresses qualms concerning our first property; on the contrary, in the brief introductory sketch to the *Theory of the Heavens* referred to in note 16 above, he speaks of it as following in an "indubitable fashion" from Kepler's third law (1, 244.18–24).

27. "In a system of several bodies A, B, C, D, etc., if any one of those bodies, as A, attract all the rest, B, C, D, etc., with accelerative forces that are inversely as the squares of the distances from the attracting body; and another body, as B, attracts also all the rest, A, C, D, etc., with forces that are inversely as the squares of the distances from the attracting body; the absolute forces of the attracting bodies A and B will be to each other as those very bodies A and B to which these forces belong" ([82], pp. 296–297; [83], p. 191). In our

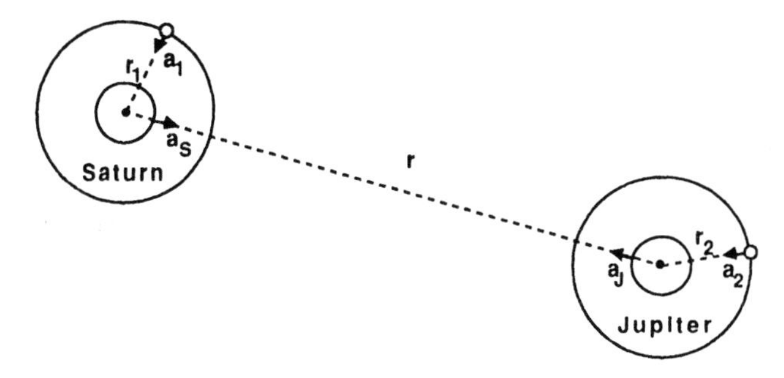

Figure 7

which can be elucidated as follows. Consider a system of bodies consisting of two planets and their respective satellites, for example, the systems of Jupiter and Saturn (see Figure 7). We know, by the first property of gravitational acceleration, that the acceleration-field on Saturn's moons is given by $a_1 = k_S/r_1^2$ and the acceleration-field on Jupiter's moons is given by $a_2 = k_J/r_2^2$. We want to show that when $r_1 = r_2$, $a_1/a_2 = k_S/k_J = m_S/m_J$, where m_S and m_J are the masses of Saturn and Jupiter respectively. To do so, we assume that the acceleration-fields of our two planets extend far beyond their respective satellites, so that we also have an acceleration $a_J = k_S/r^2$ of Jupiter and an acceleration $a_S = -k_J/r^2$ of Saturn, where r is now the distance between the two planets. But, according to the third law of motion, $m_J a_J = -m_S a_S$. Therefore, we have $m_S/m_J = -a_J/a_S = k_S/k_J$, as desired. We are now—and only now—in a position to compare the masses of Jupiter and Saturn by reference to the acceleration-fields on their respective satellites, that is, by reference to k_J and k_S.[28]

What matters here are two general features of the argument. First, we need to extend the acceleration-fields of the two planets far beyond the regions of their respective satellite-systems and suppose that they affect the two planets as well. In other words, we need to assume that gravitational attraction takes place not only between primary bodies and their satellites,

illustration *A* is Saturn, *B* is Jupiter, and *C*, *D* are their respective satellites. (In reference to the issue raised in note 17 above, suppose that *A* is the earth, *B* is the sun, *C* is the moon, and *D* is Venus. From the point of view of the Tychonic system, the moon and the sun will not be governed by the *same* acceleration-field, and neither will the earth and Venus.)

28. Using just the third law of motion and our first property of gravitational acceleration discussed above we can show the following: for any two gravitationally interacting masses m_A and m_B, there is a constant G_{AB} such that $F_{AB} = -F_{BA} = G_{AB}m_A m_B/r^2$. Thus for these two masses we have $k_A = G_{AB}m_A$ and $k_B = G_{AB}m_B$. We need the *universality* of gravitational interaction, however, to conclude that G_{AB} is the same constant for every such pair; that is, that G is a "universal constant." It now—and only now—follows that, universally, $k_A = Gm_A$.

but also among the primary bodies themselves. This is a first—and absolutely crucial—step on the way towards *universal* gravitational attraction. Second, we need to apply Newton's third law of motion, the equality of action and reaction, to the gravitational interactions among the primary bodies: in our case, to the interaction between Jupiter and Saturn. For only so do the masses of the primary bodies enter directly into our calculations at all.

Suppose now that we seriously entertain the possibility of an explanation of gravitational attraction by the action of an Aetherial Medium. This would cause overwhelming difficulties for the above derivation. First, on this kind of model of gravitational attraction, one would have no reason to extend the attractions in question beyond the regions of the respective satellite-systems. Indeed, one would naturally expect the attractions to be limited to precisely these regions, and, in any case, it would be extremely difficult coherently to incorporate *universal* attraction into the model.[29] Second, and even more fundamental, however, one would have no license whatever for applying the third law of motion—applying the equality of action and reaction to the accelerations of Jupiter and Saturn, for example. For, on an aether model, even if one could somehow contrive to incorporate accelerations of Jupiter and Saturn towards one another, there would still be absolutely no warrant for applying the equality of action and reaction directly to these two accelerations themselves: on the contrary, this principle is to be applied to the interactions among Jupiter, Saturn, *and the intervening medium*—conservation of momentum will not hold, in other words, if we consider merely the accelerations of Jupiter and Saturn alone.

This last point is in fact explicitly proposed to Newton in a well-known letter (of [March] 18, 1712/13) by Roger Cotes—a most acute Newtonian who is of course himself strongly tempted to conceive gravity as an essential property of matter.[30] Cotes argues that the third law of motion can be applied only "when the Attraction may properly be so called," but not in cases of merely apparent attraction effected by pressure (he illustrates the point by considering one globe rotating about another on a table in

29. Stein [104], pp. 178–180, points out that on a vortex-model of gravitational attraction one would certainly not expect the gravitational acceleration-fields of the planets to extend beyond their respective satellite-systems, and that Newton's inference to *universal* gravitation was accordingly disputed by intelligent critics such as Huygens and Leibniz (who both, in particular, fully accepted the inverse-square law). See also Koyré [61], chap. III, appendix A.

30. Newton [87], pp. 391–393. Cotes asserts that gravity is an essential property of matter in a draft of his Preface to the second edition of *Principia;* in response to criticism from Clarke he changed "essential" to "primary" in the final version. For discussion of the Newton-Cotes correspondence see Koyré [61], chap. VII.

virtue of being pushed by an "invisible Hand"). Cotes even remarks that neglect of this point causes difficulties with the argument of Proposition VII of Book III. Kant, as I read him, is entirely in agreement with this fundamental observation of Cotes. Kant goes further, however, by also stressing the importance of the assumption of universality in the argument of Proposition VII. We must presuppose, according to Kant, that the action of gravitation is *immediate:* effected by no intervening medium. We must also presuppose that the action of gravitation is *universal:* acting between each body in the solar system and each other body. Together, these two presuppositions constitute the content of Kant's own Propositions 7 and 8 of the Dynamics of the *Metaphysical Foundations;* and the two together amount, in Kant's eyes, to the claim that gravitational attraction is an essential property of matter.

It is worth reminding ourselves of precisely what this claim means for Kant. We need to presuppose the immediacy and universality of gravitational attraction in order to develop a rigorous method for comparing the masses of the primary bodies in the solar system.[31] We need such a method, in turn, in order rigorously to determine the center of mass of the solar system. This, in turn, is necessary for rigorously determining a privileged frame of reference and thus for giving objective meaning, in experience, to the distinction between true and apparent (absolute and relative) motion.[32] This, finally, is necessary if matter, as the movable in space (Definition 1 of the Phoronomy: 480.5–10), is to be itself possible as an object of experience. Hence an essential—that is, immediate and universal—attraction is necessary to matter as an object of experience. It follows, for Kant, that the immediacy and universality of gravitational

31. That universal gravitation provides a generally applicable measure of mass is emphasized in the Observation to Proposition 1 of the Mechanics (541.14–26); interestingly enough, this idea is reemphasized at the very end of the Phenomenology (564.29–30). That universal gravitation provides the *only* generally applicable measure of mass is an important theme of the *Opus postumum:* see, for example, 21, 406.22–25, and especially 21, 403.26–29, where Kant speaks of the two "original forces" of attraction and repulsion. The latter is characterized as "that without which no space would be filled," the former as "that without which no quantity of matter is cognizable: gravitation."

32. It has been objected—notably by DiSalle [20]—that Kant's claim here ignores the obvious possibility of determining masses via collisions, in which case we assume only the third law of motion but nothing whatsoever about the properties of gravitation. This possibility is of course undeniable, but its force is mitigated, I believe, by two important considerations. First, Kant's aim is to reconstruct the logic of Newton's actual construction of a privileged frame of reference in Book III, and universal gravitation certainly plays the essential role in this particular argument. Second, as emphasized in note 11 above, Kant aims to construct a *single global* privileged frame of reference embracing simultaneously all the phenomena of nature, and here universal gravitation—which defines a simultaneous comparison of all masses across arbitrary distances to infinity—indeed appears as the primary method.

attraction must be viewed—like the laws of motion themselves—as in an important sense a priori. These two properties cannot be straightforwardly obtained from our experience of matter and its motions—by some sort of inductive arguement, say—for they are necessarily presupposed in making an objective experience of matter and its motions possible in the first place.[33]

Now this last Kantian claim—that the immediacy and universality of gravitational attraction must be viewed as in an important sense a priori—may seem obviously and outrageously false, and this is especially so in the case of the universality of gravitation. Is not this property subject to inductive confirmation of the most straightforward kind: namely, by observations of the planetary perturbations? Does not Newton, in the second edition of *Principia,* appeal to observations of precisely such perturbations in the orbits of Jupiter and Saturn?[34] Does not Kant's correspondent J. H. Lambert himself make an important contribution in 1773–1776 to the theory of the so-called Great Inequality of Jupiter and Saturn by providing tables—*obtained purely empirically*—that suggest for the first time that the anomaly in question is periodic?[35]

These facts are undeniable, and they do show that the assumption of universality has indeed been corroborated by observational data. Kant's problem, however, concerns the *interpretation* of such observations. For what do such observational data—Lambert's, for example—actually show? They show that in a given frame of reference—perhaps a frame of reference fixed at the center of the sun and in which the fixed stars are at rest—certain motions that we call planetary perturbations occur. Yet such

33. In the *Principia* itself the remarkable extrapolation to *universal* gravitation is of course supported principally by quasi-inductive arguments (compare note 18 above). Corollary II to Proposition VI—which Kant singles out for special criticism, it will be recalled—is typical: "Universally, all bodies about the earth gravitate towards the earth; and the weights of all, at equal distances from the earth's center, are as the quantities of matter which they severally contain. This is the quality of all bodies within the reach of our experiments; and therefore (by Rule III) to be affirmed of all bodies whatsoever" ([82], p. 574; [83], p. 413). The moon and all sublunary bodies experience an inverse-square acceleration towards the earth; therefore, all bodies whatsoever—no matter how distant—must also experience an inverse-square acceleration towards the earth! Such an extrapolation can certainly be questioned by reasonable men, as it was by intelligent critics such as Huygens and Leibniz. The irony is that Kant, in acknowledging the force of the criticisms of Newton's quasi-inductive arguments for universal gravitation put forward by Huygens and Leibniz, responds by giving an a priori foundation for precisely what they feared most: immediate action at a distance (to infinity) across empty space.

34. See Stein [104], pp. 180–181. Such evidence of mutual planetary attractions was just beginning to be observed by the time of the second edition of *Principia* (1713).

35. I am indebted to Judson Webb for emphasizing the importance of this problem to me. See Wilson [122].

data, by themselves, do not and cannot show that the motions in question can be interpreted as true or absolute motions: that *absolute* gravitational acceleration is universal. Kant's point is that the notion of true or absolute motion does not even have objective meaning or content unless we employ Newton's procedure for determining the center of mass of the solar system and hence *presuppose* that absolute gravitational acceleration is in fact universal. To be sure, observations of planetary perturbations turn out, fortunately, to provide corroboration for this whole scheme; but consider what our situation would have been if such perturbations had not been observed. We would not then merely be in the position of having disconfirmed an empirical hypothesis, the hypothesis that gravitational accelerations are universal; rather, we would be left with no coherent notion of true or absolute motion at all. For the spatio-temporal framework of Newtonian theory—which, for Kant, can alone make such an objective notion of true or absolute motion first possible—would itself lack all objective meaning.

· IV ·

The above considerations seem to me to lend considerable plausibility to the present reconstruction. In particular, whether or not one accepts my attempt to find an allusion to the moon test—and hence to *Principia,* Book III—in the General Observation to Phenomenology, it is clear that fundamental questions surrounding Newton's argument in Book III are absolutely central to Kant's concerns in the *Metaphysical Foundations.* Assuming that this is correct, then, let us conclude by indicating briefly how Kant's reinterpretation of the *Principia,* so construed, is to serve as a realization or "example *in concreto*" of the abstract transcendental principles of the first *Critique.*[36]

The Phenomenology of the *Metaphysical Foundations* depicts a constructive procedure for "transforming appearance [Erscheinung] into experience [Erfahrung]" whose goal is a description of all true motions in the universe. This procedure therefore seeks to construct an objective spatio-temporal framework (described from the point of view of a privileged frame of reference) within which the objective alterations of state are accelerations and the underlying "natural" states, relative to which such alterations or events are defined, are states of inertial (uniform) motion.[37] Each such alteration or event is to have an objective or determinate

36. This idea is nicely illustrated, with respect to the three analogies of experience in particular, in Martin [73], §11.

37. Compare A207/B252n: "One should carefully note that I do not speak of the alter-

spatio-temporal position relative to every other alteration or event. The result is a unified and thoroughly interconnected spatio-temporal representation of all objective motions.

This constructive procedure then realizes or instantiates the procedure of the postulates of empirical thought in the first *Critique*.[38] The schema of this group of categories is "time itself, as the correlate of the determination of whether and how an object belongs in time" (A145/B184), and the postulates themselves depict a procedure—modeled on the constructive procedure expressed in the Postulates of Euclidean geometry (A233–235/B285–287)—for constructing such a representation of time.[39] The essential difference is that, whereas the Postulates of Euclidean geometry proceed wholly a priori, the postulates of empirical thought involve a "material" element derived from "sensation" or "perception." In other words, the starting point or "input" of our constructive procedure consists of a posteriori given data of experience.

Such data are to be brought into agreement with—or perhaps better: are to be operated upon or transformed by—the "formal conditions of experience" (A218/B266) and, in particular, the analogies of experience:

> They are nothing but principles of the determination of the existence of appearances in time, according to all its three modes: relation to time itself as a magnitude (the magnitude of existence, that is, duration), relations in time as a series (successively), and finally relations in time as a sum of all existents (simultaneously). This unity of time-determination is through and through dynamical: that is, time is not viewed as that wherein experience immediately determines a place for every existent—which is impossible, since absolute time is no object of perception by means of which appearances could be held together—rather, it is the rule of the understanding, through which alone the existence of appearances can acquire synthetic unity according to time-relations, that determines the place of each existent—and this determination is therefore a priori, and valid for each and every time. (A215/B262)

One should observe that the "time" Kant aims to construct here is actually what we would now call *space-time*. Kant's temporal relations essentially include a *simultaneity relation*, and simultaneous alterations or events are necessarily contained in a single, instantaneous, three-dimensional Euclid-

ation of certain relations in general, but of alterations of state. Thus, when a body moves uniformly, it thereby alters its state (of motion) not at all—but only when its motion increases or decreases."

38. The importance of the Phenomenology of the *Metaphysical Foundations* as a realization of the postulates of empirical thought is well emphasized by Schäfer [103], §§17, 18; see, in particular, n. 66 on pp. 128–129.

39. See Chapter 1 above for an attempt to elucidate the significance of the constructive procedure expressed in Euclid's Postulates for Kant's own conception of geometry.

ean space for Kant (see the principle of the third analogy in the *second* edition: B256). Hence the succession of "times" in question here is really a succession of such instantaneous, three-dimensional Euclidean spaces,[40] and the object of our construction is best viewed—from a modern point of view—as a four-dimensional (Newtonian) space-time.[41]

How does the construction now proceed? First of all, the abstract principles of the first *Critique*—in particular, the analogies of experience—are to be further specified by means of Kant's articulation of the concept of matter in the *Metaphysical Foundations,* which renders this concept "a priori suitable for application to outer experience." The result is a realization or instantiation of the analogies of experience according to which the conservation of mass realizes the first analogy (Prop. 2 of the Mechanics), the law of inertia realizes the second analogy (Prop. 3 of the Mechanics), and the equality of action and reaction realizes the third analogy (Prop.

40. As Manley Thompson has emphasized to me, the status of time in general is considerably more complicated than this, for there is no doubt that time as the *form of inner sense* is one-dimensional, having to do "neither with figure [Gestalt] nor place [Lage]" (A33/B49–50). Yet it is necessary to distinguish time as the *form* of inner sense from time as itself an *object* of intuition. The latter, I suggest, is only constructed in the Transcendental Analytic; and it is this "time" which is necessarily four-dimensional. Compare A33/B50 with B160 (distinction between "form of intuition" and "formal intuition"). See also the Preface to the *Metaphysical Foundations* (471.11–21), and Reflexion 13 at 14, 55.4–7, where Kant distinguishes between "Zeitform" and "Zeitgröße"—the latter essentially involves space. For further discussion of this issue see Chapter 2 above.

41. For the notion of Newtonian space-time, see Stein [104]. Parsons [93] emphasizes that—especially in the second edition of the first *Critique*—the categories appear to require schematization in terms of both time and space, and he introduces the useful notion of a "second schematization of the category in terms of space" ([93], p. 226). If the present point is correct, however, it follows that Parsons's "second schematization" is already built into the schematization in terms of time itself—in so far as the latter essentially involves a simultaneity relation. In other words, schematization in general necessarily involves what we now call *space-time.* Moreover, Kant himself comes very close to making the idea of space-time explicit in a remarkable footnote to §14.5 of the *Inaugural Dissertation,* appended to a complaint that the Leibnizean view of time "completely neglects *simultaneity,** the most important consequence of time": "**Simultaneous* [events] are not so in virtue of not being successive. For removing succession certainly annuls a conjunction due to the series of time, but *another* true relation—as is the conjunction of all at the same moment—does not immediately arise thereby. For simultaneous [events] are connected through the same moment of time, just as successive [events] are through different moments. Therefore, although time has only one dimension, the *ubiquity* of time (to speak with Newton), whereby all that is sensitively thinkable is at *some time,* adds another dimension to the quantum of actual [events], in so far as they hang, so to speak, from the same moment. For if one designates time by means of a straight line produced to infinity and simultaneous [events] at any point by means of ordinate lines, then the surface thereby generated represents the phenomenal world—both with respect to substance and with respect to accidents" (2, 401). Stein [105], p. 13, cites this passage in the same connection.

4 of the Mechanics). Inertial motion, then, realizes (provides an example for) the category of causality (B292); matter, and the quantity thereof, realizes the category of substantiality (B278, B291); mutual interaction in space, with respect to a common center of mass frame, realizes the category of community or simultaneity.[42]

Note, however, that it is really somewhat artificial to consider our realized categories as literally divided up in this way. For, in order to apply the law of inertia, we need to set up (at least approximately) a privileged frame of reference. This in turn requires application of the equality of action and reaction so as to determine the relevant center of mass. This in turn requires that we be able rigorously to estimate the masses of the bodies in our system and thus, as we have seen, that we have a universal "fundamental force" of attraction. The concepts of causality, interaction, substantiality, and force are therefore inextricably linked (A204–206/B249). Thus, for example, once we have set up (at least approximately) a privileged frame of reference by means of the equality of action and reaction (and hence have also invoked the concept of mass), it is then the law of inertia that enables us to consider time as a magnitude: equal times are those during which a freely moving body would traverse equal distances.[43]

It remains to specify the initial a posteriori data or "input" of our constructive procedure. I suggest we take them to be the observable, purely relative motions in the solar system described in §II above: the Galilean phenomena of free fall, projectile motion, etc., relative to the earth and the Keplerian-Tychonic orbital motions of the sun, the planets, and their satellites. In other words, our initial data are just Newton's "phenomena" of *Principia*, Book III.[44] As we have seen at length above,

42. Kant's example of community or simultaneity at B257 is the earth-moon system; his second example at A213/B260 is the "heavenly bodies [Weltkörper]."

43. Compare B293, A215/B262, A183/B226. See also *Inaugural Dissertation*, §14.5, immediately following the passage cited in note 41 above, where Kant accuses the Leibnizean view of "confusing all use of sound reason" and "entirely abolishing all certainty of rules," "because it does not postulate the determination of the laws of motion according to the measure of time, [namely motion]" (2, 401.5, [514]). (In a Newtonian space-time consisting of a succession of instantaneous Euclidean three-spaces, the relation of temporal congruence is completely determined by the Euclidean geometry of the three-spaces together with the inertial trajectories or "affine structure.") Compare also Newton's remarks on the determination of true or "absolute" time in the Scholium to the Definitions of *Principia* ([82], p. 48; [83], pp. 7–8).

44. Again, Kant does not explicitly refer to Newton's derivation of his theory of the solar system from our Galilean-Keplerian initial data in either the *Metaphysical Foundations* or the first *Critique*. He does, however, explicitly describe Newton's derivation of the law of universal gravitation from Kepler's laws in the above-cited sketch from the *Theory of the Heavens:* see note 16 above. (As Robert Butts has emphasized to me, A662–A664/B690–692 contains an unmistakable allusion to the argument of the *Theory of the Heavens*.) Moreover,

these initial data, when fed into our constructive procedure, yield precisely the law of universal gravitation and a frame of reference that is privileged to a high degree of approximation: the center of mass frame of the solar system. With respect to this frame of reference the equality of action and reaction holds necessarily, simultaneous events (accelerations) are those connected by an (instantaneous) gravitational interaction, successive events (accelerations) are those lying in succeeding instantaneous spaces or "simultaneity-slices" in this (deterministic) universe, and the magnitude of temporal duration is determined by the law of inertia in the manner just indicated. Our representation of time—more precisely, of space-time—is, to a high degree of approximation, complete.

Yet as we have also stressed above, this representation is not fully complete, for the frame of reference thereby constructed is not *exactly* privileged. Indeed, not only is our constructive procedure strictly non-terminating in principle for Kant, but even the limited success achieved so far—the construction of an *approximately* privileged frame of reference—depends entirely on fortunate, and contingent, facts about our initial a posteriori data: in this case, on the relative accessibility of the Galilean-Keplerian "phenomena" and the relative isolation of the solar system. The third law of motion, for example, is in this sense merely regulative: it supplies no guarantee that a privileged frame of reference actually exists, but merely "a rule for seeking it in experience, and a mark whereby it can be detected" (A179–180/B222). Thus, unlike *mathematical* principles, which are "constitutive with respect to intuition" (I can, for example, instantiate the category of quantity a priori in pure intuition by means of geometrical construction), *dynamical* principles of pure understanding are "merely regulative principles of intuition" (A664/B692). In other words, there can be no intuition realizing the dynamical principles given completely a priori.

The dynamical principles of pure understanding, and their specifications given in the *Metaphysical Foundations* (that is, the laws of motion), are nonetheless "constitutive with respect to experience" (A664/B692), of course; but what exactly does this mean? Kant explains at A721/B749:

> Accordingly, transcendental propositions can never be given through construction of concepts, but only in accordance with a priori concepts. They contain merely the rule according to which a certain synthetic unity of that

---

the fundamental importance of this derivation is noted explicitly at several points in Kant's unpublished *Nachlaß*. Thus, for example, Reflexion 5414 (1776–1780) begins: "One can certainly discover rules empirically, but not laws—as Kepler in comparison to Newton—for to the latter belong necessity; and therefore, that they can be cognized a priori" (18, 176.19–21). (This Reflexion is cited in Buchdahl [13], n. 6 on p. 148.) For further discussion and references see Chapters 4 and 5 below.

> which cannot be intuitively represented a priori, namely perception, is to be sought empirically Yet these transcendental propositions cannot exhibit any one of their concepts a priori in any instance whatsoever, but do this only a posteriori by means of experience—which is first possible only in accordance with these synthetic principles.[45]

The dynamical principles of pure understanding, and their specifications given in the *Metaphysical Foundations,* are rules which govern a procedure for constructing objective experience from given perceptions *if such experience is possible at all.* There can be no a priori guarantee, however, that the proper object of pure understanding, namely, objective experience, is in fact constructible. In the end, only the utterly remarkable success of Newton's *Principia* itself shows that—and how—objectivity is realized.

45. The importance of this passage is stressed by Plaass [97], p. 60.

· CHAPTER 4 ·

Space, the Understanding, and the Law of Gravitation: *Prolegomena* §38

Section 38 of Kant's *Prolegomena to Any Future Metaphysics* (1783) is of great interest. For Kant there attempts, uncharacteristically, to illustrate one of the central claims of his exceedingly abstract and general transcendental philosophy by means of a concrete example. The claim in question is stated as the conclusion of §36:

> *The understanding does not extract its laws* (a priori) *from, but prescribes them to, nature.* (4, 320.11–13)

This same claim figures prominently in the conclusion to the central argument of the *Critique of Pure Reason,* the transcendental deduction of the categories, and it is as obscure—and apparently also as outlandish and bizarre—as it is striking.[1] Here in the *Prolegomena* Kant promises to illuminate this peculiar claim by means of an actual example, which is announced in §37:

> We shall illustrate this seemingly bold proposition through an example, which is to show: that laws, which we discover in objects of sensible intuition, especially if they are cognized as necessary, are indeed held by us to be such as the understanding has placed there, although they are equally similar otherwise in all respects to natural laws that we ascribe to experience. (320.15–20)

The example provided in §38 is the law of gravitation: "a physical law of mutual attraction diffusing over all of material nature, whose rule is that it diminishes inversely as the square of the distances from each attracting point" (321.13–16).

1. See A126–128, B159–160, B163–165. Kant himself is well aware of the apparently bizarre nature of his claim: "As exaggerated and as contrary to common sense as it may sound, to say: the understanding is itself the source of the laws of nature; such an assertion is nevertheless equally correct and appropriate to the object, namely experience" (A127).

Unfortunately, the discussion of this example in §38 is also extremely compressed and obscure, and it raises at least as many questions as it answers. Is Kant saying that the law of gravitation itself is knowable a priori? Such a view is certainly suggested by the text, for immediately after identifying the law of gravitation Kant describes its "sources" as "resting merely on the ratio [Verhältnis] of spherical surfaces of various radii" (321.18–20). In other words, Kant employs a representation of the inverse-square law in terms of the diffusion of a quantum of attractive force from a central point uniformly on concentric spherical surfaces (which of course stand to one another as the squares of their radii) and seems to suggest that the inverse-square law is knowable a priori for precisely this reason—as a geometrical consequence of the three-dimensionality of space, as it were.[2] If this is in fact the view of §38, however, then it becomes hard to see how it is to illustrate the claim at issue: namely, that the laws of nature are prescribed by the *understanding*. For the three-dimensionality of space is presumably a feature of the faculty of *sensibility* for Kant (see, for example, #3 of A24, B41) and, accordingly, does not derive from the faculty of understanding. Moreover, Kant emphasizes the distinction between these two faculties in a passage following in §38 itself:

> The mere universal form of intuition called space is thus indeed the substratum of all intuitions determinable into particular objects, and certainly the conditions of the possibility and manifoldness of the latter lie in it; but the unity of the objects is nevertheless determined solely through the understanding, and in fact according to conditions that lie in its own nature; and so the understanding is the origin of the universal order of nature . . . . (322.1–8)[3]

On the other hand, if Kant is not propounding a purely geometrical deriva-

2. In *Thoughts on the True Estimation of Living Forces* (1747), Kant suggests, in §10, that the three-dimensionality of space is derived from the fact that the fundamental law binding together substances in our world is an inverse-square law; if it were instead an inverse-cube law, say, then space would have more than three dimensions. Here in the *Prolegomena* Kant appears to reverse this inference and to argue from three-dimensionality to the inverse-square law.

3. Indeed, §38 is the concluding section of a subheading of Part Two of the *Prolegomena* entitled "How Is Nature Itself Possible?" The subheading begins, in §36, by distinguishing between "nature in its *material* meaning, namely, according to intuition, as the totality of appearances . . . space, time, and that which fills both, the object of sensation" and "nature in its formal meaning, as the totality of rules, under which all appearances must stand, if they are to be thought in an experience as connected." The possibility of the first is explained "by means of the constitution of our sensibility," and this is said to be accomplished in the Transcendental Aesthetic in the *Critique* and in the First Part of the *Prolegomena*. The possibility of the second is explained "by means of the constitution of our understanding," and this is said to be accomplished in the Transcendental Logic in the *Critique* and in the Second Part of the *Prolegomena* (318.7–27).

tion of the inverse-square law in §38, then why does he put such emphasis on a representation in terms of concentric spherical surfaces—thereby strongly suggesting such a geometrical derivation? Moreover, if the law of gravitation is not to be geometrically derived in this way, how then is it to be derived? And how, in particular, is it to illustrate the claim that the laws of nature are prescribed by the understanding?

I will approach these questions by first articulating a view of the law of gravitation based on Kantian texts other than §38 of the *Prolegomena.* According to this interpretation, the law of gravitation is not itself an a priori law for Kant, and, accordingly, it does not derive a priori from either the understanding or sensibility. Rather, the law of gravitation has a particular kind of mixed status: it is derived from a combination of a priori laws of the understanding (the analogies of experience), a priori laws of sensibility (Euclidean geometry), *and* a posteriori given data of experience. Because of its essential dependence on the latter, the law of gravitation is an empirical law and hence, strictly speaking, is neither a priori nor necessary. Nevertheless, because of its equally essential dependence on the above-mentioned a priori laws, the law of gravitation still enjoys a particular kind of "empirical" or "material" necessity in virtue of which it is more firmly established and secure relative to the a posteriori given data than any mere inductive generalization or hypothesis. After articulating and defending this interpretation of the status of the law of gravitation for Kant, I will then return to §38 of the *Prolegomena.* I hope to show that it confirms the interpretation to be developed in a most illuminating fashion and, in fact, that its role is to emphasize precisely the priority of the understanding over sensibility in the particular kind of mixed procedure for establishing the law of gravitation I will describe below. In this way, §38 does after all shed light on the claim it is intended to illustrate.

· I ·

The *Metaphysical Foundations of Natural Science,* published in 1786 (three years after the publication of the *Prolegomena* and one year before the appearance of the second edition of the *Critique*), represents Kant's most sustained attempt to depict the relationship between the system of the *Critique*—in particular, the transcendental principles of the understanding—and natural science—in particular, the science of Newton's *Principia.* In the Preface to the *Metaphysical Foundations* Kant distinguishes between general metaphysics or the *transcendental* part of the metaphysics of nature and *special* (besondere) metaphysics. The former comprises the transcendental principles of the understanding; the latter applies these principles to a particular species of objects of our senses: in this case to objects of outer sense or *matter*. And, although the concept of matter is said to be an empirical concept, the principles of special metaphysics (the metaphysics of corporeal nature) are just as a priori as are the principles of general metaphysics (transcendental philosophy). Indeed, the principles of special metaphysics constitute an indispensable complement to those of general metaphysics, without which the latter would remain "a mere form of thought" without "sense and meaning."[4]

Of special importance, in the present context, are the principles of special metaphysics that correspond to, and thus exemplify and realize, the analogies of experience. These are given in Chapter 3 or Mechanics of the *Metaphysical Foundations* as the three "laws of mechanics": (1) the principle of the conservation of mass or quantity of matter, (2) the law of inertia, (3) the principle of the equality of action and reaction. Each is derived by an a priori "proof" which proceeds by starting with the corresponding principle from the *Critique* and substituting in, as it were, the empirical concept of matter.[5] These three laws of mechanics, it is clear, correspond closely to Newton's three laws of motion;[6] but what is even

4. See, in particular, paragraphs six and sixteen of the Preface (4, 469.26–470.12, 477.14–478.20); and compare also, and especially, B291–294 of the *Critique* (the General Note to the System of the Principles). Special metaphysics is a priori because "it occupies itself with a particular [besondere] nature of this or that kind of thing, of which an empirical concept [in this case *matter*] is given, but nevertheless in such a way that, besides what lies in the concept, no other empirical principle [Prinzip] is required for its cognition" (470.1–4). Compare also A847–848/B875–876 of the *Critique:* "how can I expect to have knowledge a priori (and therefore a metaphysics) of objects in so far as they are given to our senses, that is, given in an a posteriori manner? . . . The answer is this: we take nothing more from experience than is required to give us an object of outer . . . sense. The object of outer sense we obtain through the mere concept of matter (impenetrable, lifeless, extension)." I do not pretend that the sense of these passages is immediately clear, however.

5. For example, the proof of the first law of mechanics begins as follows: "From general metaphysics we take as our basis the proposition that in all alterations of nature no substance either arises or perishes, and here it is only shown what substance amounts to in matter" (541.32–35).

6. The most obvious difference, of course, is the absence, in Kant's enumeration, of Newton's second law: in modern terms, $F = ma$. However, since Kant formulates the equality of action and reaction in terms of equal and opposite *momenta,* and states that a "dynamical law" of moving *forces* corresponds to his "mechanical law" of the community of *motions* (Note 2 to Proposition 4 of the Mechanics: 548.13–29), Newton's second law appears to be contained in Kant's third law: in modern terms, Kant's third law states that any two interacting masses m_A, m_B have accelerations a_A, a_B such that $m_A a_A = -m_B a_B$; Kant's "dynamical law" then states that $F_{AB} = -F_{BA}$, where F_{AB}, F_{BA} are the forces exerted by A on B and by B on A respectively. It appears, then, that Kant is presupposing $F_{BA} = m_A a_A$ and $F_{AB} = m_B a_B$. The fact that Kant gives special prominence, in his first law, to the conservation of mass or quantity of matter is due, I think, to the special role of mass or quantity of matter in defining an objective spatio-temporal framework for Kant; this will, I hope, become clear below.

more striking is the way in which they are employed in the overall argument of the *Metaphysical Foundations*.

The *Metaphysical Foundations* begins, in the first chapter or Phoronomy, by emphatically rejecting Newtonian absolute space. Such a space "can be no object of experience," and so "absolute space is *in itself* therefore nothing and indeed no object at all" (4, 481.28–33). It follows that "all motion that is an object of experience is merely relative; the space in which it is observed is a relative space" or, as Kant also puts it, an "*empirical space*" which "is characterized through that which can be sensed" (481.14–22). Kant does not hold, however, that all such relative spaces or empirical spaces (in modern terms, reference frames) are equally suitable for representing motion. On the contrary, in the fourth chapter or Phenomenology he explicitly follows Newton in sharply distinguishing between "true" and "apparent" motion. The problem is to understand how, in the absence of Newtonian absolute space, Kant can possibly implement this distinction.

The fourth chapter or Phenomenology is intended to exemplify or realize the postulates of empirical thought of the *Critique* (that is, the principles corresponding to the modal categories of possiblity, actuality, and necessity), and it has as its aim the transformation of *appearance* (Erscheinung) into *experience* (Erfahrung). More specifically, its aim is to transform *apparent motions* into *true motions* (Observation to the Definition with which the Phenomenology begins: 554–555). The former are defined relative to an arbitrarily chosen reference frame (by taking some arbitrarily chosen body to be at rest); the latter, however, are the outcome of a procedure for "reducing all motion and rest to absolute space"—where the latter is conceived not as a pre-existing object, as it were, but as nothing more or less than the goal of this procedure: in Kant's terms, the notion of absolute space is a "necessary concept of reason" or "mere *idea*" (559.7–560.7). The procedure itself unfolds in three steps or stages. In the first stage we assume some arbitrarily chosen body (Kant's example is the earth) to be at rest and describe all motions relative to it. These relative motions are merely *possible*. In the second stage we apply the law of inertia (Kant's second law of mechanics), together with "active dynamical influences given through experience" to determine states of true or *actual* rotation.[7] In the third stage we apply the equality of action and reaction (Kant's third law of mechanics) to declare as *necessary*

7. In his description of this stage Kant refers explicitly to Newton's Scholium to the Definitions of *Principia* and gives as examples the axial rotation of the earth and the orbital rotation around a common center of gravity of a two-body system (557.33–558.3, 562.32–40); as examples of "active dynamical influences" he gives "gravity, or a tensed cord" (552.12–13).

equal and opposite contrary motions for any state of true or actual motion.

Kant's discussion of this procedure (the way in which he deploys the laws of motion, in particular) strongly suggests that he takes the argument of Book III of *Principia* as his model.[8] Newton begins the argument with "Phenomena," cases of observable, so far merely relative motions in the solar system: the relative motions of the satellites of Jupiter and Saturn with respect to their primary bodies, the relative motions of the planets with respect to the sun and the fixed stars, and so on. (It is thus particularly noteworthy that, in his statement of Phenomenon IV, Newton carefully and deliberately leaves it open whether the earth rotates around the sun or vice versa.) Newton then applies his first and second laws of motion to these relative motions to conclude that there is an inverse-square force directed towards the center of each primary body in question, and, moreover, that this force is in fact identical to the force of terrestrial gravity. Next, Newton applies the third law of motion to conclude first, that the forces in question are mutual—equal and opposite—and second, that *any* two bodies in the solar system interact through such forces. This last conclusion then yields the result that the forces in question are also directly proportional to the masses of the two interacting bodies, and, on this basis, Newton is able to develop a method for rigorously estimating the masses of the various bodies in the solar system. It follows, in particular, that the common center of mass of the solar system is located sometimes within, sometimes without the surface of the sun, but never very far from the sun's center; and Newton is thus finally in a position to settle the issue between Copernicus and Ptolemy (more precisely, between Kepler and Tycho) in favor of heliocentrism. Such, then, is Newton's method for implementing the distinction between true and apparent motion: for "obtaining the true motions from their causes, effects, and apparent differences" (Scholium to the Definitions: [82], p. 53; [83], p. 12).

Kant, I have suggested, takes this Newtonian argument as the model for his procedure for distinguishing true from apparent motions, but he also views the procedure in his own characteristic way. Whereas Newton presents it as the determination of a fact, the fact that the sun is (approximately) in a true state of rest, Kant views the very same procedure as a necessary condition for giving the distinction between true and apparent motion objective meaning in the first place. In particular, since for Kant there is no pre-existing absolute space relative to which states of true motion are automatically well defined, it literally makes no sense independently of a method for actually determining the common center of mass of the solar system to assert that the sun is either in a true state of rest or

8. An attempt to provide detailed support for this idea can be found in Chapter 3 above.

that it is in a true state of motion. Rather, the true motions just are motions defined relative to the common center of mass of the system of interacting bodies in question; hence, since the center of mass of the solar system turns out to be very close to the center of the sun, the sun is (approximately) in a true state of rest in the context of this system. Of course we are here assuming that the solar system is an effectively isolated system, which is not strictly the case. It follows that the procedure must be pushed far beyond the argument of Book III: for Kant, to the center of mass of the Milky Way galaxy, from there to a common center of mass of a system of such galaxies, from there to a common center of mass of a system of such systems, and so on.[9] In the end, questions of true motion and rest are decided only with respect to "the common center of gravity of all matter" (563.4–5), and this point, for Kant, is forever beyond our reach. It is in this sense that he views absolute space as no object at all but rather an idea of reason: that is, as a rule for constructing an infinite sequence of better and better *approximations* to the distinction between true and apparent motion—a sequence which converges but never actually terminates.

It is important to note, at this point, that Kant's characteristic perspective on the Newtonian procedure for determining the true motions also involves a characteristically Kantian perspective on the laws of motion. Thus, Newton presents the laws of motion as facts, as it were, about a notion of true motion that is antecedently well defined. Accordingly, he attempts to provide empirical evidence for their truth—especially in the case of the third law. For Kant, on the other hand, since there is no such antecedently well-defined notion of true motion, the laws of motions are not facts but rather conditions under which alone the notion of true motion first has objective meaning. And, as we have seen, the third law is particularly important in this regard, for the true motions are defined relative to the common center of mass of the system of interacting bodies in question: in other words, true motions are just those satisfying the third law. In is within this context, then, that Kant, in the third chapter or Mechanics of the *Metaphysical Foundations,* casts doubt on Newton's attempt to invoke experience in support of the third law (Observation 1 to Proposition 4: 549.6–9) and holds, on the contrary, that all the laws of motion must be proved a priori: the laws of motion are not empirical facts about true motions but a priori conditions of the possibility of such motions—just as the analogies of experience, which the laws of motion are intended to instantiate or realize, are not facts of objective experience but a priori conditions of the possibility of such experience.

9. This conception of an infinity of ever-widening rotating systems is developed in Kant's *Theory of the Heavens* of 1755. See Chapter 3 above.

But what about the law of gravitation? What is the status of this law in the context of our argument? The first point to notice is that the Newtonian procedure for determining the true motions in the solar system—for determining its center of mass—is also, and at the same time, a procedure for establishing the law of gravitation. We begin with "phenomena": the observed relative motions of various systems of satellites with respect to their primary bodies (Jupiter and its moons, Saturn and its moons, the sun and its planets, and so on). These relative motions are described, in all cases, by Kepler's three laws: each orbit is an ellipse with the primary body at one focus; each orbit satisfies the area law; and any two concentric orbits satisfy the harmonic law. We then apply the first two laws of motion, together with the provisional assumption that the relative motions in question closely approximate to true motions, to derive mathematically the inverse-square law. In particular, it follows via Kepler's first two laws that each orbit is governed by an inverse-square force; and it follows via Kepler's third law that the same proportionality holds between any two orbits: that is, not only is the acceleration within any single orbit inversely proportional to the square of the distance from the focus of that orbit, but the accelerations governing any two concentric orbits are also inversely proportional to the squares of their respective distances from their common focus.[10] This last result is actually extremely important, for it shows that the acceleration of any satellite depends only on its distance from the primary body and is entirely independent, therefore, of any facts about the constitution of the satellite. In Howard Stein's very useful terminology, we have an "acceleration-field" around each primary body which determines the acceleration of any object that may be placed therein solely as a function of the distance.[11] In particular, then, the acceleration of any satellite is entirely independent of its mass, and it follows that the latter is directly proportional, at a given distance, to the satellite's weight.

So much for the inverse-square law. We are still not in a position to assert the full law of gravitation, however. For this law goes far beyond the existence of an inverse-square acceleration-field around each primary body and states that *any* two bodies in the solar system *mutually* attract one another according to the same inverse-square force, which, in addi-

10. It follows from Kepler's first two laws that if r is the distance from the focus to an arbitrary point of any single orbit and a is the acceleration (directed towards the focus) at that point we have $a = k/r^2$. This leaves it open that in two concentric orbits we have $a_1 = k_1/r_1^2$, $a_2 = k_2/r_2^2$, $k_1 \neq k_2$. But Kepler's third law states that for any two such orbits, if r_1, r_2 are the respective mean distances and T_1, T_2 the respective periods, then $T_1^2/T_2^2 = r_1^3/r_2^3$. Assuming circular orbits for simplicity, we have $a = v^2/r$ (where v is the uniform linear speed), $v = 2\pi r/T$, and therefore $a_1/a_2 = r_2^2/r_1^2$; *hence* $k_1 = k_2$.

11. See Stein's seminal paper [104], to which I am greatly indebted throughout my discussion of the Newtonian argument.

tion, is also directly proportional to the masses of the two interacting bodies in question. Moreover, this latter property (proportionality to mass) in fact depends on the former properties (universality and mutuality) in the Newtonian argument. The key point is that, in order to derive proportionality to mass, we need to extend the acceleration-fields of the two bodies in question (Jupiter and Saturn, for example) to incorporate not only any satellites they may have but also each of the two primary bodies mutually; we then apply the third law of motion directly to this last interaction to infer proportionality to mass.[12] Now, that gravitational attraction is mutual of course also follows from the third law of motion, but how do we establish the universality of such mutual attractions? Such universal mutual accelerations are of course in no way given in the observable "phenomena" to which we have appealed so far. Newton himself, at this crucial point in the argument, appeals to inductive considerations: all bodies "within the reach of our experiments" gravitate towards the earth with an inverse-square acceleration, therefore all bodies whatever so gravitate (Corollary III to Prop. VI); moreover, the acceleration-fields of the other primary bodies in the solar system are "appearances of the same sort" as the acceleration-field of the earth, therefore all bodies whatever gravitate towards every primary body (Prop. V); and so on.[13] With the universality of gravitational attraction thus accepted, we are now, and only now, in a position to assert the full law of gravitation. Accordingly, we can now, and only now, rigorously estimate the masses of the various bodies in the solar system, rigorously determine the center of mass of the solar system, and rigorously describe the true motions in the solar system. In this way, we are also in a position to discharge the provisional assumption with which we began: namely, that the relative motions constituting our initial "phenomena" closely approximate to true motions.

Such, in outline, is the argument by which the law of gravitation is established in Book III. What I want to suggest, of course, is that this argument is also Kant's model for how the law of gravitation is to be established. And, in this connection, it is especially noteworthy that Kant discusses the central, and especially problematic, steps of the argument in

12. This is established in Proposition VII of Book III, which depends on the previously asserted universality of gravitational attraction (Prop. VI) and Proposition LXIX of Book I. See Chapter 3 above for an attempt to depict some of the intricacies of this argument: one particularly subtle point, which sufaces explicitly in the correspondence between Newton and Cotes, is that we must apply the third law of motion *directly* to the gravitational interactions in question and hence assume that gravity acts immediately at a distance (rather than through the action of an intervening medium, say).

13. Of course these inductive arguments for universality are greatly strengthened by subsequent observations of the planetary perturbations. Newton himself, in the second (1713) edition of *Principia,* adds a remark on the observed perturbation of Saturn by Jupiter to Proposition XIII (compare also Corollary III to Proposition V).

Propositions 7 and 8 of the second chapter or Dynamics of the *Metaphysical Foundations.* There Kant offers the one explicit criticism of Newton to be found in his book: namely, that Newton cannot leave it open whether gravity is an essential property of matter without fatally compromising the proof that gravitational attraction is directly proportional to mass. By "essential" here Kant means immediate and universal, and he holds that Newton can neither remain agnostic about the first nor allow the second to be supported merely inductively: on the contrary, both the immediacy and the universality of "the fundamental force of attraction" are to be postulated a priori as belonging to the very concept of matter. The underlying idea, once again, is that neither the immediacy nor the universality of gravitational attractions can be viewed simply as an empirical property which holds (or fails to hold) of the true motions in the solar system, for without these properties we are unable to define the true motions in the first place: the true motions are determined relative to the center of mass of the solar system; the determination of this center of mass presupposes a determination of the masses of the various bodies in the solar system; this latter presupposes proportionality to mass of gravitational attraction; and this, as we have seen, presupposes both immediacy and universality. In this way these properties of gravitational attraction, like the laws of motion themselves, are viewed as necessary conditions of the possibility of an objective notion of true motion; and they are in this sense a priori for Kant.[14]

The significance of this last point is now that Kant, in the context of his fundamental reinterpretation of the Newtonian argument for the law of gravitation, has eliminated all of its explicitly inductive steps. We begin, to be sure, with Kepler's three laws as empirical descriptions of the observable "phenomena": the empirically given relative motions in the solar system. From this point on, however, we do not proceed inductively. For Kant, the laws of motion we subsequently apply are fixed a priori, and so are the crucial properties of immediacy and universality (as well as the Euclidean geometry of space, of course). It follows that the law of gravitation is determined (or perhaps better, "constructed") from Kepler's laws

14. Thus the immediacy and universality of gravitational attraction are part of the very concept of matter (as conditions of its possibility), because the concept of matter is that of "the movable in space" (Definition 1 of the Phoronomy: 481.6–10): in order to apply this concept to experience we thus need an objective notion of true motion. In reference to note 13 above, Kant's point holds also for observations of planetary perturbations: these too can only be regarded as *true* motions if we follow Newton's procedure for determining the center of mass of the solar system and hence presuppose universality. Of course if such perturbations were not observed the entire structure would collapse, and we would be left with no notion of true motion at all. This is one important reason for understanding the concept of matter Kant articulates as an *empirical* concept.

by means of a procedure that admits no room for inductive uncertainty: there can be no question, in particular, of alternative possible explanations of the same given data. In this sense, the law of gravitation does not count as a mere *hypothesis* for Kant, and he is therefore in a position to give a very strong interpretation to Newton's notion of "deduction from the phenomena."[15]

In this way, Kant is also in a position to ascribe a particular kind of "empirical" or "material" necessity to the law of gravitation. The relevant notion of necessity here is in fact just Kant's official category of necessity, as this category is explained in the Postulates of Empirical Thought at A218–219/B265–266:[16]

> 1. That which agrees with the formal conditions of experience (according to intuition and concepts), is *possible.*
> 2. That which connects with the material conditions of experience (sensation), is *actual.*
> 3. That whose connection with the actual is determined in accordance with universal conditions of experience, is (exists as) *necessary.*

Now recall that in the *Metaphysical Foundations,* Chapter 4 or Phenomenology realizes or instantiates the postulates of empirical thought. Recall also that this chapter applies the category of possibility to merely relative motions, the category of actuality to true rotational motions, and the category of necessity to equal and opposite motions satisfying the third law of motion. The argument for the law of gravitation then corresponds to the postulates of empirical thought in the following way. In the first stage we begin with the purely relative motions described by Kepler's three laws. In the second stage we assert that these motions are due to an

15. Compare the well-known concluding Scholium to *Principia:* "whatever is not deduced from the phenomena is to be called an hypothesis; and hypotheses, whether metaphysical or physical, whether of occult qualities or mechanical, have no place in experimental philosophy. In this philosophy particular propositions are inferred from the phenomena, and afterwards rendered general by induction. Thus it was that the impenetrability, the mobility, and the impulsive force of bodies, and the laws of motion and of gravitation, were discovered" ([82], p. 764; [83], p. 547). Because Newton explicitly includes induction in his method, and especially because he places the laws of motion and the law of gravitation on the same level, he himself is not in a position to give a strong interpretation to "deduction from the phenomena." On the contrary, I do not see how Newton's method can, in the end, be distinguished from the hypothetico-deductive method: in the end, therefore, he does not and cannot avoid hypotheses. (Of course in the end this Newtonian predicament remains our predicament as well, for we certainly cannot embrace Kantian apriorism in the context of contemporary physics.)

16. For a discussion of this notion of "empirical" or "material" necessity, also in the context of Kant's perspective on the Newtonian argument for the law of gravitation, see Harper [44].

inverse-square *force* and are therefore actual or true motions.[17] Finally, in the third stage, we apply the third law of motion in the rather intricate fashion sketched above to conclude that such inverse-square forces are universal, everywhere mutual, and directly proportional to mass. The center of mass of the solar system is determined as necessary for describing the true motions therein (that is, the determination of this point is necessary for the possibility of such true motions), and the law of gravitation, which describes these true motions completely, is itself necessary in precisely the sense of the third postulate: it is determined in its connection with the actual in accordance with universal conditions of experience—namely, the laws of motion and hence the analogies of experience.

Further confirmation for this reading is provided by Kant's remarks on the postulates of empirical thought in §25 of the *Prolegomena:*

> . . . their relation to experience in general [involves] possibility, actuality and necessity according to universal natural laws, which would constitute the physical doctrine of method (distinction of truth and hypotheses and the limits of reliability of the latter). (307.33–308.5)

And this is particularly so when the above passage is read in conjunction with the following important footnote in the Preface to the second edition of the *Critique:*

> Thus the central laws [Zentralgesetze] of the motion of the heavenly bodies provided established certainty to what *Copernicus* had first assumed only as an hypothesis and at the same time established [bewiesen] the invisible force holding the universe together (the *Newtonian* attraction), which would have remained forever undiscovered if Copernicus had not dared—in a manner contrary to common sense but yet true—to seek the observed motions not in the objects of the heavens but in their observer. (Bxxii)[18]

Note that Kant here gives clear recognition to the fact that the law of gravitation is established at the same time, and by the same argument, as is the Newtonian determination of the true motions in the solar system.

The most explicit statement I have yet encountered, however, is found in an unpublished Reflexion written somewhere between 1776 and the early 1780s (R. 5414):

17. In the *Metaphysical Foundations,* Chapter 2 or Dynamics, where the "fundamental forces" of attraction and repulsion are introduced, corresponds to the categories of quality in the *Critique,* where *sensation* is introduced.

18. The expression "central laws [Zentralgesetze]" is a peculiar one, and presumably refers to the laws of *central forces* deployed in *Principia.* These include the various measures of centripetal force, the inverse-square law, and so on—and also, of course, the laws of motion. Compare the use of this expression in Observation 7 in Part Two of *The Only Possible Basis* (1763) (2, 147.8–149.28), where it is interchangeable with "rules of central forces [Regeln der Centralkräfte]" (149.19–20).

> Empirically one can certainly discover rules, but not laws—as Kepler in comparison with Newton—for to the latter belongs necessity, and hence that they are cognized a priori. Yet one always supposes that rules of nature are necessary—for on that account it is nature—and that they can be comprehended a priori; therefore one calls them laws by way of anticipation. The understanding is the ground of empirical laws, and thus of an empirical necessity, where the ground of law-governedness can in fact be comprehended a priori: e.g., the law of causality, but not the ground of the determinate law. All metaphysical principles [Prinzipien] of nature are only grounds of law-governedness. (18, 176.19–28)

Of particular interest is the sliding usage of "a priori" and "necessity" in this passage. At first Kant appears to be saying that Newton's law of gravitation, in contradistinction to the Keplerian laws from which it is in fact derived, is itself an a priori law. It turns out, however, that only the grounds of this law in the understanding—which are identified with grounds of law-governedness in general such as the law of causality—are known a priori.[19] The point, I think, is that, whereas Kepler's laws themselves are merely inductive generalizations and are as such not yet grounded in a priori laws, the law of gravitation is obtained precisely by applying a priori laws such as the laws of motion to these Keplerian generalizations—it thereby obtains "an empirical necessity." Thus, the sliding usage of "a priori" and "necessity" here perfectly illustrates the particular kind of mixed status that the law of gravitation has on our interpretation.

· II ·

We are now in a position to state the problem posed by §38 of the *Prolegomena* more precisely and also to appreciate some of the wider implications of this problem in the general context of Kant's transcendental philosophy.

First, we have seen above that one can develop a rather plausible and attractive interpretation of Kant's view of the law of gravitation, according to which this law is not an a priori but an empirical law. It is empirical because it depends on Kepler's three laws governing the observed relative motions of satellites with respect to the primary bodies in the solar system, and these Keplerian laws are themselves mere inductive generalizations. Nevertheless, despite this undoubtedly empirical starting point, the argument for the law of gravitation, on a Kantian account,

19. Another example of the same phenomenon can be found in Kant's discussion of the necessity found in particular causal judgements in the footnote to §22 of the *Prolegomena:* 305.23–35.

then proceeds in an a priori and non-inductive fashion. In particular, the argument then invokes a priori laws of the understanding, that is, the analogies of experience, instantiates these laws by means of the empirical concept of matter to obtain the equally a priori laws of mechanics or laws of motion, and finally applies the latter (together with the a priori laws of Euclidean geometry, of course) to our initial a posteriori given data in such a way as to determine a unique conclusion: there are inverse-square mutual accelerations, directly proportional to mass or quantity of matter, between each body in the solar system and each other body. In this way, although the law of gravitation is not itself a priori, it is *grounded* in a priori laws and thereby acquires a particular kind of mixed or "empirical" necessity. This procedure of grounding empirical laws by a priori laws of the understanding then provides us with a plausible interpretation of the claim that §38 of the *Prolegomena* is intended to illustrate: namely, that the understanding prescribes laws to nature and is thus the source of its law-governedness.

Moreover, the procedure of grounding empirical laws by a priori laws of the understanding illustrated by the argument for the law of gravitation, as articulated above, also illuminates Kant's remarks on the relationship between empirical laws and a priori laws of the understanding generally. Thus, in the first edition transcendental deduction we have:

> Although we learn many laws through experience, these are nonetheless only special determinations [besondere Bestimmungen] of yet higher laws, among which the highest (under which all the others stand) originate a priori in the understanding itself, and are not borrowed from experience but rather provide appearances with their law-governedness, and precisely thereby make experience possible. (A126)

Note that only the highest-level laws are said to "originate a priori in the understanding itself," a point which is emphasized even more strongly in the second edition deduction:

> Nature, considered merely as nature in general, is dependent on these categories, as the original ground of its law-governedness (as nature viewed formally). Pure understanding is not, however, in a position, through mere categories, to prescribe to appearances any a priori laws other than those which are involved in a *nature in general*, that is, in the law-governedness of all appearances in space and time. Special [besondere] laws, because they concern empirically determined appearances, can *not be completely derived* therefrom [können davon *nicht vollständig abgeleitet* werden], although they one and all stand under them. (B165)[20]

20. Kemp Smith renders "können davon *nicht vollständig abgeleitet* werden" as "cannot in their specific character be *derived* from them," which is certainly acceptable. However, this rendering leads Harman, for one, to quote the passage as "cannot . . . be *derived* from

Finally, this same conception of the relationship between empirical laws and a priori laws of the understanding is found in the concluding paragraph of §36 of the *Prolegomena:*

> We must, however, distinguish empirical laws of nature, which always presuppose particular perceptions, from the pure or universal natural laws, which, without being based on particular perceptions, contain merely the conditions of their necessary uniting in an experience—and with regard to the latter nature and possible experience are entirely and absolutely one and the same; and, since in nature law-governedness rests on the necessary connection of appearances in an experience (without which we could cognize absolutely no object of the sensible world at all)—and therefore rests on the original laws of the understanding—it thus at first indeed sounds strange, but is nonetheless certainly true, if with regard to the latter I say: *The understanding does not extract its laws* (a priori) *from, but prescribes them to, nature.* (320.1–13)[21]

Thus it is particularly noteworthy that even here in the *Prolegomena* Kant explicitly restricts his claim of an a priori prescription by the understanding to "pure or universal natural laws" or "original laws of the understanding"—among which the law of gravitation itself is certainly not to be found.[22]

Yet, as we have also seen, Kant goes on to illustrate this claim of a priori prescription by the understanding, in §38 immediately following, by precisely the law of gravitation. This in itself would not necessarily be problematic, of course, for the illustration might simply be intended to show how empirical laws, although not themselves prescribed a priori

---

them," and to use it in support of the idea that there are no deductive links at all between the transcendental principles of the *Critique* and the special laws of natural science proper—not even between the analogies and the laws of motion: see [43], chap. IV, §§1, 6 (the quotation occurs on p. 59). Compare also A127–128: "To be sure, empirical laws as such can in no way derive their origin from pure understanding—no more than the immeasurable manifold of appearances can be adequately comprehended from the pure form of sensibility. Yet all empirical laws are only special determinations of the pure laws of the understanding, under which and in accordance with the norm of which they first become possible, and the appearances take on a lawful form—just as all appearances, notwithstanding the variety of their empirical form, nonetheless also must always be in accordance with the conditions of the pure form of sensibility."

21. The translations following Carus unaccountably omit the second occurrence of "with regard to the latter" in this passage (although Lucas retains it), thereby omitting Kant's emphasis that *only* the original laws of the understanding are prescribed a priori to nature.

22. In §15 of the *Prolegomena* Kant even hesitates to allow that the principles of the special metaphysics of corporeal nature are part of "pure science of nature" or "universal science of nature," because they depend on the empirical concept of matter. On the contrary, only the transcendental principles of the understanding "have the required generality"—as instances Kant gives the first two analogies of experience and says: "These are actual universal natural laws, which hold completely a priori" (294.32–295.22).

by the understanding, are nonetheless grounded in a priori laws of the understanding as outlined above. And, if this were the case, the illustration would then fit in perfectly with all the ideas we have been considering so far. The problem, rather, is that Kant goes on to suggest a purely a priori derivation of the law of gravitation, based on the geometrical properties of a force diffusing from a central point uniformly on concentric spherical surfaces, and, moreover, he does not explicitly invoke any of the factors we have been emphasizing above: neither Kepler's laws, nor the immediacy and universality of gravitational attraction, nor even the laws of motion. Section 38 therefore confronts us with two pressing problems. First, how can Kant, in the context of all the ideas we have been considering, nevertheless entertain a purely a priori derivation of the law of gravitation? Second, what would such a derivation, given its purely geometrical character, have to do with the principles of the *understanding* in any case? For the principles of the understanding—in particular, the analogies of experience and the laws of motion that are to instantiate them—appear to play no role whatsoever in the purely geometrical derivation Kant envisions.

This last point can be sharpened and put into perspective in the context of one of the central distinctions of Kant's transcendental philosophy: namely, the distinction between the *mathematical* and the *dynamical*. In the *Critique* the distinction is first introduced as a division of the table of categories: the categories under the first two headings, of quantity and quality, are entitled mathematical; those under the second two headings, of relation and modality, are entitled dynamical (B110). The division is then carried over into the table of principles (A160–162/B199–200) and explained most fully in the proof of the general principle of the analogies of experience at A178–181/B220–224. Mathematical principles are concerned with "appearances [Erscheinungen], and the synthesis of their empirical intuition." They "extend to appearances according to their mere possibility" and teach us how appearances "can be generated according to rules of a mathematical synthesis" so that we can "determine appearance as magnitude." As such, the mathematical principles are "constitutive." The dynamical principles, on the other hand, are concerned with "the *existence* of such appearances and their *relation* to one another in respect of their existence." And "since existence cannot be constructed," the dynamical principles "can yield only *regulative* principles [Prinzipien]." In particular, the analogies of experience, unlike mathematical analogies, do not actually give us the fourth term of a proportion:

> But in philosophy the analogy is not the equality of two *quantitative* but rather *qualitative* relations [Verhältnisse], where from three given terms I can cognize and give a priori only the *relation* [*Verhältnis*] to a fourth, but not

> this very fourth *term* itself; but I do nonetheless have a rule for seeking it in experience and a mark for discovering it therein. An analogy of experience is thus only to be a rule according to which unity of experience (not as perception itself, as empirical intuition in general) may arise from perceptions, and it is to hold of the objects (the appearances) as a principle not *constitutively* but merely *regulatively*. (A180/B222–223)

Similarly with the postulates of empirical thought: they too are regulative, not constitutive, and are to be distinguished from mathematical principles in terms of "the manner of their evidence [Evidenz], that is, what is intuitive therein (and therefore also of their demonstration [Demonstration])" (A180/B223).[23]

Now it is not immediately clear what Kant has in mind here, of course, but the distinction between mathematical and dynamical principles can be substantially clarified, in the present context, through comparison with a kindred distinction Kant makes in the General Observation to Chapter 2 or Dynamics of the *Metaphysical Foundations:* namely, between the "mathematical-mechanical" and "metaphysical-dynamical" explanatory schemes, that is, between what Kant calls the "mechanical natural philosophy" and the "dynamical natural philosophy." The former, as its name suggests, is identified with a mechanical, purely geometrical physics of the Cartesian type (532.34–533.4). It attempts to explain physical interac-

23. As the language here indicates, the distinction between mathematical and dynamical principles is related to—although certainly not identical with—the distinction between mathematical and philosophical reasoning developed in the Methodology of the *Critique* at A712–738/B740–766. "*Philosophical* knowledge is *rational knowledge from concepts*, mathematical knowledge is rational knowledge from the *construction* of concepts" (A713/B741). There is consequently a corresponding distinction between philosophical and mathematical proof: "Even from a priori concepts (in discursive knowledge) there can never arise intuitive certainty, i.e., evidence [Evidenz]," so that "Only mathematics therefore contains demonstrations [Demonstrationen], because it derives knowledge not from concepts, but from the construction of concepts: that is, from the intuition that can be a priori given corresponding to the concepts" (A734/B762). And the explanation for this is that philosophical reasoning, "the employment of reason in accordance with concepts," concerns not only "the form of intuition (space and time)" but also "the matter (the physical), or the content, which signifies something that can be met with in space and time, and therefore contains an existent and corresponds to sensation." In this employment of reason "we can do nothing more than bring appearances, according to their real content, under concepts, which can be determined thereupon in no other way than empirically, that is, a posteriori (but in accordance with these concepts as rules of an empirical synthesis)" (A723/B751). In mathematical reasoning, "the employment of reason through the construction of concepts," on the other hand, "there is absolutely no question of existence at all, but only of the properties of the objects in themselves [Gegenstände an sich selbst], solely in so far as these properties are connected with the concepts of the objects" (A719/B747). Hence, "since the concepts after all relate to an a priori intuition, they can for this very reason be given determined a priori and without any empirical data in pure intuition" (A724/B752).

tions and material constitution solely in terms of the sizes, motions, and figures of elementary particles distributed in empty space, and it accordingly has the advantage that "the possibility of the figures as well as of the empty intermediate spaces can be shown with mathematical evidence [Evidenz]" (525.5–7). In the "metaphysical-dynamical" scheme, on the other hand, we attempt to explain physical interactions and material constitution in terms of the "fundamental forces" of attraction and repulsion, and we eschew empty space. What is of primary importance here is the circumstance that "it is in general beyond the horizon of our reason to comprehend original forces a priori according to their possibility" (534.20–22), so that

> . . . if the material itself is transformed into fundamental forces (whose laws we are not capable of determining a priori—and still less can we reliably specify a priori a manifold of such forces that suffices to explain the specific variety of matter), then all means escape us for *constructing* this concept of matter and for presenting as possible in intuition what we think universally. (527.7–12)[24]

Nevertheless, despite the advantage of a "merely mathematical physics" with respect to a priori insight into the possibility of its basic concepts, a "dynamical explanatory scheme [dynamische Erklärungsart]" is still

> . . . much more favorable and suitable to experimental philosophy, in that it leads directly to the discovery of the moving forces proper to matter and their laws, and, on the other hand, limits the freedom to assume empty intermediate spaces and fundamental corpuscles of determinate figures, neither of which can be determined or discovered in any experiment. (533.21–26)

These last remarks, together with Kant's complaint that the mathematical-mechanical scheme "gives far too much freedom to the imagination to compensate for the lack of inner natural knowledge through feigning [Erdichtung]" (532.13–15), inevitably call to mind Newton's well-known implicit criticism of the mechanical philosophy at the conclusion of the General Scholium to *Principia* (see note 15 above).

In the context of our account of the empirical determination and discovery of the law of gravitation in §I above, we can then illustrate the sense in which the dynamical principles of the understanding—in particular, the analogies of experience—are regulative rather than constitutive. Grav-

24. Compare paragraph six of the Preface: "*Properly* so-called natural science first presupposes metaphysics of nature; for laws, that is principles [Prinzipien] of the necessity of that which belongs to the *existence* of a thing, are concerned with a concept which cannot be constructed, because existence can be presented in no a priori intuition" (469.25–30; compare also note 17 above on the connection between *force* and *sensation*).

itational force is of course identical to the "fundamental force of attraction" for Kant (Note 2 to Proposition 8: 518.17–19). As such, its laws cannot be determined a priori, but only a posteriori:

> . . . no law of attractive nor of repulsive force may be risked on a priori conjectures, but everything, even universal attraction, as the cause of gravity, must, together with its laws, be inferred from data of experience. (534.15–18)[25]

(And, on our account, the "data of experience" here are just the observed relative motions in the solar system.) As we have seen, this a posteriori determination of the law of gravitation is grounded (and thereby given a more than merely hypothetical or inductive status) by laws originating a priori in the understanding: above all, by the third law of motion, which is to realize or instantiate the third analogy of experience. Yet the third law of motion in no way supplies or guarantees the *existence* of a system in which it is satisfied: in this case, the solar system as described from the point of view of its center of mass. The third law merely supplies us with "a rule for seeking" such a system in experience and "a mark for discovering it therein"—the system itself (in this case the solar system) can only be given a posteriori, that is, through perception.[26] In this sense the third law of motion, together with the third analogy of experience it is intended to realize, is a regulative rather than a constitutive principle.

Be this as it may, however, the point I principally wish to emphasize here is the following. When Kant claims that the understanding prescribes laws a priori to nature, and that the understanding is accordingly the ground of the law-governedness of nature, he has in mind primarily the *dynamical* principles of the understanding, and, in fact, the analogies of experience. This is clear in the *Critique*, where, in his concluding remarks on the three analogies taken together, Kant says:

> By nature (in the empirical sense) we understand the connection of appearances according to their existence, in accordance with necessary laws, that is, in accordance with rules. There are thus certain laws, in fact a priori laws, that first make a nature possible. Empirical laws can obtain, and be

25. Thus gravity is an empirical concept (A173/B215), and as such its "possibility must either be known a posteriori and empirically, or it cannot be known at all" (A222/B269–270). Compare also A770/B798—*retaining* the reference to "a force of attraction without any contact"!

26. Compare A180/B222–223, and also A720–721/B748–749: "Therefore transcendental propositions can never be a priori given, through construction of concepts, but only in accordance with a priori concepts. They contain merely the rule according to which a certain synthetic unity of that which is incapable of a priori intuitive representation (perceptions) is to be empirically sought. Yet they cannot present a priori any one of their concepts in any instance whatsoever, but do this only a posteriori, by means of experience, which is itself first possible only in accordance with these synthetic principles."

> discovered, only by means of experience, and indeed in virtue of these original laws through which experience itself first becomes possible. Our analogies therefore properly present the unity of nature in the connection of all appearances under certain exponents, which express nothing other than the relation of time (in so far as it comprehends all existence within it) to the unity of apperception, which can only take place in the synthesis according to rules. Taken together they thus state: all appearances lie in a nature and must lie therein, because, without this a priori unity, no unity of experience, and thus no determination of objects therein, would be possible. (A216/B263)[27]

Thus, the unity of nature here depends first and foremost on the analogies, and, as Kant expresses it immediately before the above passage: "this unity of time-determination is through and through dynamical" (A215/B262).

Moreover, this emphasis on the dynamical principles of the understanding, and, in particular, on the analogies of experience, is just as clear in Part Two of the *Prolegomena*. Thus, for example, Kant illustrates the laws of "pure science of nature" in §15 by precisely the analogies of experience (see note 22 above). And, when summarizing the results of the first two Parts in §40, Kant again takes pains to contrast mathematics and pure science of nature:

> Pure mathematics and pure natural science had, *for the sake of their own security* and certainty, no need for the kind of deduction that we have made of both. For the former rests on its own evidence [Evidenz]; and the latter, however, although originated from pure sources of the understanding, nevertheless rests on experience and its thoroughgoing confirmation—which latter testimony it cannot wholly renounce and dispense with, because, despite all of its certainty, as philosophy it can never imitate mathematics [es der Mathematik niemals gleich tun kann]. (327.5–13)

Thus, using precisely the language of the mathematical/dynamical distinction, Kant thereby reemphasizes the need for a posteriori imput in pure natural science and hence its dynamical character.

Even more fundamentally, however, we should recall that the general theme of Part Two of the *Prolegomena* is the transformation of *judgements of perception* into *judgements of experience:* that is, of *appearance* (Erscheinung) into *experience* (Erfahrung) (compare the distinction be-

27. Compare A159/B198: "Even natural laws, when they are considered as principles of the empirical employment of the understanding, at the same time carry with themselves an expression of necessity and thus at least the suggestion of a determination from grounds that hold a priori and antecedently to all experience. Yet all laws of nature without distinction stand under higher principles of the understanding, in that they merely apply these to special cases [besondere Fälle] of appearance. These principles alone therefore give the concept that contains the condition, and as it were the exponent, of a rule in general; but experience gives the case that stands under the rule."

tween the material and the formal sense of nature in §36: see note 3 above). Now, according to §22, "Experience consists in the synthetic connection of appearances (perceptions) in a consciousness, in so far as this connection is necessary" (305.8–9); and, according to §25, it is the analogies of experience which first make this possible:

> With regard to the relation of appearances, and indeed simply in view of their existence, the determination of this relation is not mathematical but dynamical, and is never objectively valid and hence suitable for an experience unless it stands under a priori principles which make knowledge of experience with respect to appearances first possible. Therefore appearances must be subsumed under the concept of substance—which is the basis for all determination of existence as a concept of a thing itself—or second, in so far as a time-sequence (that is, an event) is met with among the appearances, under the concept of effect in relation to cause—or, in so far as simultaneity is to be known objectively (that is, through a judgement of experience), under the concept of community (reciprocity). And thus a priori principles are the basis for objectively valid, although empirical judgements: that is, of the possibility of experience in so far as it is to connect objects in nature according to their existence. These principles are the proper natural laws, which can be called dynamical. (307.14–30)

When these passages are compared with the general principle of the analogies in the *Critique*—namely, "Experience is only possible through the representation of a necessary connection of perceptions" (B218)[28]—it is clear beyond the shadow of a doubt, I think, that the analogies constitute the mechanism for transforming mere appearance into objective experience.[29]

We can finally state the problem posed by §38 of the *Prolegomena* in a more explicit and sharper form: What does the a priori derivation of the law of gravitation from the geometry of a system of concentric spheri-

28. In A the principle runs: "All appearances [Erscheinungen] stand, according to their existence, a priori under rules of the determination of their relation to one another in time" (A176–177).

29. This of course fits in very well with the procedure, discussed in §I, by which the fourth chapter or Phenomenology of the *Metaphysical Foundations* transforms appearance into experience: namely, by applying the laws of motion to observed cases of so far merely relative motion. And, in this connection, it is noteworthy that the *Prolegomena* itself gives the following example of "mere appearance": "The senses represent to us the movement of the planets as now progressive, now retrogressive; and herein is neither falsehood nor truth, because, so long as one acquiesces for the moment in this being only appearance, one absolutely does not yet judge concerning the objective character of their motion. Yet, because a false judgement can easily arise if the understanding has not taken sufficient care to prevent this subjective mode of representation from being taken for objective, one says: they present a semblance [sie scheinen] of retrogression. However, the illusion [Schein] here is not to be charged to the senses but to the understanding, whose province alone it is to make an objective judgement from the appearances" (291.2–11).

cal surfaces Kant envisions there have to do with the *dynamical* principles of the understanding? Such a geometrical derivation would in any case still involve the mathematical categories and principles of the understanding—concepts and principles of quantity, for example[30]—and one might hope thereby to preserve a formal consistency between the example of §38 and the thesis it is intended to illustrate: namely, that the understanding prescribes laws a priori to nature. Yet such consistency would be *merely* formal nonetheless, for the example of §38 has still not been shown to involve the dynamical principles of the understanding, and, as we have just seen, the latter, above all, are what Kant is concerned with here.

· III ·

Section 38 begins, somewhat unexpectedly, with an example from pure mathematics, presented together with a question concerning the role of the understanding in this example:

> For example, two lines that cut one another and also the circle—however they are randomly drawn—still always divide according to the rule: that the rectangle from the segments of one of the lines is equal to that of the other. Now I ask: "does this law lie in the circle, or does it lie in the understanding," i.e., does this figure [Figur], independently of the understanding, contain the ground of this law in itself, or does the understanding—in that it has constructed the figure itself according to its concepts (namely the equality of the radii)—at the same time inject into [hineinlegen] the figure the law of the chords intersecting one another in geometrical proportion? (320.25–30)

The example is Proposition 35 of Book III of Euclid: if two straight lines intersect one another within a circle at point E, and meet the circle at A, C and B, D respectively, then $AE \times EC = BE \times ED$.[31] The answer to

30. Kant is perfectly explicit in the *Prolegomena* itself that even judgements of pure mathematics involve the categories, and hence the understanding: "Even the judgements of pure mathematics in its simplest axioms are not exempt from this condition [of subsumption under a category]. The principle: the straight line is the shortest between two points, presupposes that the line is subsumed under the category of magnitude, which is certainly no mere intuition but has its seat solely in the understanding and serves to determine the intuition (of the line) with regard to the judgements that may be made concerning it with respect to their quantity" (301.20–27).

31. Kant appears to have had a particular fondness for this property of the circle. Thus, he also discusses it in #1 of Observation One of Part Two of *The Only Possible Basis*, as an illustration of "The unity in the manifold of the essence [Wesen] of things exhibited in the properties of space" (2, 93.9–10). There he also discusses the extension of this property to a point outside the circle—taken together, the two properties can be expressed generally as follows: if two straight lines intersect one another at point E (whether inside or outside the circle), and also intersect the circle at A, C and B, D respectively, then $AE \times EC =$

Kant's question, of course, is that the understanding is responsible for the law: "One will soon become aware, if one attends to the proof of this law, that it could only be derived from the condition on which the understanding bases the construction of this figure: namely, the equality of the radii" (320.34–321.3)—a point which is crystal clear in Euclid's proof of the Proposition. The same point is also made in the well-known remarks on mathematical method in the Preface to the second edition of the *Critique:*

> A new light dawned on the first man (whether he may be Thales or whoever) who demonstrated the *isosceles triangle;* for he found that he must not inspect what he saw in the figure [Figur], or even in the mere concept of it, and as it were learn its properties therefrom, but he must rather bring forth what he himself has injected in thought [hineindachte] and presented (through construction) according to concepts, and that, in order to know something a priori, he must attribute nothing to the thing except that which follows necessarily from what he himself has placed in it in accordance with his concept. (Bxi–xii)

Thus the idea here is simply that mathematical propositions are not established by the inspection of figures that hover before the mind's eye, as it were, but rather by the construction of concepts and the rigorous (Euclidean) proofs resulting thereby (the example at Bxi–xii is Proposition I.5 of Euclid).[32]

The significance of these considerations, in the present context, is, I think, to be found in the first sentence of §38 by which Kant introduces the above example:

> If one considers the properties of the circle, through which this figure [Figur] unites so many optional determinations of space in it immediately in a universal rule, then one cannot help attributing a nature to this geometrical thing [so kann man nicht umhin, diesem geometrischen Dinge eine Natur beizulegen]. (320.22–25)

Kant's language here is significant, for, according to the Preface to the *Metaphysical Foundations,* it is a mistake to attribute a nature to a geometrical figure. Kant there defines *nature* (in the "formal sense") as "the

---

$BE \times ED$. As a special case of the second property we then have Proposition III.36 of Euclid: if D is a point outside the circle such that DB is tangent to the circle at B, and if a second straight line DA intersects the circle at C and at A respectively, then $DB^2 = DC \times DA$. It is also of interest that Kant next proceeds to discuss Proposition VI of "On Naturally Accelerated Motion" from the Third Day of Galileo's *Two New Sciences,* which depends on Euclid III.35.

32. Hintikka has long emphasized that Kant's model for mathematical method here is Euclid: see especially "Kant's 'New Method of Thought' and His Theory of Mathematics" (1965)—reprinted in [53]. For further development of the point see Chapter 1 above.

first inner principle [Prinzip] of all that belongs to the existence of a thing*" (467.2–4), and remarks in the footnote:

> *Essence [Wesen] is the first inner principle of all that belongs to the possibility of a thing. Therefore one can attribute only an essence but not a nature to geometrical figures [Figuren] (since in their concept nothing is thought which expresses an existence). (467.22–25)

And this fits together perfectly with Kant's insistence, in the Methodology of the *Critique,* that in mathematical problems "there is absolutely no question of existence at all, but only of the properties of the objects in themselves, solely in so far as these properties are connected with the concepts of the objects" (A719/B747: see note 23 above).[33]

It appears, then, that Kant begins §38 by reminding us of the distinction between mathematical and philosophical reasoning (the theme of the Methodology passages) and, at the same time, of the distinction between the mathematical and the dynamical.[34] The notion of nature pertains to the dynamical categories and principles because it explicitly involves the

33. Of course the *Metaphysical Foundations* was published three years after the *Prolegomena,* and one can therefore not immediately assume that Kant takes such attribution of a nature to a "geometrical thing" to be mistaken already in the *Prolegomena.* Indeed, one might view the footnote in the *Metaphysical Foundations* as a correction to §38 of the *Prolegomena.* This suggestion, however, appears to be most implausible. For, in a footnote to the *Critique* at A419/B446 (which is therefore found even in the first edition) Kant defines *nature* in a closely related fashion: "Nature, taken adjectivally (formally) signifies the connection [Zusammenhang] of the determinations of a thing according to an inner principle [Prinzip] of causality"—and of course both *existence* and *causality* pertain to the dynamical, as opposed to the mathematical, categories. Moreover, §14 of the *Prolegomena* begins: "*Nature* is the *existence* of things, in so far as it is determined by universal laws" (294.7–8). Finally, in a series of unpublished Reflexionen from approximately 1776–1783 (therefore predating the *Prolegomena*) Kant clearly distinguishes *nature* and *essence* along the same lines as in the *Metaphysical Foundations:* R. 5406, "The first inner (determining) principle [Principium] of that which belongs to the concept of a thing is essence. But the first inner (determining) principle of that which belongs to existence is called nature* (*materialiter:* the totality of appearances, *formaliter:* the nexus of determinations)" (18, 174.16–20); R. 5409, "Nature is the inner causal principle [*principium causale*] according to constant laws: *Leges stabiles.* Substances have nature" (18, 175.9–10). To be sure, Kant does not here explicitly associate geometrical figures with essence rather than nature, yet he does, nonetheless, explicitly associate this distinction with the mathematical/dynamical distinction: R. 5412, "The mathematical antinomies extend to the essence of appearances [Erscheinungen], the dynamical to nature. The former to appearances determined merely through space and time, the latter determined through real grounds [realgründe]" (18, 175.23–25). In view of these passages it is hard to see how Kant could fail to have been aware of the mistake in attributing a nature to a "geometrical thing" in 1783; indeed, his language here ("one cannot help attributing a nature") suggests just such a mistake.

34. It is noteworthy that the Preamble to the *Prolegomena* explicitly refers not once but twice to the distinction between mathematical and philosophical reasoning developed in the Methodology of the *Critique* (266.3–7, 272.9–16).

concept of existence. And, since the latter concept cannot be constructed, the notion of nature belongs to metaphysics or philosophy rather than to mathematics. It therefore appears that the point of the example from pure mathematics with which §38 begins may be precisely to distinguish the role of the understanding in pure mathematics from its role in universal physics or pure natural science.

This suggestion can be developed and, I think, illuminated if we attend to the distinction, central to Kant's philosophy of mathematics, between *images* and *schemata:*

> In fact our pure sensible concepts are not based on images of objects but on schemata. To the concept of a triangle in general absolutely no image thereof would ever be adequate. For it would not achieve the universality of the concept, which makes it valid for all, right-angled or obtuse-angled, etc., but would be limited always to only a part of this sphere. The schema of the triangle can never exist anywhere but in thought and signifies a rule of synthesis of the imagination with respect to pure figures [Gestalten] in space. (A140–141/B180)

Thus the schema of the concept of a triangle is the universal rule for constructing any and all particular instances of this concept, whereas an image of the concept of a triangle is one particular such instance thereby constructed. Kant's point is that the universality of geometrical proof depends on operating with schemata, with constructions of concepts, rather than with particular images or figures.[35]

Moreover, since images of geometrical concepts, unlike their schemata, are particular individual figures, there is an important sense in which such images, unlike the corresponding schemata, belong to *empirical* intuition: "the *image* is a product of the empirical capacity of the productive imagination, the *schema* of sensible concepts (such as figures [Figuren] in space) is a product, and as it were a monogram, of the pure a priori imagination" (A141–142/B181). In this way, images of geometrical concepts serve to relate such concepts to empirical (and not just pure) intuition and thereby to secure their objective meaning:

> Therefore all concepts, and with them all principles, even those that are possible a priori, still relate to empirical intuitions, that is, to data for possible experience. . . . One may take only the concepts of mathematics as an exam-

35. Comparing the above passage with A164–165/B205, it appears that the schema of the concept of a triangle—the relevant "rule of synthesis of the imagination"—is just the Euclidean construction of the triangle (Prop. I.22); similarly, it appears from A234/B287 that the schema of the concept of a circle is the Euclidean construction of the circle (Postulate 3); and so on. This reinforces the point that Euclidean proof is Kant's model for mathematical method here (compare note 32 above). See Chapter 2 above for further development of this point.

> ple, and, in fact, first of all in their pure intuitions. Space has three dimensions, between two points there can be only one straight line, etc. Although all these principles, and the representation of the object with which this science occupies itself, are generated in the mind completely a priori, they would still mean absolutely nothing if we could not always present their meaning in appearances (empirical objects). . . . Mathematics fulfills this demand through the construction of the figure [Gestalt], which is an appearance present to the senses (although accomplished a priori). (A239–240/B298–299)

Thus, whereas the universality of geometrical schemata facilitates geometrical reasoning and proof, the particularity of the images thereby constructed meets the demand for actual individual objects corresponding to geometrical concepts.[36]

Images of geometrical concepts are therefore actual individual objects in the world of appearances—in the empirical world. If, for example, in perceiving a house "I draw as it were its figure [ich zeichne gleichsam seine Gestalt]" (B162), this figure of the house has a definite size and a definite location in space. It also exists throughout a definite interval of time and is accordingly a suitable object of causal explanation. In other words, the dynamical categories and principles apply to such empirical figures, and they, unlike their corresponding geometrical schemata, consequently fall under the concept of *nature*. It follows that to attribute a nature to a "geometrical thing," such as a circle, is to focus on the empirical images corresponding to the geometrical concept rather than on the associated universal schema—the construction of the concept. One thereby mistakenly attempts to derive geometrical laws by inspecting individual figures and misses the all-important role of the understanding in the construction of geometrical concepts. One therefore misconceives the character of geometrical reasoning and proof, and this, as we have seen, is precisely the point of the first example of §38.

The second example of §38 generalizes the above-mentioned property of the circle to all conic sections:

36. See also A713–714/B741–742: "Thus I construct a triangle, in that I present the object corresponding to this concept, either through mere imagination in pure intuition, or in accordance therewith also on paper in empirical intuition—but in both cases completely a priori, without having borrowed the pattern for it from any experience. The individual drawn figure [einzelne hingezeichnete Figur] is empirical, and nonetheless serves to express the concept without detriment to its universality; for in this empirical intuition we consider always only the act of the construction of the concept—for which many determinations, e.g., the magnitude, the sides, and the angles, are entirely indifferent—and thus we abstract from these differences, which do not alter the concept of the triangle." Note that the individual drawn figure is an empirical object whether it is produced in mere imagination or on paper.

> If we now extend this concept, in order to pursue still further the unity of manifold properties of geometrical figures under common laws, and we consider the circle as a conic section—which thus stands with other conic sections under precisely the same fundamental conditions of construction—we then find that all chords that intersect within the conic sections (the ellipse, the parabola and hyperbola) are always such that the rectangles from their segments are not indeed equal, but still always stand to one another in constant ratios. (321.3–11)

This property of conic sections is the natural generalization of Propositions 35–36 of Book III of Euclid and follows from Propositions III.16–17 of Apollonius's *Conics:* suppose that the members of two given pairs of straight lines intersect one another at E, E' respectively and meet the conic at A, C and B, D and at A', C' and B', D' respectively; then, if the lines AC, $A'C'$ and BD, $B'D'$ are respectively parallel to one another, $(AE \times EC)/(BE \times ED) = (A'E' \times E'C')/(B'E' \times E'D')$.[37]

How does Kant now wish to employ this example? An answer is suggested in the following sentence:

> If we proceed from here still futher—namely to the basic doctrines of physical astronomy—then a physical law of mutual attraction diffusing over all of material nature emerges, whose rule is that it diminishes inversely as the square of the distances from each attracting point . . . . (321.11–15)

Thus Kant is here suggesting a route from our property of conic sections to the inverse-square law. In this way, he also suggests a transition from the purely mathematical examples with which §38 begins to the physical example from natural science which is apparently the principal point of that section.[38]

Now there is of course a close relationship between conic sections and the inverse-square law, more precisely, between conic sections, Kepler's law of areas, and the inverse-square law: if a body moves in a conic section in such a way that it satisfies Kepler's law of areas with respect to

37. See Heath [45]: Proposition 59 in Heath's numbering. Kant was certainly acquainted with Apollonius—see, for example, Section One of *On a Discovery According to Which Any New Critique of Pure Reason Has Been Made Superfluous by an Earlier One* (8, 191–192). Note that the translations following Carus obscure the property of the conics at issue here by rendering "in gleichen Verhältnissen" as "in a constant ratio" rather than "in constant ratios" (Lucas, however, correctly has the plural). This makes it look as if we generalized the property of the circle in question simply by substituting $AE \times EC = k(BE \times ED)$ for $AE \times EC = BE \times ED$, and this is nonsense: for we can certainly choose *one* pair of lines that intersect each other and the conic such that $AE \times EC = BE \times ED$, and it would thereby follow that $k = 1$ for *all* such pairs.

38. Again, the translations unfortunately obscure this suggestion by unaccountably omitting "from here" in their renderings of the first clause [Gehen wir von da noch weiter].

a focus of the conic, then its total acceleration is directed to that focus and is inversely proportional to the square of its distance from that point. Indeed, the relationship here is an equivalence: if the total acceleration of a body with respect to a given point is inversely proportional to the square of the distance, then the body moves in a conic section with that point as focus and satisfies Kepler's law of areas with respect to it. (Note also that this equivalence is so far a purely mathematical result concerning the kinematics of relative motion and does not depend, in particular, on the Newtonian laws of motion.)[39]

Yet Kant has something even more specific in mind here, I think, for Newton appeals explicitly to a special case of the particular property of the conics to which §38 refers in his derivation of the inverse-square law in Propositions X–XIII of Book I of *Principia.* The special case in question is as follows. Let PG and DK be two conjugate diameters of the conic (that is, each bisects every chord parallel to the other) intersecting (and therefore bisecting) one another at C and meeting the conic at P, G and D, K respectively; and let QV be an ordinate of the diameter PG (that is, QV is one-half of a chord in a system of parallel chords bisected by PG) meeting the conic at Q (and hence the diameter PG at V); then $QV^2/(PV \times VG) = CD^2/PC^2$ (see Figure 8).[40] This property plays a crucial role in Newton's derivation of the inverse-square law from motion in a conic section, and it appears extremely likely, therefore, that Kant is suggesting precisely that Newtonian derivation here.[41]

39. Newton appeals to the first and second laws of motion in his proof of this equivalence and related Propositions in Book I of *Principia.* But such appeals are dispensable in principle here, for what is actually used is simply the *definition* of accelerated motion. Thus, neither force, nor mass, nor anything more than merely *relative* accelerations need be mentioned here. See Chapter 5, §I, below for further discussion of this point.

40. This property, a case of Propositions I.11–19 of Apollonius (Propositions 1–6 in Heath's numbering), can in fact be applied only to central conics, that is, to the ellipse and the hyperbola, for the parabola does not have conjugate diameters. Thus in proceeding to the inverse-square law for the parabola one has either to consider the parabola as a limiting case of an ellipse with center at infinity (as Newton does in the Scholium to Proposition X, for example) or apply a closely related property (also a case of Apollonius I.11–19) involving an ordinate and the *latus rectum* associated with the point P (as Newton does in Proposition XIII).

41. In the Newtonian derviation of the inverse-square law our property functions essentially as a means for the construction of an *osculating circle* to the conic (a circle with the same radius of curvature as that of the conic). Although this is not entirely evident in the exposition of *Principia* itself, it emerges clearly and explicitly in manuscripts from the early 1690s in which Newton attempted to revise the text of *Principia* so as to allow the essentials of the argument to stand out in sharper relief; these manuscripts are collected and annotated in Newton [86], ed. Whiteside, vol. VI: pp. 568–593. The basic point is that our property enables us to prove that the osculating circle at P intersects the diameter PG at M such that PM = the *latus rectum* of the diameter = DK^2/PG (Lemma XII in [86], vol. VI, pp. 583–585), and this then enables us to apply the measure of centripetal acceleration of

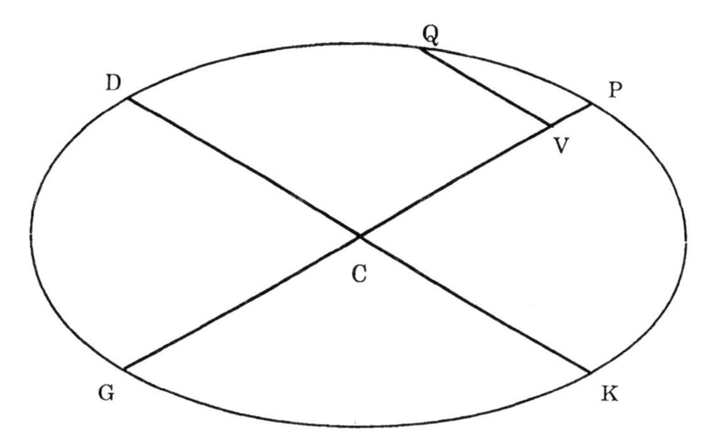

Figure 8

Hence, despite first impressions to the contrary, it appears extremely likely that §38 is concerned with Newton's "deduction from the phenomena" of the law of gravitation from Kepler's laws after all.[42] And this derivation proceeds a posteriori, from the empirically given "phenomena" of observed relative motions in the solar system: in this case, from the "phenomenon" that satellites of primary bodies orbit in ellipses with the primary body at one focus (and, of course, satisfy the law of areas in such orbits)—relative to the primary body in question and the fixed stars. By Newton's argument in Proposition XI of Book I it then follows purely mathematically that the total acceleration of any such satellite—in the frame of reference in question—is directed towards the primary body and satisfies the inverse-square law. Note, however, that the use of geometry here is essentially different from that of the purely mathematical example with which §38 begins. Rather than focusing on the universal schemata

Corollary III of Proposition VI of Book I—by which the acceleration in any orbit is reduced to that in a corresponding osculating circle—so as to derive the law of centripetal acceleration for the conic (Proposition X in [86], vol. VI, pp. 585–589; compare also "the same otherwise" to Propositions X–XII of *Principia*, Book I, and [86], vol. VI, pp. 594–595). I have no idea whether Kant himself thought of the matter in this way, yet the manner in which §38 arrives at the inverse-square law via the circle and the above property of the conics is highly suggestive nonetheless.

42. I am therefore in agreement here with Brittan's remarks concerning §38 in [10], pp. 140–142, and, in particular, with his contention that the law of gravitation is not itself synthetic a priori for Kant. Yet I cannot follow Brittan when he asserts that "if the [law of gravitation] is not synthetic *a priori*, then it is difficult to see how the laws of motion could be" ([10], pp. 141–142). On the contrary, as we have seen above, the Newtonian argument becomes a true "deduction from the phenomena" only if we *presuppose* the laws of motion as synthetic a priori: see, in particular, note 15 above.

of geometrical concepts, we are here focusing on particular individual figures (what Kant calls images): namely, on the actual individual ellipses (which have definite sizes, locations in space, and so on) described by the motions of the heavenly bodies under consideration. These particular ellipses are empirical objects given a posteriori, which are then subsumed under a pure geometrical concept in such a way that further properties (in this case the inverse-square law) can be mathematically derived. At this point, then, we are on the verge of a transition from the mathematical concepts and principles to the dynamical concepts and principles, from the realm of *essence* to that of *nature*.[43]

Yet such a transition is apparently rudely interrupted by what comes next, for it is at precisely this point that Kant inserts the reference to a purely a priori, purely geometrical derivation of the inverse-square law:

> . . . a physical law of mutual attraction diffusing over all of material nature emerges, whose rule is that it diminishes inversely as the square of the distances from each attracting point, just as the spherical surfaces in which this force diffuses increase—which seems to lie necessarily in the nature of the things themselves and is therefore customarily expounded as cognizable a priori. (321.12–18)

We are therefore faced, at precisely this point, with our old uncomfortable question: How, in view of all the evidence to the contrary (which must now also include the apparently clear reference to Newton's a posteriori derivation from Kepler's laws immediately preceding this passage), could Kant possibly envision such a purely geometrical derivation?

We must clearly proceed very cautiously here. It is significant, first of all, that Kant does not explicitly say that the inverse-square law is a priori. Rather, he says that it *seems* to lie in the nature of the things themselves and that, for this reason, it is *customarily* expounded as a priori. However, if a law is a priori in virtue of "lying necessarily in the nature of the things themselves [als notwendig in der Natur der Dinge selbst zu liegen]," then this would seem to contradict Kant's basic point in §§36–38: namely, that a priori knowledge of nature is possible only of appearances and not of things in themselves. Indeed, Kant reemphasizes this point in the next (and final) paragraph of §38: "we do not have to do with the nature *of the things in themselves* [*Dinge an sich selbst*]" (322.11–12).[44]

43. We are therefore also on the verge of a transition from *appearance* (Erscheinung) to *experience* (Erfahrung) or, in the terminology of §36, from "nature in its *material* meaning" to "nature in its formal meaning": compare note 3 above.

44. The translations following Carus obscure the tension here by rendering "in der Natur der Dinge selbst" in the first passage as "in the very nature of things" (while Lucas has "in the nature of things themselves"). And, although we certainly cannot immediately equate "things themselves [Dinge selbst]" with "things in themselves [Dinge an sich selbst]" in Kant, the *Prolegomena* exhibits a tendency to use these two expressions almost interchangeably.

It is not immediately clear, therefore, that Kant himself actually endorses the a priori derivation envisioned here. He says only that the law of gravitation is thereby *customarily* expounded as knowable a priori, and the natural question is: By whom? In the *Physical Monadology* of 1756—where, as far as I know, Kant first entertains the geometrical derivation under consideration—it is attributed to John Keill.[45] The reference appears to be to Keill [56], first published in 1702, where we find the following general "Theorem" proved in Lecture I: "Every Quality or Virtue that is propagated every way in right Lines from a Center, is diminished in a duplicate Proportion of the Distance from that Center"—the application to gravity is briefly mentioned two pages later.[46] Keill was a rather extreme Newtonian of the first generation, and his book is otherwise noteworthy for its zealous defense of a predominantly geometrical approach to physics. In particular, he defends action-at-a-distance against

Thus, for example, §14 asserts: "Should nature signify the existence of things *in themselves* [Dinge *an sich selbst*], then we could cognize it neither a priori nor a posteriori. Not a priori, for how do we hope to know what belongs to the things in themselves [Dingen an sich selbst], since this can never take place through analysis of our concepts (analytic propositions)—because I do not want to know what may be contained in my concept of a thing (for that belongs to its logical essence), but what is added to this concept in the actuality of the thing and by which the thing itself [Ding selbst] may be determined in its existence outside of my concept. My understanding, and the conditions under which alone it can connect the determinations of the things in their existence, prescribes no rule to the things themselves [Dingen selbst]" (294.8–18). Beck renders "Ding selbst" as "thing itself" and "Dingen selbst" as "things [in] themselves"; Ellington drops the brackets and leaves us with "things in themselves" (Carus and Lucas both have "things themselves").

45. Scholium to Proposition X of Section II (1, 484.22–39). Note that Kant does not explicitly endorse the derivation here either: he presents the construction only as one possible way to think about the laws of attractive (and repulsive) force. Moreover, the same is true of the analogous construction sketched in Observation 1 to Proposition 8 of the Dynamics of the *Metaphysical Foundations:* Kant presents it not as a result or theorem but as "a small preliminary suggestion on behalf of the attempt at such a perhaps possible construction" (4, 518.33–34)—where the construction in question is not that of the law of gravitation itself but of the dynamical concept of matter (from the interplay of attractive and repulsive force). In addition, Kant expresses considerable diffidence about the prospects for this latter construction, which, in any case, does not belong to metaphysics (compare Note 1 and Observation 2: 517–523). Indeed, in view of Kant's explicit assertion in the General Observation to Dynamics that the dynamical concept of matter *cannot* be constructed (527. 7–12, cited in §II above), such diffidence is entirely appropriate.

46. Keill's book was first published in Latin as *Introductio ad veram physicam: lectiones physicae.* I quote from the English version of 1726 [56], pp. 5–7. In fairness to Keill, I should point out that the application to gravity is quite incidental here: "the Intensions of Light, Heat, Cold, Odors, and the like Qualities, will be reciprocally as the Squares of their Distances from the Point whence they proceed. Hence also may be compared amongst themselves, the Action of the Sun on different Planets: but this is not the Business of our present Design" ([56], pp. 6–7). Indeed, Keill's book discusses only terrestrial gravity and does not develop the theory of universal gravitation at all.

the Cartesians (Preface, pp. iv–vii), infinite divisibility against the "philosophers" (Lectures III, IV), Newtonian absolute space, time, and motion (Lectures VI, VII), and void space (Lecture II)—together with the explanation of differences of density in terms of interspersed "pores" of void space (Lecture X). Now, whereas Kant strongly agrees with Keill concerning the first two of these ideas, he just as strongly disagrees concerning the last two. As far as we know, then, it is quite possible that Kant also disagrees with the idea of an a priori, purely geometrical derivation of the inverse-square law.

It is possible, in other words, that Kant intends in §38 *to contrast* an a posteriori derivation from "phenomena" of the Newtonian type with the kind of a priori geometrical derivation that is here said to be "customarily expounded": his intention, that is, may be to accept the former and reject the latter. Moreover, this possibility appears to be reinforced by what Kant says next:

> Now as simple as the sources of this law are, in that they rest merely on the ratio [Verhältnis] of spherical surfaces of various radii, the consequence thereof is nevertheless so splendid with respect to the manifold of its harmony and its regularity, that it not only follows that all possible orbits of the heavenly bodies are conic sections, but also such a ratio [Verhältnis] of the orbits to one another results that no other law of attraction except that of the inverse-square ratio [umgekehrten Quadratverhältnis] of the distances can be devised as suitable for a cosmic system [Weltsystem]. (321.18–26)

Thus, although one can undoubtedly *represent* the inverse-square law geometrically in terms of a quantum of attractive force diffusing uniformly on concentric spherical surfaces (and Kant is clearly intrigued by this geometrical representation), what is truly decisive here is the totality of the consequences of such a representation in the empirical realm of the actual orbital motions in the solar system. In particular, it is because of a certain feature of the orbits of the heavenly bodies (their ratio to one another), and not simply because of our geometrical representation, that "no other law of attraction . . . can be devised."

We shall have to return to the question of precisely what feature of the orbits of the heavenly bodies Kant has in mind here—and precisely why, in particular, the inverse-square law is thereby privileged. First, however, it will perhaps be helpful to attempt to gain some perspective on the matter by looking at what Kant goes on to say in the next (and final) paragraph of §38. Just as he does in the discussion of the purely mathematical example with which §38 begins, Kant proceeds to ask a disjunctive question about the physical example from gravitational astronomy:

> Here is therefore nature which rests on laws that the understanding cognizes a priori, and, to be sure, chiefly from universal principles of the determi-

nation of space [aus allgemeinen Prinzipien der Bestimmung des Raums]. Now I ask: do these natural laws lie in space, and does the understanding learn them—in that it merely seeks to investigate the copious significance [reichhaltigen Sinn] that lies in space—or do they lie in the understanding and the mode in which it determines space according to the conditions of synthetic unity towards which its concepts all proceed [darauf seine Begriffe ingesamt auslaufen]? (321.27–33)

This passage raises several important questions. What are the "laws that the understanding cognizes a priori" here? Is the law of gravitation itself to be included among them? What are "universal principles [Prinzipien] of the determination of space"? Are these just the laws of geometry, for example? (Answering the second and fourth questions affirmatively supports the idea that Kant is endorsing a purely geometrical derivation of the inverse-square law.) Most important, perhaps: What is the point of the disjunctive question Kant poses? What kind of view, in particular, holds that the laws of nature "lie in space"?

Kant answers his disjunctive question as follows:

Space is something so uniform and with respect to all particular properties so indeterminate that one will certainly not seek a wealth of natural laws in it. On the contrary, that which determines space to the figure of a circle, the figure of a cone and a sphere is the understanding, in so far as it contains the ground of the unity of their construction. The mere universal form of intuition called space is thus indeed the substratum of all intuitions determinable into particular objects, and certainly the condition of the possibility and manifoldness of the latter lies in it; but the unity of the objects is nevertheless determined solely through the understanding, and in fact according to conditions that lie in its own nature. . . . (321.3–322.7)

Thus the laws of nature at issue here lie in the understanding rather than in space. For the latter, considered independently of the understanding, as it were—considered as "mere universal form of intuition"—is "uniform" and "indeterminate" and does not support "the unity of the objects."

What we have here, I suggest, is a precursor of the important distinction, which is only made explicit in the second edition of the *Critique*, between space as *form of intuition* and as *formal intuition:*[47]

However space and time are not represented merely as *forms* of sensible intuition, but as *intuitions* themselves (which contain a manifold), and are

47. I am certainly not alone in suggesting such a connection between §38 of the *Prolegomena* and the distinction between form of intuition and formal intuition: see, in particular, Buchdahl [12], pp. 572–587; Brittan [10], p. 142 and pp. 96–99; and Torretti [112], pp. 32–33. My disagreement with these authors will emerge in note 50 below.

> thus represented with the determination of the *unity* of this manifold in them (see the Transcendental Aesthetic*). (B160)

The footnote then begins:

> *Space, represented as *object* (as one actually needs to do in geometry), contains more than mere form of intuition: namely, *uniting* [*Zusammenfassung*] of the manifold in accordance with the given form of sensibility in an *intuitive* representation; so that the *form of intuition* yields only a manifold, but the *formal intuition* yields unity of representation.

Moreover, it is by thus uniting the mere form of intuition that we acquire *determinate* combination and *determinate* objects:

> Thus the mere form of sensible intuition, space, is still absolutely no cognition; it yields only the manifold of a priori intuition for a possible cognition. To cognize anything at all in space, e.g., a line, I must *draw* it, and thus synthetically achieve a determinate combination of the given manifold, in such a way that the unity of this act is simultaneously the unity of consciousness (in the concept of a line), and thereby an object (a determinate space) is first cognized. (B137–138)

The kinship between these passages and our last passage from *Prolegomena* §38 is therefore evident.[48]

But how exactly are we to understand the distinction between space as *form of intuition* and as *formal intuition?* My suggestion is the following. To consider space merely as form of intuition is to consider only the circumstance that each individual representation of outer sense includes a spatial structure, a structure which can be considered in abstraction from the contributions of the understanding and of sensation:

> Thus, if I abstract from the representation of a body that which the understanding thinks therein—substance, force, divisibility, etc.—and likewise what belongs to sensation therein—impenetrability, hardness, color, etc.—something still remains over from this empirical intuition: namely, extension and figure. These belong to pure intuition, which, even without an actual object of the senses or sensation, occurs a priori in the mind as a pure form of intuition. (A20–21/B35)

However, when one considers only this feature of each individual representation of outer sense one does not yet consider how different such representations are related to one another or combined: one does not yet

48. As the first passage from B160 suggests, an analogous distinction of course holds for time: "Inner sense, on the other hand, contains the mere *form* of intuition, but without combination of the manifold in it . . . and therefore as yet contains absolutely no *determinate* intuition, which is only possible through the consciousness of the determination of the manifold through the transcendental action of the imagination (synthetic influence of the understanding on inner sense) which I have called figurative synthesis" (B154).

consider a "*uniting* of the manifold in accordance with the given form of intuition." In particular, one does not yet consider how different such individual representations are related to one another in *time*—the form of inner sense in which all representations whatsoever take place.[49] I am suggesting, in other words, that to consider space as a formal intuition is to consider precisely the relations between different individual representations of outer sense occurring at different times in inner sense: it is to consider the relations subsisting in a collection or manifold of spatial structures that is itself temporally ordered.[50]

49. See A34/B50–51 and especially A98–99: "However our representations may arise, whether effected through the influence of outer things or through inner causes, whether originating a priori or empirically as appearances, they nevertheless belong to inner sense as modifications of the mind; and as such all our cognitions are still in the end subject to the formal condition of inner sense, namely time, as that in which they must all be ordered, connected, and brought into relation."

50. In modern terms the point can perhaps be expressed as follows. Space and time as mere forms of intuition comprise a one-dimensional sequence of three-dimensional Euclidean spaces: a four-dimensional *space-time* structure comprising a one-dimensional ordering of three-dimensional "planes of simultaneity." (This four-dimensional style of conceiving spatio-temporal structure is not entirely foreign to Kant himself: compare the footnote to §14 of the *Inaugural Dissertation* at 2, 401.28–38—cited above in Chapter 3, §IV.) When one considers space and time as formal intuitions, however, one considers the relations between different such instantaneous Euclidean three-spaces. Thus, for example, one may introduce a "rigging" or relation of being-at-the-same-spatial-location defined between pairs of points on different such instantaneous three-spaces; alternatively, one may introduce an "affine structure," which yields only a relation of lying-on-the-same-inertial-trajectory defined between triples of points on different such instantaneous three-spaces; and so on. It is in this way of understanding the distinction between space as form of intuition and as formal intuition that I disagree with the authors mentioned in note 47 above. For these authors focus exclusively on *spatial* relations rather than on *spatio-temporal* relations here and suggest, in fact, that space as form of intuition is merely topological or "metrically amorphous": space acquires metrical relations only from the understanding, which then "injects" a particular metric into an "indeterminate," as yet merely topological manifold so as to first make possible "determinate" space as a formal intuition (see Buchdahl [12], pp. 606–615—on p. 647, however, there is an allusion to a conception closer to the present one; Brittan [10], pp. 96–99; Torretti [112], p. 33). On the present conception, by contrast, space as formal intuition is not concerned with the transition from topological to metrical spatial structure, but rather with spatio-temporal structure. To be sure, Kant certainly holds that the structure of Euclidean geometry requires the action of the understanding on the form of sensibility. But it does not follow that the mere form of sensibility subsisting prior to the determination by the understanding can be conceived as topological or "metrically amorphous" space (if only because topological structure is just as conceptually articulated as is metrical structure); nor does it follow that the understanding has any choice, as it were, as to which metrical structure to "inject" (for Kant, the full structure of Euclidean space is the only spatial structure that could possibly be in question here). Moreover, on the present interpretation, space as formal intuition necessarily involves *more* than the Euclidean structure of pure geometry: since the act by which the understanding determines sensibility ("figurative synthesis") involves *both* time and space, we are led in the end to the dynamical structure of *physical* space. See below, and compare also Chapter 2, §V, above.

Now how are such relations, which lead to a uniting of the spatial manifolds at different times, produced? The footnote to B160 quoted above continues as follows:

> This unity I had reckoned merely under sensibility in the Aesthetic in order only to observe that it precedes all concepts, although it in fact presupposes a synthesis which does not belong to the senses but through which all concepts of space and time first become possible. For since through it (in that the understanding determines the sensibility) space or time are first *given* as intuitions, the unity of this a priori intuition belongs to space and time, and not to the concept of the understanding. (§24)

And, when we turn to §24, we find that Kant distinguishes *intellectual synthesis,* which is thought by the understanding merely with respect to the categories, from what he calls *figurative synthesis* or *transcendental synthesis of the imagination*. The former synthesis, as purely intellectual, does not belong to the senses, but the latter is both intellectual and sensible. Since it involves our particular (spatio-temporal) form of sensibility, figurative synthesis, "owing to the subjective condition under which alone it can give to the concept of the understanding a corresponding intuition, belongs to sensibility." On the other hand, since it is "determining and not, like sense, merely determinable, and therefore can determine sense a priori according to its form in accordance with the unity of apperception," figurative synthesis also belongs to the understanding. Indeed, this synthesis is "an action of the understanding on sensibility and the first application of the understanding (at the same time the ground of all the rest) to objects of our possible intuition" (B151–152). It follows, then, that it is this figurative synthesis—this "action of the understanding on sensibility"—which makes the consideration of space as a formal intuition first possible.[51]

Thus far, however, we may appear to be in danger of explaining the obscure by reference to the still more obscure; so it is necessary, at this point, that we attempt to achieve a more concrete understanding of Kant's exceedingly abstract notion of figurative synthesis. Fortunately, Kant himself illustrates his meaning as follows:

> This we also always observe in ourselves. We can think no line without *drawing* it in thought, no circle without *describing* it. We can absolutely not represent the three dimensions of space without *setting* three lines at right-angles to one another from the same point. And even time we cannot represent without attending in the *drawing* of a straight line (which is to be the outer figurative representation of time) merely to the action of synthesis

51. This same figurative synthesis first makes the consideration of *time* as a formal intuition possible as well: see note 48 above.

> of the manifold, by which we successively determine inner sense—and thereby attend to the succession of this determination in it. Motion, as action of the subject (not as determination of an object*), and thus the synthesis of the manifold in space—if we abstract from the latter and attend merely to the action by which we determine *inner* sense according to its form—even produces the concept of succession in the first place. (B154–155)

Thus it is *motion*—"as the describing of a space" (B155n)—for example, the motion of a mathematical point in the successive synthesis underlying geometrical construction—which exemplifies the abstract notion of figurative synthesis here. And this makes sense, for the idea of motion—which unites time and space (A41/B58)—induces just the kind of relations between different particular spatial manifolds occurring at different times that were suggested above to constitute space as a formal intuition.

There are in fact two distinguishable aspects of this concrete realization of the notion of figurative synthesis. On the one hand, as the first examples in the above passage indicate, motion is considered as underlying the constructive procedures of pure geometry. Although what really matters here are the spaces thereby described, the constructive activity itself is still necessarily spatio-temporal for Kant. This is why Kant asserts, in the footnote to our passsage, that motion so considered belongs to geometry (B155n).[52] As such, we are here dealing with a merely *mathematical* synthesis.[53] On the other hand, this very same figurative synthesis is not solely mathematical, for it also produces the concept of *succession*. And this latter concept, in turn, is of course closely connected with the category of *causality*.[54] In this way, we are at the same time dealing with a *dynamical* synthesis, and this, I suggest, is why Kant also asserts that "motion, as the *describing* of a space, is a pure act of successive synthesis of the

52. Compare the Observation to Definition 5 of the Phoronomy of the *Metaphysical Foundations*, where Kant compares geometry and phoronomy (kinematics) in this respect: "In phoronomy, since I am acquainted with matter through no other property than its movability, and therefore it may be considered only as a point, motion is considered only as *describing of a space*, but still in such a way that I do not merely attend, as in geometry, to the space described, but also to the time in which, and thus the speed with which, a point describes the space" (489.6–11).

53. See the important footnote at B201 for the distinction between *mathematical* synthesis ("composition [Zusammensetzung] of the homogeneous") and *dynamical* synthesis ("connection [Verknüpfung] of the inhomogeneous").

54. Indeed, the *schema* of the category of causality "consists in the succession of the manifold, in so far as it is subject to a rule" (A144/B183). Compare the General Observation to the System of Principles (added in B): "In order to present *alteration*, as the intuition corresponding to the concept of *causality*, we must take motion, as alteration in space, as our example—indeed, we can even make alteration, whose possibility can be comprehended by no pure understanding, intuitive to ourselves in no other way" (B291).

manifold in outer intuition in general through productive imagination, and belongs not only to geometry but even to transcendental philosophy" (B155n).[55]

Given this background, I now want to suggest that the "universal principles [Prinzipien] of the determination of space" to which Kant refers in the first sentence of the ultimate paragraph of §38 of the *Prolegomena* are not to be equated with the laws of geometry. They comprise, rather, the general principles that first make possible the consideration of space as a formal intuition: that is, the consideration of spatio-temporal relations subsisting between different spatial manifolds occurring at different times. Such principles underly the possibility of both mathematical and dynamical synthesis and therefore belong to transcendental philosophy.[56] It follows that the principles in question serve to make possible the twofold consideration of the concept of motion discussed above: motion as a mathematical concept belonging to pure geometry and motion as a dynamical concept realizing the category of causality.

We have now reached the central point I wish to emphasize here: namely, that the consideration of motion in the latter sense—as a dynamical concept—unavoidably requires that we be able to apply this concept to our perceptual experience and hence, for Kant, that we can distinguish between *true* and *apparent* motion. In the terminology of the fourth chap-

55. It is central to Kant's transcendental philosophy that these two distinguishable aspects of figurative synthesis are at bottom identical: "precisely the same image-forming [bildende] synthesis, by which we construct a triangle in the imagination, is entirely identical with that which we exercise in the apprehension of an appearance, so as to make an empirical concept thereof" (A224/B271). Compare A301/B357: "the principle [of causality] shows how one is able to obtain a determinate empirical concept of that which happens in the first place." Thus the empirical concept at issue in the first passage is not the concept of a triangle (for this is a pure a priori concept) but rather the concept of a particular kind of appearance—of "that which happens." We obtain such an empirical concept through a *dynamical* synthesis. Compare also A664/B692 and A206–207/B252.

56. In this connection, it is to be emphasized that the term *Prinzipien* here has a special meaning in Kant. Kant defines *Prinzipien* as "synthetic cognitions from concepts" (A301/B357) and thereby *contrasts* cognition from *Prinzipien* with cognition of mathematical axioms (A300/B356–357). Indeed, Kant never, as far as I know, calls mathematical propositions *Prinzipien* and instead consistently uses the term *Grundsätze* for the latter; moreover, he explicitly distinguishes mathematical axioms from his own axioms of intuition by calling the latter "the principle [Prinzipium] of the possibility of axioms in general and itself only a principle [Grundsatz] from concepts" (A733/B761). Kant even hesitates to apply the term *Prinzipien* to the principles [Grundsätze] of pure understanding (A301/B357–358) and asserts that *reason,* not understanding, is "the faculty of principles [das Vermögen der Prinzipien]" (A299/B356). Accordingly, Kant tends to reserve the term for regulative principles (Prinzipien) of reason. Yet he also explicitly calls the dynamical principles of pure understanding "mere *regulative* principles [Prinzipien]" (A179/B222) and asserts that they, in contradistinction to the mathematical principles, are "mere regulative principles [Prinzipien] of *intuition*" (A664/B692).

ter or Phenomenology of the *Metaphysical Foundations,* we require a procedure for transforming appearance (Erscheinung), which involves merely relative motions, into experience (Erfahrung), which involves true or absolute motions. But this procedure, as we have seen, is effected through the analogies of experience and, in particular, by the laws of motion or laws of mechanics which concretely realize them. My final suggestion, then, is this: When Kant, in the same passage from §38, says that nature as described by gravitational astronomy "rests on laws that the understanding cognizes a priori, and, to be sure, chiefly from universal principles of the determination of space," the laws primarily in question here are just these laws of motion or laws of mechanics. They, as we have seen, are indeed cognized a priori (as necessary conditions for the determination of true motions), and, in fact, they are cognized a priori from principles of the understanding—and not simply from space itself, as it were, considered as mere form of intuition.

I will illustrate, and attempt to corroborate these points by now returning to Kant's discussion of the law of gravitation in the first paragraph of §38. There, as argued above, Kant appears to be suggesting Newton's "deduction from the phenomena" of the law of gravitation from Kepler's laws: in particular, from motion in an ellipse (satisfying the law of areas) to the inverse-square law. At this point, then, we have subsumed the observable relative motions in the solar system under a mathematical concept (that of a conic section), and, accordingly, we have subjected the appearances (Erscheinungen) to a mathematical synthesis; from there we can proceed purely mathematically to the inverse-square law.[57] We have not yet arrived at the law of gravitation itself, however, and in order to achieve this we need to proceed still further to a dynamical synthesis. In particular, all we have so far is a purely kinematical description of relative accelerations. If we are to go beyond kinematics and conclude that these accelerations manifest a *force,* we need to assume that the relative motions in question closely approximate to true motions (in modern terms, that the reference frames in which we record the relative motions of the various satellites closely approximate, for this purpose, to inertial frames).[58] And,

57. Kant also appears to be suggesting this procedure in the ultimate paragraph of §38: "that which determines space to the figure of a circle, the figure of a cone and a sphere is the understanding, in so far as it contains the ground of the unity of their construction"—a passage which should be compared with A662–663/B690–691, where Kant describes how we are led by experience to consider first circular orbits, then orbits in conic sections, and thence to "unity in the kinds [Gattungen] of these orbits in their figure [Gestalt], and thereby unity in the cause of all laws of their motion (gravitation)." The idea, presumably, is that the understanding—guided by experience—is responsible for the particular mathematical representations we choose to employ in our description of nature.

58. From a modern point of view, then, we need (in the terminology of note 50 above) to introduce an "affine structure" into space-time—which is, in effect, implicitly defined by

as we have seen, there is no way to substantiate this assumption—or even, in fact, to give it objective meaning in the first place—without explicitly appealing to the laws of motion.

Moreover, as we have also seen above, it is at precisely this point that Kant interrupts the required transition from a merely mathematical to a truly dynamical synthesis by inserting the reference to a possible purely mathematical derivation of the law of gravitation from the geometry of a system of concentric spherical surfaces. We are now in a position finally to appreciate the fundamental weakness in any such derivation—which may be elucidated as follows. A system of concentric spherical surfaces is a purely geometrical object. As such, it is an essentially static representation and has no implications whatsoever concerning the dynamical relations between different spatial manifolds occurring at different moments of time. If we say that a system of spherical surfaces represents the gravitational force, on the other hand, we necessarily presuppose just such dynamical relations. In particular, we presuppose a given *spatio-temporal* frame of reference—a frame of reference at least approximating an inertial frame. Thus, for example, I can certainly set up a system of concentric spherical surfaces in a frame of reference fixed at the center of the earth and in which the moon experiences no rotation (hence the frame itself is rigidly rotating). Yet in this frame of reference the moon will of course not experience an inverse-square acceleration towards the earth and will therefore, as far as this frame of reference is concerned, fail to manifest an inverse-square centripetal force. It follows that we can use a system of spherical surfaces to represent the gravitational force only if we have previously set up a privileged, spatio-temporal frame of reference (in modern terms, an inertial frame of reference). But this in turn requires the laws of motion, and the purely geometrical derivation under consideration makes no use whatever of the laws of motion.

Now, as we also know, Kant himself does not actually endorse such a purely geometrical derivation. Kant says only that the law of gravitation is thereby *customarily* expounded as knowable a priori and appears to have in mind, in particular, the views of John Keill, who sketches such a geometrical derivation in the Preface to his *Introduction to the True Physics* of 1702. Moreover, according to the rather extreme form of Newtonianism expounded in Keill's book, absolute space and time are treated

the laws of motion. As I suggested above in Chapter 3, §II, however, Kant himself follows Newton in defining the true motions relative to an ideal "absolute space," and thus (in the terminology of note 50) in terms of a "rigging" of space-time—which, for Kant, is defined by the laws of motion *together with* the procedure for approximating the center of mass frame of the universe made possible by the theory of universal gravitation.

simply as given (Lecture VI), and, accordingly, the distinction between true and apparent (absolute and relative) motion is assumed to make sense entirely independently of the laws of motion (Lecture VII)—indeed, Keill does not even formulate the laws of motion until much later in the book (Lectures XI, XII). On such a view a purely geometrical understanding of the law of gravitation could thus make sense after all, for we simply assume, without question, that the space in which we set up our concentric spherical surfaces is uniquely given as the one absolute space. But this kind of view is also fundamentally opposed to Kant's, of course; for Kant explicitly rejects such an absolute space and, in the *Metaphysical Foundations,* deliberately refrains from discussing the distinction between true and apparent motion until the fourth chapter or Phenomenology—that is, *after* he has formulated the laws of motion in the third chapter or Mechanics.

It is nevertheless certainly still possible, on a Kantian conception of space, time, and motion, *to represent* the inverse-square law by a system of concentric spherical surfaces—understood as set up in an appropriate frame of reference, of course. Indeed, this representation has an important advantage over the previously discussed derivation from motion in conic sections (satisfying the area law). For, if we represent the inverse-square law by a system of concentric spherical surfaces, we thereby suggest that the inverse-square proportion holds not only within each individual orbit but also from orbit to orbit in a given system of concentrically orbiting satellites. In this way, we emphasize the very significant difference, noted in §I above, between the derivation of the inverse-square law from Kepler's first and second laws and the additional content provided by Kepler's third (harmonic) law. In particular, it is only Kepler's third law that yields the result that the motions of the various satellites are governed by an inverse-square acceleration-field and, accordingly, that such accelerations are entirely independent of the masses of the satellites.[59] And this last result is significant indeed; for it alone makes it possible, in conjunction with the third law of motion and the universality and immediacy of gravitational attraction, to assert that gravitational acceleration is also proportional to mass and to proceed thereby to the construction of the

59. Note that the geometrical representation in terms of a system of concentric spherical surfaces does not actually *entail* this result either: for it does not tell us *what* field diffuses uniformly on the spherical surfaces in question. Thus, for example, it could have turned out that the gravitational *force* field obeys the inverse-square proportion (and thus diffuses uniformly on concentric spherical surfaces) while gravitational *acceleration* nevertheless did not. That the gravitational field has this latter property is a unique and very remarkable circumstance, which is by no means common to inverse-square forces generally: it certainly does not hold for the electro-static field, for example.

center of mass frame of the solar system. In other words, it is only Kepler's third law which allows us to apply the laws of motion so as actually to determine the desired distinction between true and apparent motion.[60]

Kant appears to be making precisely this point in the final sentence of the first paragraph of §38 (cited above), where he describes the consequences of the inverse-square law:

> . . . it not only follows that all possible orbits of the heavenly bodies are conic sections, but also such a ratio [Verhältnis] of the orbits to one another results that no other law of attraction except that of the inverse-square ratio of the distances can be devised as suitable for a cosmic system [Weltsystem].

It appears, in other words, that Kant is here emphasizing precisely the difference between Kepler's first two laws and Kepler's third law noted above. Not only does the inverse-square proportion hold within each possible single orbit (which is therefore a conic section), but it also holds from orbit to orbit. And it is this latter property, and only this latter property, which allows us to construct the center of mass frame of the solar system so as rigorously to distinguish true and apparent motion therein.[61] The indispensable a priori conditions for this construction are of course the laws of motion.

On the present interpretation, the final paragraph of §38 then proceeds

60. In this connection, it is noteworthy that when Newton actually derives the inverse-square proportion for gravitation in Book III, he explicitly appeals not to Kepler's first and second laws but to Kepler's third law (Props. I–II). (Some of the empirical motivations for this choice are discussed in a detailed and subtle study by Wilson [121].) Kant also proceeds in this way when he actually presents the Newtonian derivation: for example, in a short introductory section to the First Part of the *Theory of the Heavens* (1, 244.1–24). An especially intriguing instance (from 1800–1803) is found in the *Opus postumum* (21, 68.14–20): "*Kepler* had already become acquainted with the temporal periods through observation but *Newton* still had to appear in order for the proposition $v:V = (d/t^2):(D/T^2)$ to be discovered.—But this nevertheless already lay in the Keplerian formula: namely, if $d^3:D^3 = t^2:T^2$, then also $v:V = (d/d^2):(D/D^2)$ [which should obviously be $(d/d^3):(D/D^3)$] $= (1/d^2):(1/D^2)$. Huygens had already discoursed on the central forces of bodies moving in circles." Here Kant clearly has in mind the derivation of the inverse-square law from Kepler's third law and Huygens's formula for centrifugal force in uniform circular motion (presented above in note 10). This derivation also appears to exemplify §37 of the *Prolegomena,* where Kant says that his example (the law of gravitation) is "otherwise similar in all respects to natural laws that we ascribe to experience" (320.12–20). The latter natural laws are Kepler's laws, and the point, I think, is that the law of gravitation acquires an additional *necessity* in virtue of the application of the laws of motion to the above, so far merely kinematical, result.

61. I am thus interpreting "cosmic system [Weltsytem]" here as the type of system our solar system exemplifies: a system of bodies (truly!) rotating about a common central point. In *The Only Possible Basis* the term is used in precisely this sense: see especially the second footnote to #2 of Observation Four of Part Two (2, 110.30; compare also 137.13, 140.19–20).

to diagnose the fundamental mistake in attempting to employ the geometrical representation of the inverse-square law by a system of concentric spherical surfaces as an a priori proof of that law. Such an attempt necessarily operates with a naive conception of absolute space as an a priori given object, as it were, and ignores the crucial role of the laws of motion in making an objective notion of true motion first possible. In other words, the attempt at an a priori, purely geometrical derivation under consideration (such as that suggested by Keill) necessarily misses the distinction between space as form of intuition and as formal intuition (and thereby misses the need for special principles of the understanding in constituting the latter):

> . . . space, because it originally makes possible all figures [Gestalten], which are simply various limitations of space, although it is equally only a principle [Prinzipium] of sensibility, is nonetheless taken for precisely that reason for an absolutely necessary thing subsisting for itself [für sich bestehendes Etwas] and an object given a priori in itself [a priori an sich selbst gegebenen Gegenstand]. (A619/B647)

Space as mere form of sensible intuition indeed appears as an infinite receptacle for all possible figures and all particular objects, and it can therefore easily be confused with Newtonian absolute space.[62] Yet this is of course a fundamental mistake, for in considering space as mere form of intuition we are ignoring precisely the dynamical relations between different particular spatial manifolds occurring at different times that are necessarily involved in the notion of absolute space—and are also necessarily involved, for Kant, in the notion of space as itself an object of intuition or formal intuition.[63] In thus hypostatizing the mere form of

62. Compare A22/B36, where space and time are described as "two pure forms of sensible intuition, as principles [Prinzipien] of a priori knowledge," and A39/B56, where the "mathematical students of nature" are said to be committed to "two eternal and infinite non-entities subsisting for themselves [für such bestehende Undinge]."

63. It is thus significant that when Kant, in the *First Grounds of the Differentiation of Regions in Space* of 1768, argues that "absolute space has its own reality independently of the existence of all matter," he explicitly distances himself from Euler's argument ([27], [28]) for absolute space based on the laws of motion, and asserts that the latter argument "leaves untouched the no less significant difficulties that remain in the application of the aforesaid laws [i.e., the laws of motion] if one wishes to represent them *in concreto* in accordance with the concept of absolute space" (2, 378.9–27). Accordingly, Kant's own argument (from "incongruent counterparts") is based on geometry rather than mechanics, and it therefore concerns only the essentially static "space of the geometers." Kant's argument here, in other words, pertains only to what he will later call space as mere form of intuition, and it therefore has nothing whatever to do with the dynamical structure underlying Newtonian absolute space, time, and motion. Indeed, this is entirely to be expected, for Kant had previously broken decisively with Newtonian absolute space, time, and motion in the *New System of Motion and Rest* of 1758, where we already find the idea (although in a more primitive

sensible intuition we presume that we are thereby given a priori an object in itself, and this explains why Kant says that when the inverse-square law is conceived as derived purely geometrically, it "seems to lie necessarily in the nature of the things themselves."

On Kant's own view, by contrast, the space in which we are to set up the distinction between true and apparent motion is always an empirical space constructed as the outcome of an a posteriori procedure—the very same procedure through which the law of universal gravitation is itself determined. Moreover, in this procedure the distinction between space as mere form of intuition and as formal intuition or object of intuition is emphasized rather than obscured, and the indispensable role of the understanding as the source of nature's law-governedness through the analogies of experience is thereby highlighted as well. It becomes clear, finally, why the laws of nature in no way lie in the nature of things in themselves:

> . . . and so the understanding is the origin of the universal order of nature, in that it comprehends all appearances under its own laws and thereby first brings about experience (according to its form), by means of which all that is to be known only through experience is necessarily subject to its laws. For we do not have to do with the nature *of the things in themselves*—which is just as independent of conditions of our sensibility as of the understanding—but rather with nature as an object of possible experience; and here the understanding, in that it makes experience possible, at the same time brings it about that the sensible world is either no object of experience at all or else a nature. (322.7–17)

The sensible world becomes a nature, in other words, only as a result of the successful application of the analogies of experience to spatio-temporal data given a posteriori—that is, their application to "phenomena" in precisely the Newtonian sense.

From the present point of view, therefore, §38 of the *Prolegomena* is to be read in the context of Kant's lifelong struggle to articulate and define the respective roles of geometry and metaphysics in the foundations of natural science. This struggle begins with Kant's first published work, and the underlying problem is expressed with particular clarity and definiteness in the Preface to the *Physical Monadology* (whose full title is *The Use in Natural Philosophy of Metaphysics Combined with Geometry*):

---

form) that true motions are to be defined relative to the center of mass frame of the system of interacting bodies in question. (In the critical period space as mere form of intuition serves as a receptacle, as it were, for geometrical construction, but the latter also requires the action of the understanding on the form of sensibility. Thus even the "space of the geometers" leads ultimately to the dynamical structure of space as an *object* of intuition or formal intuition.)

> But how in this business can metaphysics be reconciled with geometry, where it appears easier to be able to unite griffins with horses than transcendental philosophy with geometry. The former precipitously denies that space is infinitely divisible, while the latter asserts this with its customary certitude. The latter contends that void space is necessary for free motion, the former denies it. The latter shows most exactly that attraction or universal gravitation is hardly to be explained by mechanical causes but rather from forces innate in bodies that are active at rest and at a distance, the former reckons this among the empty playthings of the imagination. (1, 475.22–476.2)

Here the side of geometry is clearly represented by the Newtonian natural philosophy, that is, by the "mathematical students of nature" to which the Aesthetic refers at A39/B56–57.[64] The side of metaphysics is of course represented by the "monadologists" or Leibnizean-Wolffians, that is, by the "metaphysical students of nature" to which the Aesthetic refers at the same place.[65]

Now Kant never had any doubt that the Newtonian natural philosophy comprises the true physics,[66] but he always insisted that this physics nonetheless requires a metaphysical foundation. The question was what precisely does such a metaphysical foundation for physics involve. In the critical period Kant answers this question in terms of the doctrine of the *schematization* of the pure concepts of the understanding: metaphysical concepts—concepts of substance, causality, and community—provide a foundation for natural knowledge in virtue of, and only in virtue of, their application to our spatio-temporal form of sensibility, "which realizes the understanding in that it simultaneously restricts it" (A147/B187). Such metaphysical concepts are thereby seen as the most general rules and conditions of "time-determination." In the present context, I suggest, this

64. It appears that in the *Physical Monadology* itself the primary representative of the side of geometry is none other than John Keill. Keill [56] zealously defends all of the above assertions of "geometry," and, as we observed in note 45 above, it is this book to which Kant implicitly refers in the Scholium to Proposition X. Moreover, the proof of infinite divisibility in Proposition III is clearly taken from Lecture III of Keill's book—as Kant explicitly states in *On a Discovery* (8, 202.16–21).

65. As is indicated by the emphasis on the question of infinite divisibility versus simple substances in the *Physical Monadology,* Kant is here responding to the controversy between the Newtonians and the Wolffians officially addressed by the Royal Berlin Academy of Sciences in 1745–1747. See the Introduction above for further discussion.

66. As we know, this statement requires qualification in the case of the second assertion of "geometry" in the above passage: that is, the necessity of void space for free motion. Kant consistently denies the existence of an *exactly* void space and holds instead that free motion requires only that the density of interplanetary space be vanishingly small in comparison with that of the heavenly bodies: see, for example, the conclusion to the Phenomenology of the *Metaphysical Foundations* (564.27–33). This issue will be extensively discussed in Chapter 5 below.

doctrine is to be understood as pointing towards the essential limitations of an exclusively geometrical approach to the foundations of physics. What is missing from a *merely* geometrical approach, that is, is a recognition of the unique and indispensable role of dynamical structure: the need for an explicit construction of the spatio-*temporal* relations subsisting in a collection of different particular spatial manifolds occurring at different moments of time. Newton's fundamental contribution was to fashion the appropriate dynamical concepts—concepts of mass, force, and interaction—which, together with the laws of motion governing these concepts, first make it possible to employ pure geometry in the construction of such a dynamical, spatio-temporal structure. Kant's fundamental contribution, however, was to see further than anyone else into the philosophical implications of this Newtonian achievement—its implications, in particular, for the nature and future of metaphysics.

· PART TWO ·

The *Opus postumum*

· CHAPTER 5 ·

Transition from the Metaphysical Foundations of Natural Science to Physics

Kant's *Opus postumum* (approximately 1796–1803) includes a succession of attempts to draft a work to be entitled *Transition from the Metaphysical Foundations of Natural Science to Physics.* The work was never successfully brought into publishable form, however, and what has come down to us instead is a bewildering collection of fragments, outlines, and sketches.[1] This ungainly and fragmentary collection, in spite of—and even in some ways because of—its highly unfinished character, nevertheless contributes an extremely illuminating perspective on Kant's philosophy of natural science, and also on the critical philosophy as a whole. A preliminary sense of why this may be so can be conveyed by noting some the

1. The *Opus postumum* comprises volumes 21 and 22 of the Akademie edition of *Kant's gesammelte Schriften.* Gerhard Lehmann's Introduction on pp. 751–789 of volume 22 explains some of the background to the form in which we now find the material. It is especially important to note that the order in which the manuscripts are printed in the Akademie edition is *not* their chronological order of composition: Adickes's generally accepted reconstruction of this chronology is found in a table printed at the end of volume 22. In presenting multiple citations from the *Opus* I follow the practice of listing them always in their chronological order. In quotations I enclose marginal insertions, as well as insertions in the main text made later by Kant, in angle brackets thus ⟨ . . . ⟩—I do not distinguish insertions in the main text made contemporaneously. Although some of the earliest loose-leafs are actually written before 1796, it is generally agreed that serious work on the project begins with the so-called Octavio-outline (Oktaventwurf) of 1796–97. An English translation of parts of the *Opus* by Eckart Förster and Michael Rosen is being published by Cambridge University Press (1992). I am grateful to Eckart Förster for allowing me to see pre-publication drafts of this translation, and I note that Förster's Introduction will contain the most complete history of the manuscripts yet available. It should be emphasized, finally, that I will focus exclusively on the project of the proposed *Transition* here—and, moreover, on that part of the *Transition* concerning physical science. Thus I shall not discuss Kant's new attempts to comprehend biological science (living organisms), or his attempts—especially prominent in the later fascicles of the *Opus*—to embed the project of the *Transition* into a revised conception of transcendental philosophy in general.

most obvious divergences between the *Metaphysical Foundations of Natural Science* of 1786 (thus written, of course, at the height of the critical period) and ideas to be found in the proposed *Transition.*

First of all, in the Preface to the *Metaphysical Foundations* of 1786 Kant apparently claims to have accomplished, or at least to have indicated, all that philosophy can possibly contribute to the "pure doctrine of nature"—the complete a priori foundations of natural science. Thus, in paragraph thirteen of this Preface, Kant contasts his philosophical undertaking with both pure mathematics and empirical natural science:

> . . . as in the metaphysics of nature in general, thus also here the completeness of the metaphysics of corporeal nature can be confidently expected—whereof the reason is that in metaphysics the object is considered only as it must be represented according to the universal laws of thought, but in other sciences according to data of intuition (pure as well as empirical). Therefore the former, because in it the object must always be compared with *all* the necessary laws of thought, must yield a determinate number of cognitions which can be completely exhausted; whereas the latter, because they offer an infinite manifold of intuitions (pure or empirical) and thus objects of thought, never attain absolute completeness but can be extended to infinity—as pure mathematics and empirical doctrine of nature. (4, 473.19–31)

As the following paragraph makes clear, the relevant "determinate number of cognitions" is to be enumerated according to the table of categories, and Kant is quite confident that he has substantially accomplished this philosophical task: "More is not to be done, to be discovered, or to be added here, except to improve it in any case where clarity or rigor may be lacking" (476.4–6); "therefore I believe that I have completely exhausted this metaphysical doctrine of body as far as it may extend—without thinking, however, that I have thereby accomplished any great work" (473.31–34).

In the *Opus* of 1796–1803, however, Kant is now convinced that a new a priori science, the *Transition,* must be added to the *Metaphysical Foundations;* and he claims, moreover, that without this new philosophical undertaking the "pure doctrine of nature" remains *incomplete:* "⟨NB. That this treatise is directed towards filling what is still a gap in the pure doctrine of nature and in general in the system from a priori principles—and thus towards accomplishing completely my metaphysical task⟩" (21, 626.7–11); "through the latter [the *Transition*] a gap in the system of pure natural science *(philosophia naturalis pura)* is now filled, and the circle of all that belongs to a priori cognition is closed" (640.4–6). Further, as these passages suggest, the "gap" that the *Transition* is to fill is not merely a local defect involving natural science and the philosophy of nature; it extends, in fact, to the "system from a priori principles" and to "all that belongs to a priori cognition." In other words, the "gap in the pure doctrine of nature" that Kant has now uncovered extends to the critical system as a whole.

That the new problem of the *Transition* generalizes in this way is explicitly stated by Kant in two letters from the fall of 1798 (during which the above two passages were also probably written).[2] In a letter to Christian Garve of September 21, 1798, Kant complains about the inadequacy of his mental powers for the difficult task on which he has embarked (he is now 74):

> [I am] as paralyzed for intellectual works, with otherwise tolerable physical good health: to see the complete balancing of my accounts, in matters that concern the whole of philosophy (relating as much to ends as to means), lying before me and yet still not completed—although I am conscious of the feasibility of this task—is a pain like that of Tantalus, which is thus still not hopeless.—The problem with which I am now occupied concerns the "Transition from the Metaphysical Foundations of Natural Science to Physics." It must be solved; for otherwise there would be a gap in the system of the critical philosophy. The demands of reason here do not weaken, nor does the consciousness of the capacity thereto; but the satisfaction of reason is postponed up to the limits of patience, if not be the complete paralysis of my vital force then by its continual diminution. (12, 257.2–15)

And Kant says substantially the same in a letter to J. G. C. Kiesewetter of October 14, 1798 (12, 258.19–27). Thus the incompleteness in the "pure doctrine of nature" with which Kant is grappling in 1798 belies not only the presumed completeness claimed in the Preface to the *Metaphysical Foundations* but also Kant's famous statement in the Preface to the *Critique of Judgement* of 1790 that "hereby [with the *Critique of Judgement*] I bring my entire critical undertaking to a close" (5, 170.20).

Second, when one compares the contents of the *Metaphysical Foundations* with those of the *Transition* a striking divergence is observed in the scientific problems constituting the primary focus of philosophical inquiry. The *Metaphysical Foundations* focuses on fundamental questions concerning space, time, and motion—and, accordingly, concentrates on rational mechanics (namely, kinematics and the Newtonian laws of motion) and the application thereof to impact and the theory of universal gravitation. We are therefore concerned only with "a priori comprehensible universal characteristics of matter" such as impenetrability (which for Kant rests on repulsive force and thus "original elasticity") and weight (resting on the universal attractive force of gravitation). Problems involving the "specific

2. The importance of the letters in question—together with the above-cited passages—has been emphasized especially in Förster [31].

variety" of matter—the difference between matter in general and determinate bodies; the explanation of cohesion, fluidity, and rigidity; expansive elasticity (that is, air); chemical interactions—do not fall within our properly metaphysical discussion (the question of "specific variety" and the indicated physical problems are only touched on, parenthetically as it were, in the General Observation to Dynamics: 4, 525.20–535.10). For only strictly *universal* properties of matter belong to metaphysics:

> Thus [original elasticity] and weight constitute the only a priori comprehensible universal characteristics of matter, the former internally, the latter in outer relations; for the possibility of matter itself rests on both: *cohesion,* if it is explained as the mutual attraction of matter that is limited simply to the condition of contact, does not belong to the possibility of matter in general and can therefore not be cognized a priori as bound up with this. This property would thus not be metaphysical but rather physical, and would therefore not belong to our present considerations. (518.21–31)

The line between metaphysical question, which are capable of philosophical treatment, and physical questions, which are best left to the progress of empirical science itself, is therefore sharply drawn (compare also 517.26–518.2, 522.39–523.4, 525.7–12, 534.31–36, 563.32–564.27).

Yet it is precisely the problems consigned to a merely physical treatment in the *Metaphysical Foundations* (and therefore touched on only in the General Observation to Dynamics)—problems involving the formation of bodies, cohesion, state of aggregation or phase (solid, liquid, gaseous), and above all heat[3]—that are central to the *Transition* project. This can be seen by glancing at the first few pages of the very first complete outline-draft, the Octavio-outline of 1796–97 (21, 373ff.), and the same problems remain central throughout the *Opus.* The fundamental divergence from the *Metaphysical Foundations* here can be articulated in a closely related but somewhat more general fashion by noting that, in the *Transition* project, the physics for which philosophy is to provide an a priori foundation or explanation includes *chemistry.* Thus, to pick two representative passages at random: "⟨Chemistry is a part of physics but not of the mere transition from the meta. to physics.—The latter contains merely the conditions of the possibility of instituting experiences⟩" (316.20–22); "⟨The whole of chemistry belongs to physics—but in the Topic [the *Transition*] only the transition to it is in question⟩" (288.5–6). The *Transition* project is not, of course, concerned with problems of chemistry as such, but rather with an explanation of the possibility of that science—that is, with a

3. This contrasts sharply with even the General Observation to Dynamics, where heat is mentioned only in passing: see 529.34–36, 530.1–3, 532.3–5. The significance of this will be explored in detail below.

"transition" from the *Metaphysical Foundations* to the latter (compare 22, 263.1–6; 21, 362.28–363.9; 22, 141.18–24; 21, 623.4–10, 625.12–20, 633.3–16).

In the *Metaphysical Foundations* itself, by contrast, chemistry is clearly not a part of the natural science for which philosophy is to provide an a priori foundation. Indeed, chemistry is not a "*proper* natural science" at all:

> A whole of cognition that is systematic can already, for precisely that reason, be called *science,* and, if the connection of cognition in this system is a complex of grounds and consequences, even a *rational* science. But if these grounds or principles in it, as, e.g., in chemistry, are still in the end merely empirical, and the laws from which the given facts are explained through reason are mere empirical laws, then they carry no consciousness of their *necessity* (are not apodictically certain) and consequently the whole [of cognition] does not deserve the name of a science in the strict sense—and therefore chemistry should rather be called systematic art than science. (4, 462.19–29)

It follows that the *possibility* of chemistry cannot be explained a priori:

> . . . chemistry can be no more than systematic art or experimental doctrine but never a proper science, because its principles are merely empirical and permit no a priori presentation in intuition. Consequently the principles of chemical appearances cannot in the least be made comprehensible according to their possibility, because they are unsuitable for the application of mathematics. (471.5–10)

Note that chemistry is here said to be incapable of a priori philosophical foundation or explanation because—unlike rational mechanics and the theory of universal gravitation, for example—it is not suitable for mathematical treatment. As we shall see, the relationship between philosophical foundations and mathematics is subject to intense reexamination in the *Transition* project.

Finally, there is an extraordinary and dramatic divergence between the *Metaphysical Foundations* and the *Transition* project with respect to a single, very particular question of natural philosophy: Is it possible to prove a priori the existence of an everywhere distributed, space-filling medium or "aether"—or, rather, does such an aether have the status of a mere physical hypothesis? In the *Metaphysical Foundations* the aether clearly has a merely hypothetical status. To be sure, Kant appears to be convinced of its existence, and he accordingly vigorously opposes a natural philosophy based on atoms and the void. Yet he also firmly asserts that it is not possible to rule out such an atomistic philosophy a priori—the philosophical task, rather, is only to oppose the aether hypothesis to the atomistic hypothesis so as to show that the latter is itself not provable a priori.[4]

Thus, in the General Observation to Dynamics of the *Metaphysical Foundations* Kant argues against the "mechanical natural philosophy," that is, against "*atoms* and the *void*" (4, 532.20–535.10). In order to combat this philosophy "it is absolutely not necessary to fabricate new hypotheses but only to oppose the postulate of the mere mechanical mode of explanation—*that it is impossible to think a specific difference of density of matters without introduction of void space*—through the mere exposition of a mode in which it may be thought without contradiction" (533.26–31). The mode in question, of course, consists simply in imagining space as filled with types of matter of different intrinsic densities (intensities), of which the space-filling aether is the smallest possible: "In this mode one would not find it impossible to think a matter (as one perhaps represents the aether) that fills its space completely without any vacuum—and yet with incomparably smaller quantity of matter at the same volume than any body that can be subject to our experiments" (534.5–9). This clearly does not amount to a proof of the existence of the aether, however; rather, the latter is something "we assume merely for the reason that *it can be thought,* simply to oppose an hypothesis (of void space) that rests only on the presumption that [difference in density] *cannot be thought* without void spaces" (534.12–15). We therefore cannot prove the impossibility of void space.[5]

4. This basically negative approach to the question of aether versus void space is of course entirely consistent with the approach taken in the Anticipations of Perception at A172–175/B213–216. See especially A173–174/B215–216: "[Atomists] assume that the *real* in space (I may here not name it impenetrability or weight, for these are empirical concepts), is *everywhere the same* and can only differ according to extensive magnitude, i.e., multiplicity [Menge]. To this presupposition, for which there can be no ground in experience and is therefore merely metaphysical, I oppose a transcendental proof—which, to be sure, is not to explain the difference in the filling of space, but which nevertheless completely annuls the supposed necessity of the presupposition that the difference in question can be explained not *otherwise but* through assumed void spaces, and has the advantage of at least freeing the understanding for thinking this variety in other ways if natural explanation should make any such hypothesis necessary."

5. Compare 534.37–535.10: "The well-known question of the admissibility of void spaces within the world can be resolved. The *possibility* thereof cannot be denied. For space is required for all forces of matter and, since the former also contains the conditions of the laws of dispersion of the latter, is necessarily presupposed before all matter. Thus attractive force is attributed to matter in so far as it *occupies* a space around itself through attraction, but without *filling* this space—which therefore can be thought as empty even where matter is active, because it is not active there through repulsive force and does not therefore fill this space. However, to assume void spaces as *actual* can be justified by no experience or conclusion from experience, nor by any necessary hypothesis to explain experience. For experience yields only comparatively-empty space for our cognition, which can be completely explained,

The status of the aether as a physical hypothesis stands out even more clearly in the General Observation to Phenomenology, where Kant takes up the question of void space once again (563.10–564.33). In particular, Kant considers the possibility of "void space *within* the world" (for example, interspersed pores of void space within bodies) and explicitly links this possibility with the physical question of cohesion:

> . . . that the assumption of void space [within the world] is not *necessary* is already shown in the General Observation to Dynamics; but that it is *impossible* can in no way be shown from its concept according to the principle of contradiction. Nevertheless, if there is here no merely logical ground to be found for discarding it, there could still be a universal physical gound for banishing it from the doctrine of nature: namely, that arising from the possibility of the compounding of a matter in general, if one only had better insight into the latter. For if the *attraction* that one assumes for the explanation of cohesion of matter should be only apparent, not true attraction—perhaps merely the action of a *compression* through external matter distributed throughout the universe (the aether), which is itself brought to this pressure only by a universal and original attraction, namely gravitation (which opinion has many grounds)—then void space within matters would be, if not logically, yet dynamically and thus physically impossible. Because any matter would itself expand into the void space assumed within it (since nothing resists its expansive force there), and the latter would thus always remain filled. (563.32–564.9)

Void space within the world is not metaphysically, but at most physically impossible—and even this is true only on a particular hypothesis for explaining cohesion (which hypothesis itself assumes the aether). Kant therefore concludes: "One easily sees that the possibility or impossibility [of void space within the world] does not rest on metaphysical grounds, but rather on the difficult to unlock mystery of nature concerning how matter sets limits to its own expansive force" (564.24–28).[6] In sum, the question of the possibility of void space, together with the intimately related questions of cohesion and the aether, can in no way be settled by transcendental philosophy.[7]

according to any chosen degree, from the property of matter to fill its space with greater or lesser (up to infinity) expansive force—without requiring void spaces."

6. One should be careful not to confuse the question here with that already settled by the well-known "balancing" argument of Proposition 5 of the Dynamics, where Kant argues that the original force of attraction must oppose the original force of repulsion so as to explain the possibility of "matter . . . filling a space in a determinate degree" (512.22–23). As Kant makes clear at 518.25–31, Proposition 5 is not concerned with the possibility of *cohesion*—that is, with matter having a determinate *volume*—but rather with the possibility of a determinate *density* (whether or not the matter having this density is confined to a definite volume).

7. Compare A433/B461n: "One easily observes that what is here intended is that

Whereas the aether is treated only marginally in the *Metaphysical Foundations,* it is absolutely central to the *Transition* project. Here it is discussed most often in connection with phenomena involving heat and, specifically, with the expansion of matter by heat associated with the three aggregative states or phases: solid, liquid, gaseous. For Kant, these phenomena are to be explained in terms of the penetration of matter by a universally distributed, continuous, space-filling, perpetually vibrating or oscillating, imponderable expansive fluid—which he characteristically calls either "aether" or "caloric [Wärmestoff]"—and the phenomenon of cohesion in particular (for example, the formation of solid bodies) is thereby linked to the aether once again (although in a more complex fashion than in the *Metaphysical Foundations*).[8] In addition, this same expansive fluid is characteristically identified with the light-aether introduced by Euler.[9] Thus, the aether of the *Transition* project embraces physical problems involving precisely the "specific variety" of matter—problems from the theory of light, the theory of heat, and chemistry—and in this way reflects the fundamental divergence in focus from the *Metaphysical Foundations* noted above.

Nevertheless, in the early outline-drafts of the *Opus* the aether tends to retain a merely hypothetical status: its function is simply to explain the phenomena in question as well as is possible.[10] At the same time, however, various intimations and finally clear articulations of a necessary and non-

statement: *void space, in so far as it is limited by appearances,* thus that *within the world,* at least does not contradict transcendental principles, and may thus be admitted as far as the latter are concerned (although its possibility is not thereby asserted forthwith)."

8. See again, for example, the first few pages of the Octavio-outline: 21, 373–377. For the formation of solid bodies see especially 392.23–394.10. The account here is more complex than in the *Metaphysical Foundations* in two respects. First, cohesion in general, including both that of solids and that of liquids (for example, drops of water) is due not simply to the pressure of the aether but more to its vibrations or oscillations (impacts). Second, the cohesion of solids is due to a process of *crystallization* by which heterogeneous constituents of a liquid are set into varying modes of vibration by the aether so as to form a periodic structure (crystal). These matters will be explored further below.

9. See, e.g., 21, 523.5–6: "Euler's *pulsus Aetheris* are here applied not only to light but also to the motion of heat" and, in the Octavio-outline, 21, 383.31–34: "We wish to call the aether the *empyreal* expansum (fire-air) which acts in two ways (namely as light and also as heat) progressively and in oscillation, and contains and penetrates all matter constituting the universe" (we will return to the notion of "fire-air" below). (Kant mentions Euler's wave-theory of light, although in a somewhat less committed fashion, in the *Metaphysical Foundations* at 4, 520.19–41.)

10. See, e.g., in the Octavio-outline, 21, 378.15–18: "To assume such a universe filling matter is an unavoidably necessary hypothesis, because without it no cohesion such as is necessary for the formation of a physical *body* can be thought." Compare also 21, 268.13–14; 22, 259.15–20, 216.17–21, 160.26–161.2, 177.5–10, 193.3–4, 199.3, 229.13–15, 273.6–7.

hypothetical status begin to emerge (see 21, 480.24–26, 297.8–11; 22, 192.14–27, 193.17–194.15, 197.10–24; 21, 192.24–30; 22, 587.10–16, 598.5–13, 605.10–606.26). This development culminates in a central, and extraordinarily striking, series of outline-drafts written in the middle of 1799 and entitled Transition 1–14. These drafts are devoted almost entirely to the so-called aether-deduction, and the problem of the aether is now characterized as *the* problem of the *Transition* project:

> There is, however, in the transition from the metaphysical foundations of natural science to physics, a problem that is unavoidable for it: namely, whether a *material* [*Stoff*] that is thoroughly distributed in the universe (and therefore penetrating through all bodies), which one perhaps could call **caloric** [**Wärmestoff**] (without thereby bringing a certain feeling of warmth into consideration, because this concerns merely what is subjective in a representation, as perception) ⟨—⟩whether, I say, such a material, as the **basis** of all moving forces ⟨present⟩ in matter, *is* or *is not:* or whether its existence is only doubtful; ⟨in other words: whether⟩ it, as merely *hypothetical material,* is assumed by the physicists only for the explanation of certain phenomena ⟨or, rather, is to be laid down *categorically* as a postulate⟩—this question is of the greatest importance for natural science, as a system; especially since it provides the guidance from the elementary-system thereof to the cosmic-system. (22, 550.4–17)

Kant's answer to this question here is unequivocal. The aether or caloric does exist—and, moreover, as an a priori assertion or postulate of the *Trarsition* project (and thus of philosophy): "The assertion of the existence of caloric, however, belongs not to the metaphys. found. of N.S., and also not to physics, but rather merely to the *transition* from the metaphysical found. of N.S. to physics" (21, 594.6–9).[11]

We are not yet in a position to attempt to analyze the proofs Kant constructs for this assertion, but a preliminary sense of what is involved may perhaps be conveyed by a representative sample:

> Now the concept of the whole of ⟨all⟩ outer experience ⟨also⟩ presupposes all possible moving forces of matter bound together in collective unity, and in fact in full space (for void space—whether it be ⟨enclosed⟩ within bodies or external to the bodies ⟨it surrounds⟩—is no object of possible experience).* But it also presupposes a continual *motion* of all matter, which acts on the *subject* as object of the senses, for without this motion—i.e., without stimulation of the sense organs as its effect—no perception of any object of the senses whatsoever takes place, and therefore no experience—as that

11. Compare also 21, 571.1–5: "⟨The transition from the meta. found. of N.S. to physics takes place precisely through the idea of caloric, which therefore must be no merely hypothetical material but rather that material which alone governs the susceptibility to experience of all bodies in all spaces and is continuously distributed *cohesively in one experience.*⟩"

> which contains only the form belonging to the former. Therefore, a particular self-agitating material that is continuously and limitlessly distributed in space, as object of experience (although without empirical consciousness of its principle), i.e., **caloric**, is *actual*—and is not a material merely devised for the sake of the explanation of certain phenomena, but rather is provable from a universal principle of experience (not from experience) according to the principle of *identity* (analytically) and given a priori in the concepts themselves. (550.27–551.14)[12]

This extraordinary argument—whatever its precise import may be—clearly and dramatically diverges from the cautious attitude towards void space and the aether found in the *Metaphysical Foundations* (and the first *Critique*): void space, including void space within the world in particular, does conflict with transcendental principles after all; indeed, this can even be shown from the principle of contradiction alone (analytically).

I · The *Transition* Project and the *Metaphysical Foundations*

The above considerations have made it clear, I hope, that a thorough grasp of its relationship to (and, in particular, its divergences from) the *Metaphysical Foundations* of 1786 is absolutely necessary for a proper understanding of the *Transition* project of 1796–1803. This relationship has recently been subject to especially clear and penetrating treatment by Tuschling ([113], [114], [115]) and Förster ([31], [32]). Although these two authors by no means agree with one another on all particular points of interpretation, they are, I think, nonetheless in at least broad agreement on a certain conception of how the project of the *Transition* relates to the *Metaphysical Foundations.* On this conception, the *Transition* project is predicated on a fundamental rejection or refutation of the *Metaphysical Foundations,* and this rejection/refutation concerns two interrelated issues. First, the *Transition* project rejects the "theory of matter" developed in the Dynamics of the *Metaphysical Foundations:* the explanation of the possibility of matter based on the original forces of attraction and repulsion articulated in Propositions 5 and 6, where matter's "filling a space in a determinate degree" is explained in terms of a kind of "balancing" of the two original forces (4, 508–511). Second, and more fundamental, however, the *Transition* project rejects the "mathematical method" of the *Metaphysical Foundations*—as articulated in the Preface and then followed throughout that book. Whereas the *Metaphysical Foundations*

12. This passage, as well as the previous passage from 22, 550, is taken from the transcription made by an amanuensis of Transition 9 from 21, 554–581. These few sheets copied from Transition 9 are the closest that any part of the *Opus* came to attaining publishable form.

tends to conflate philosophical and mathematical methods, the *Transition* project insists on a sharp distinction between "mathematical foundations of natural science" and "philosophical foundations of natural science"—only the latter belong to philosophy and thus to the project of the *Transition.*[13] We shall here concentrate principally on the second and more fundamental issue.

As Förster has rightly emphasized, the relevant opposition between "mathematical foundations of natural science" and "philosophical foundations of natural science" initially arises in loose-leafs 6, 3/4, and 5 of the fourth fascicle of the *Opus* (21, 474–485).[14] We first find the question sharply posed: "⟨*NB:* Of the mathematical foundations of physics. Whether this too belongs to the transition⟩ " (477.21–22). The issue is then explained as follows:

> The *mathematical* foundations of natural science are those containing the moving forces that proceed from the actual motion of matter. Central force, light and sound. The *dynamical* are those containing the inner moving forces. The former locomotive forces the *mechanical,* the latter the dynamical (internal) moving forces of matter. (479.3–7)

Finally, the original question is answered decisively in the negative:

> What one calls the *mathematical* foundations of natural science *(philos. nat. principia mathematica),* as Newton delivered in his immortal work, is, as the expression already indicates, no part of *natural philosophy*—but is rather only an instrument (to be sure very necessary) for estimating the magnitude of motions and moving forces (which latter must be given in the observation of nature) and for determining their laws for physics (so that the quality of these forces in relation to the central force of a body moving in a circle, the motions of light, of sound and tone can be estimated thereby with respect to their direction and degree); so that consequently this doctrine properly constitutes no part of philosophical natural knowledge [Naturkunde]. (482.4–18)

And very similar passages, including the references to the title of Newton's

13. Tuschling develops these ideas primarily in chaps. III and V of [113], Förster in §II of [31]. With respect to the "theory of matter" developed in the Dynamics of the *Metaphysical Foundations,* both Tuschling and Förster call attention to Kant's correspondence with J. S. Beck in 1792, which discusses a *circularity* Kant now sees in the theory of the *Metaphysical Foundations:* Tuschling [113], pp. 46–56; Förster [31], pp. 547–548 (compare also Adickes's discussion: 14, 337.15–339.10). Because I concentrate on the problem of philosophical and mathematical methods here, I must leave consideration of the former problem for another occasion. In any case, however, since Kant never refers to the supposed circularity in question in the *Opus* itself, it is hard to see how this problem can have fundamental importance for the *Transition* project.

14. See [31], pp. 548ff. These pages are also emphasized, although from a slightly different point of view, by Tushling [113], e.g., pp. 93ff., 108ff.

"immortal work" (which title is now seen as self-contradictory) and to the mathematical theories of central forces, light, and sound, recur continually throughout the *Opus*.[15]

What exactly are the implications of such ideas for the *Metaphysical Foundations?* Tuschling suggests the following picture.[16] According to the new conception of the *Transition* project, it is a mistake, in philosophy, to attempt to derive original moving forces from motion. Specifically, whereas *mathematical* foundations of natural science derive moving forces from motion, *philosophical* foundations are to proceed in the opposite direction and derive motion from moving forces:

> Now one can think moving forces as given ⟨in two ways⟩ : namely, either the motion must precede and moving force is the effect of it, or, on the contrary, moving force is first taken as the basis and motion is derived from this.—The former principles contain the mathematical principles of natural science *(philosophiae naturalis principia mathematica)* as in Newton's immortal work; the latter can be called the physiological foundations of natural science *(philosophiae naturalis principia physiologica)*. (22, 164.21–29)[17]

Indeed, if one proceeds in the former, merely mathematical fashion, one is not really dealing with forces properly speaking but only with motions themselves: "⟨Mathematical foundations of natural science do not treat of the moving forces proper to matter but only of motions that are impressed and their composition. Wherein centrifugal force belongs⟩" (21, 359.7–10). And now the decisive point is this: the *Metaphysical Foundations* proceeds precisely in such a mathematical fashion, namely, by deriving moving forces from motion. It follows that the procedure of this work cannot now be viewed as truly philosophical at all, and, in addition, that it cannot now be viewed as dealing with "the moving forces proper to matter."

In particular, the *Metaphysical Foundations* attempts to derive original moving forces from motion at two crucial points. In the proof of Proposi-

15. A compilation of references to such passages is presented by Tuschling on p. 91 of [113], divided into the subtopics of what Tuschling calls "phoronomy-critique," "mathematics-polemic," "Newton-polemic," and "phenomenology-critique."

16. See [113], especially chap. V ("Phoronomy-critique"); Förster endorses Tuschling's account here on pp. 549–550 (including n. 36) of [31].

17. It is clear that the latter *physiological* foundations are to be identified with (or at least included within) *philosophical* foundations. See, for example, the characterization of the *Transition* project at 21, 638.11–18: "With respect to the tendency to physics that the metaphysics of nature carries within itself, matter is *the movable in space in so far as it* ⟨*for itself*—i.e., not through its impressed motion (as the centrifugal force of a stone twirled in a sling, say)⟩—has *moving force*.—This determination of the concept of [matter] makes it already *empirical*, i.e., dependent on perception as action on the senses, and the principles of natural science can therefore be called *physiological* . . . . "

tion 2 of the Phenomenology Kant infers the existence of a force from circular motion:

> Circular motion is (as every curvilinear motion) a continual alteration of rectilinear [motions], and since the latter is itself a continual alteration of relations with respect to the external space, circular motion is an alteration of these external relations in space and thus a continual arising of new motions. Now, since, according to the law of inertia, a motion, in so far as it arises, must have an external cause, and yet the body in every point of this circle (according to precisely the same law) is striving to proceed in the straight line tangent to the circle (which motion acts contrary to the external cause), it follows that every body in circular motion manifests [beweiset] a moving force through its motion. (4, 557.5–16)[18]

This inference appears to be directly disputed in the *Transition* project:

> There are moving forces *proper* to matter *(vires congenitae, non impressae)*—not merely those *communicated* through motion. If a body is moved in a circle then it exerts a force of striving from the center *(vis centrifuga);* this is not *proper* to the imagined body but rather a force impressed through motion—and this is also the situation with the force striving towards the center *(vis centripeta)* ⟨in circular motion⟩. There must be original moving forces given, although no motion is original ⟨but every one is imparted . . .⟩. (21, 170.6–16)[19]

Even more seriously, however, the *Metaphysical Foundations* derives an original moving force from motion in the very first Proposition of the Dynamics, where forces are initially introduced into the exposition (4, 497.15–28): here the original force of repulsion is derived from matter's property of "resisting every movable that is striving through its motion to enter into a certain space" (496.7–9). It follows, then, that if such a procedure is illegitimate—as the *Transition* project apparently holds—then the *Metaphysical Foundations* never really succeeds in incorporating moving forces at all (since all later Propositions depend on Proposition 1 of the Dynamics). Thus, the *Metaphysical Foundations* can actually contain nothing more than mere phoronomy or kinematics (namely, mere mathematical theory of *motion*): true forces and true dynamics are necessarily left out of account.[20]

18. In the Observation Kant remarks: "This Proposition determines the modality of motion with respect to *Dynamics;* for a motion that cannot take place without the influence of continually acting external moving force manifests [beweiset], directly or indirectly, original moving force of matter—whether it be attraction or repulsion" (557.28–32).

19. This passage is thus an example of what Tuschling calls "phenomenology-critique."

20. This last idea constitutes the heart of Tuschling's conception of "phoronomy-critique": "The [*Metaphysical Foundations*] does not contain, as it pretends and every reader must assume, a presentation of the moving forces (478,11ff.). It attributes to matter only the predicate of [being] the movable in space (289,18ff.). It therefore has not attained its

Yet Kant's distinction between mathematical moving forces (which proceed from motion) and dynamical or physiological moving forces (from which motion proceeds) does not, it seems to me, support Tuschling's analysis. Rather, this distinction is to be understood in the context of the eighteenth-century struggle between the corpuscular or mechanical natural philosophy and the Newtonian natural philosophy. On the former conception, force is to be understood as the *effect* of motion: moving force is an internal property possessed by a moving body in virtue of its motion, which property is communicated to a second body initially at rest through impact. And it is within this common framework of agreement, in particular, that the Cartesians and Leibnizeans can debate whether the force of a body moving with velocity v should be measured by mv or mv^2—both sides agree, that is, that the force in question is a property possesssed by the moving body in virtue of its motion. On the Newtonian conception of force, by contrast, force is the *cause* of motion: moving force is an external action exerted by one body on another body so as to change the state of motion of the latter. As such, the force exerted by a body (for example, gravitational attraction) is entirely independent of this body's state of motion—being exerted whether at motion or at rest—and only the effect of such a force involves motion (namely, change of motion of the body acted upon by the force). What happens during impact, then, is not communication of force (as the effect of motion) from one body to another, but rather action and reaction of repulsive forces (as causes of change of motion) possessed by the two bodies independently of their states of motion.[21]

---

goal in principle (164,8ff.), has supplied as a whole merely a phoronomy but no dynamics, and thus appears to provide nothing more than Newton's *Principia mathematica* does also (352,4ff.; I will enter in detail into this aspect later). Expressed in a brief formula, this critique therefore asserts that the [*Metaphysical Foundations*] is identical with its first chapter—the content of the remainder, in so far as it is still useful, reduces to phoronomy" ([113], p. 93; all references here are to volume 21). For Tuschling's specific discussion of Proposition 1 of the Dynamics, see [113], pp. 96–98. (Compare also Förster [31], p. 550.)

21. For a detailed discussion of the emergence of the Newtonian conception of force see Westfall [119]. For a discussion of the interplay between force as cause of motion and force as effect of motion in the eighteenth century see Heilbron [47], chap. I. It is noteworthy that Robert Boyle, in his well-known essay of 1674, *Excellence and Grounds of the Corpuscular or Mechanical Philosophy,* characterizes the common root of the mechanical philosophy (which, after all, has to embrace both the plenum of Descartes and the void of Boyle and Gassendi) in exactly these terms: all forces and powers of bodies arise from local motion alone, that is, by impact or pressure—the "active principles" of the Platonists, the Aristotelians, and the Paracelsans are to be entirely rejected. A nice discussion of Boyle's essay, including a discussion of how the Newtonian natural philosophy, in turn, is to be seen as a rejection of the fundamental principles of the former, can be found in the Introduction to Cantor and Hodge, eds. [16].

Kant's dynamical or physiological moving forces are forces in this second, Newtonian sense—that is, causes rather than effects of motion: "The movable in space in so far as it (itself and other matters) has moving force. What does force mean here?—that which contains the active *cause* of motion and thus also the reaction of the latter" (21, 507.28–508.2). Similarly, Kant characterizes the relevant distinction between the two types of moving forces as follows:

> Now there are two types [of moving forces]: namely 1. those that follow from actual motion, e.g., the central forces of a body rotating in a circle; or 2. those that precede motion as causes.—The former contain the mathematical ⟨(as in Newton's immortal work, *philos. nat. princ. mathem.*)⟩, the latter the physical principles of natural science. The former objects are *impressed* forces *(vires impressae),* the latter are forces belonging to ⟨the nature of⟩ matter *(vires connatae).* If the latter are given—e.g., attraction as moving force of gravitation, or the moving forces of light, sound, or fluids in general—then the mathematics applied thereto is not a particular part of the natural science of the moving forces as objects thereof, but rather a particular type of doctrine for scientifically handling them. (616.8–20)

As the reference to gravitational attraction makes clear, it is precisely force in the Newtonian sense that is paradigmatic of the dynamical, physiological, or physical forces *(vires connatae).*[22]

What here creates confusion, however, is the circumstance that Kant also mentions Newton's *Principia* as paradigmatic of merely mathematical principles of natural science—and, in particular, he mentions the Newtonian theory of central forces in this regard. If the Newtonian theory of central forces is relegated to the mathematical principles of natural science, how then can Kant, in drawing the distinction between mathematical and dynamical moving forces, possibly be endorsing the Newtonian conception of force?

As a first step towards untangling this question, we must observe that central forces can be analyzed from two very different points of view. From one point of view we consider only aspects of the orbiting body's motion: to speak of centrifugal force is to speak of the body's inertial motion (velocity) directed along the tangent, to speak of centripetal force

22. Compare 286.19–287.2: "If one still speaks nonetheless of mathematical foundations of natural science (as in Newton's *philosophiae naturalis principia mathem.*) then here the moving forces are presupposed as belonging to physics—e.g., gravitation, light-, sound- and water-moving [forces]—and it is not taught how motion arises from the moving forces but rather how certain forces arise from motion (e.g., with the central forces of bodies moved in circles); and it is therefore only a part of physics that can be handled mathematically—namely how certain moving forces bring forth motions according to determinate laws and how to determine the form of the latter."

is to speak of the body's accelerated motion directed towards the center (thus centripetal force = centripetal *acceleration* here). From a second point of view we consider also the force that causes the motion—whether it be the gravitational force due to a central body around which the moving body rotates, the elastic force in a sling used to twirl the rotating body, the pressure (directed towards the center) of an external aether, or whatever.[23] The first point of view is indeed merely mathematical; the second, however, is truly physical:

> ⟨But the mathematical foundations presuppose physical [foundations]: e.g., central force of an attracting body at the center of the circle around which a body is rotated. Therefore motions of this kind do not proceed from purely mathematical principles.⟩ The first occurs, e.g., in the doctrine of central forces where a sling-stone is set in rotation, or the moon is drawn by gravitational attraction to the center of the earth but at the same time is represented as striving to distance itself from the center along the tangent to its orbit—where then motion precedes the moving force as cause of the latter . . . . (22, 164.29–165.9)[24]

Thus, if we use "central force" to denote the (Newtonian) force responsible for the centripetal acceleration, we thereby necessarily go beyond the merely mathematical moving forces.[25]

Considered from the first point of view, moreover, the orbital motion in question is treated as a one-body problem: we consider only the body actually rotating, and the central point about which the rotation takes place is treated as a fixed, merely mathematical point. This point of view dominates most of Book I of *Principia* (namely, §§I–X), in fact, and is clearly articulated by Newton in his comments on Definition VIII:

> I likewise call attractions and impulses, in the same sense, accelerative, and motive; and use the words attraction, impulse, or propensity of any sort towards a centre, promiscuously, and indifferently, one for another;

23. Here there is of course an important asymmetry between the two aspects of motion (centrifugal and centripetal) considered from the first point of view. For only the second is a *true acceleration,* and therefore, according to precisely the Newtonian conception of force, only the second is correlated with a true external force as cause.

24. Of course in the case of centrifugal force the rotational motion literally is the cause of the *(pseudo-)* force: the tendency to flee from the center is due to nothing but the motion itself, and there is consequently no true (Newtonian) force responsible for this effect. Centripetal *acceleration,* by contrast, necessarily has an external cause: namely, a true (Newtonian) force.

25. The key point is clearly expressed at 21, 627.5–9: "The principles of the derivation of the moving forces of matter from motion ⟨e.g., those of the central forces,⟩ are *mathematical.* But those that derive motion from the moving forces are *physiological:* e.g., those of attraction through gravitation."

> considering those forces not physically, but mathematically: wherefore the reader is not to imagine that by those words I anywhere take upon me to define the kind, or the manner of any action, the causes or the physical reason thereof, or that I attribute forces, in a true and physical sense, to certain centres (which are only mathematical points); when at any time I happen to speak of centres as attracting, or as endowed with attractive powers. ([82], pp. 45–46; [83], pp. 5–6)

Thus, in these sections of Book I we are not really dealing with dynamics at all: no use is made of the concept of mass (the rotating body too can be treated as a mere mathematical point) nor therefore of the second or third laws of motion. Indeed, although Newton frequently appeals explicitly to the first law (for example, in the proof of Proposition I), such appeals are dispensable in principle, for what is actually used is simply the *definition* of accelerated motion: what is actually being shown, in other words, is that if a body moves in a certain type of orbit then its acceleration has certain properties or takes a certain form. The notion of centripetal force, therefore, is here used merely for centripetal acceleration, and the entire discussion falls within kinematics (or phoronomy).[26]

A decisive break from mere kinematics is made in §XI of Book I, which begins with a consideration of the two-body problem. Newton signals this break with his introductory remarks:

> I have hitherto been treating of the attractions of bodies towards an immovable centre; though very probably there is no such thing existent in nature. For attractions are made towards bodies, and the action of the bodies attracted and attracting are always reciprocal and equal, by Law III; so that if there are two bodies, neither the attracted nor the attracting body is truly at rest, but both (by Cor. IV of the Laws of Motion), being as it were mutually attracted, revolve about a common centre of gravity. And if there be more bodies, which either are attracted by one body, which is attracted by them again, or which all attract each other mutually, these bodies will be so moved among themselves, that their common centre of gravity will either be at rest, or move uniformly forwards in a right line. I shall therefore at present go on

26. Thus, for example, Propositions I and II together assert that a motion describes equal areas in equal times with respect to a given fixed point if and only if its total acceleration is directed towards that point; Proposition XI asserts that if an *elliptical* motion describes equal areas in equal times with respect to a focus of the ellipse, then the total acceleration is directed towards the given focus and is inversely proportional to the square of the distance from that focus; and so on. In modern terms, perhaps the best way to make the point is to note that these Propositions involve only a spatial metric and temporal metric as elements of space-time structure: we do not require, in particular, an *affine* structure. Thus, no notion of *absolute* acceleration is really involved here, but only a kinematical analysis of purely *relative* motions: see note 29 below. (I am indebted to William Harper and Robert DiSalle for discussion of this point.)

to treat of the motion of bodies attracting each other; considering the centripetal forces as attractions . . . . ([82], p. 266; [83], p. 164)[27]

Here the application of the laws of motion, the third law of motion, in particular, is absolutely essential; and here the full Newtonian conception of force—which is, in effect, defined by these laws of motion—appears for the first time. We have at this point, and only at this point, crossed the boundary between pure mathematical kinematics and true physical dynamics.

Moreover, we see precisely such a progression from mathematical kinematics to physical dynamics in the central argument of *Principia:* the argument for universal gravitation of Book III.[28] Newton begins with a purely mathematical description of the relative accelerations found in various systems of rotating bodies throughout the solar system: the relative accelerations of the moons of Jupiter and Saturn with respect to their primary bodies, of the planets with respect to the sun, and of the moon with respect to the earth. All of these accelerations are of course directed towards the central bodies in question and obey the inverse-square law. The next step is to assume that the indicated frames of reference can be considered as approximately inertial (that the laws of motion are approximately valid therein), so that the given accelerations can be correlated

27. The passage continues: "though perhaps in a physical strictness they may more truly be called impulses. But these Propositions are to be considered as purely mathematical; and therefore, laying aside all physical considerations, I make use of a familiar way of speaking, to make myself the more easily understood by the mathematical reader." Thus Newton here continues in the same agnostic, purely mathematical, vein as in the first passage—no doubt in order to forestall qualms about action-at-a-distance. Unfortunately, Newton thereby becomes entangled in a significant mistake—a mistake that becomes perfectly explicit in the Scholium to Proposition LXIX, where Newton asserts that: "I here use the word *attraction* in general for any endeavor whatever, made by bodies to approach to each other, whether the endeavor arise from the action of the bodies themselves, as tending to each other or agitating each other by spirits emitted; or whether it arises from the action of the ether or of the air, or of any medium whatever, whether corporeal or incorporeal, in any manner impelling bodies placed therein towards each other" ([82], p. 298; [83], p. 192). This is a mistake, because the third law of motion *cannot* be applied (as it is crucially in the proof of Proposition LXIX) to the "attractive" accelerations of two bodies impelled towards one another by impulse or pressure—of a surrounding medium, say—for conservation of momentum must then take into account the impulsive agent as well: applying the third law to the "attraction" of two bodies requires that this attraction be true and direct, not merely apparent and mediate. For further details here, see the references cited in note 29 below. (I am indebted to Howard Stein for pointing out the mistake in the Scholium to Proposition LXIX.)

28. Cohen, particularly in §3 of Part One of [19], sees the progression from the one-body problem (pure mathematical kinematics), through the two-body problem (third law of motion), to the many-body problem (universal gravitation), as paradigmatic of what he calls "the Newtonian style."

with (Newtonian) forces (at this stage, for example, the inverse-square acceleration of the moon is ascribed to terrestrial gravity). We then suppose that the forces in question are exerted by each primary body directly on its satellites (immediate attraction at a distance), and it thereby follows by the third law of motion that these satellites, in turn, must attract their primary body as well. Finally, we suppose that each body in the solar system attracts every other by precisely the same force (universality). It now follows, by the third law of motion once again, that the force between any two such bodies is not only inversely proportional to the square of the distance between them, but also directly proportional to the product of their masses. This last result now enables us to determine the masses of the primary bodies and thus to determine the center of mass (center of gravity) of the solar system: it turns out, of course, to be sometimes within, sometimes without the surface of the sun, but never very far from the sun's center (Prop. XII).[29]

Now, in the *Opus* there are very clear indications that Kant's effort to distinguish "mathematical foundations of natural science" from "philosophical foundations of natural science" is directed, not against Newton, but rather against precisely the mechanical natural philosophy (which philosophy constituted Newton's principal opponent). And there are clear indications, moreover, that the argument for universal gravitation of Book III of *Principia* is taken as paradigmatic of what is at stake here. Thus, in the early part of 1799 (hence immediately before the work on Transition 1–14) Kant writes:

> ⟨The system of the universe is based on universal attraction of all matter at all distances in empty space. Is this a mere empirical proposition or does it also have an a priori principle as its basis? Why had Huygens not yet proved [gelehrt] this from observation according to Kepler's rule? Newton first introduced the word *attraction*⟩. (22, 269.4–8)
>
> ⟨The *philosoph. Natur. princip. Mathematica* proceed only up to the doctrine of Huygens of the living [belebten] forces that arise from motion.

29. The best analysis of this extraordinarily subtle argument is Stein [104]. For detailed discussion of the argument in the context of Kant's treatment of space, motion, and gravitation in the *Metaphysical Foundations* and in the *Prolegomena* see Chapters 3 and 4 above, and also [35]. I attempt to read the *Metaphysical Foundations* as concerned primarily with the problem of absolute space and absolute motion, and as taking Newton's procedure in Book III as the solution to this problem. Kant reads Newton in Book III as *constructing* a privileged frame of reference—the center of mass frame of the solar system—relative to which the concept of true (as opposed to apparent) motion is first *defined*. In this way, certain crucial principles involved in the construction are understood as a priori conditions for applying the concept of true motion to experience: in particular, the laws of motion and the universality and immediacy (direct action-at-a-distance) of gravitational attraction turn out to have this a priori status.

> Newton first brought into consideration *principia dynamica* [involving] a *particular attraction* (which is not mere *phaenomenon*).〉 (608.26–609.4)

It is therefore Huygens, the great proponent of the mechanical natural philosophy, who is here said to be limited to merely mechanical principles.

Huygens, of course, provided the first correct mathematical treatment of centrifugal force.[30] Conceiving the force generated by twirling a weight attached to the end of a string as analogous to the force of gravity exerted by a weight suspended from a string, Huygens showed that the former force is directly proportional to the square of the (linear) velocity of the weight and inversely proportional to the radius of circular motion: $F_{\text{centrifugal}} \propto v^2/r$. Centrifugal force was thereby conceived as a real force present in the body and arising from its motion, which, if the body is not actually to flee from the center, requires *counterbalancing* by an opposing centripetal force striving towards the center: the physical connection to the string or, in the case of orbital (planetary) motion, the pressure exerted by the aether. This counterbalancing centripetal force must therefore also obey the formula $F_{\text{centripetal}} \propto v^2/r$. The problem to which Kant alludes in the first passage is then the following. From Kepler's laws of planetary motion (in particular, from Kepler's third or harmonic law) and Huygens's formula for centrifugal (and hence centripetal) force one can easily derive the inverse-square law.[31] Why, then, did Huygens not discover the theory of gravitation?

In the outline-drafts of the tenth and eleventh fascicles, written in late 1799 through early 1800 (hence immediately succeeding the work on Transition 1–14), Kant develops a clear, and I think extremely penetrating, answer to this question: Huygens's commitment to the mechanical natural philosophy prevented precisely the required progression from a purely mathematical to a truly dynamical treatment. Thus in the earlier drafts we find:

> Attraction and repulsion.—Without contact or in contact. (Remark: As Kepler's 3 analogies, with which [Huygens] was so near to bringing about Newton's *philos. nat. pr. mathem.* according to Huygens's mechanical principle, it was still not discovered by him—because he did not [appeal] to the dynamical (missing the *world-attraction*) but rather remained always with repulsive forces). (22, 315.15–20)

30. For Huygens's treatment of centrifugal force see, e.g., Westfall [119], chap. IV.

31. Kepler's harmonic law states that if r, R and t, T are, respectively, the mean distances and periods of two concentric orbits, then $r^3/R^3 = t^2/T^2$. Assuming circular orbits for simplicity, and using the fact that here $v = 2\pi rt$, it follows from Huygens' formula that the two centripetal forces (accelerations) satisfy $f/F = 1/r^2 : 1/R^2$. Kant poses the problem in precisely these terms at 21, 68.14–20.

⟨Galilei, Kepler, Hugenius and Newton.
Huygens's transition from the metaph. F. of N.S. to the mathematical [foundations], and Newton's to physics—merely through the concept of gravitational attraction which Kepler did not think of.⟩ (353.9–12).

The point, I think, is that Kepler's laws, although they certainly yield the inverse-square law (of centripetal acceleration) by a purely mathematical (purely kinematical) analysis, in no way suffice for a derivation of the law of universal gravitation. For this we additionally require: first, the Newtonian conception of force, as defined by the laws of motion, and second, the conception of an attraction that is universal (exerted by each body on every other body) and acts immediately at a distance (with no mediation, in particular, by the pressure of a surrounding aether).[32] Yet both of these conceptions are directly in conflict with the fundamental principles of the mechanical natural philosophy.

So long, therefore, as we remain within the mechanical natural philosophy—and thus, in the end, within mere mathematical kinematics—we can never approach the true cause (true Newtonian force) responsible for the planetary motions:

> It is, namely, a remarkable appearance in the field of science that there was a moment where its progress appeared to be terminated, where the ship lay at anchor and there was nothing further to be done for philosophy in a certain field.—*Kepler's* three analogies* had enumerated the phenomena of circular motion of the planets completely, ⟨although still only empirically,⟩ and mathematically describe them without yet [providing] ⟨an intimation of⟩ the *moving forces,* ⟨together with their law,⟩ which may be the cause thereof.
>
> Instead of Kepler's *aggregation* of motions containing empirically assembled rules, Newton created a principle of the system of moving forces from active causes. Unity[.]
>
> ⟨Hugenius had also been able to name centripetal and centrifugal forces and still Newton's theory was not discovered. Attraction belonged to it.⟩ (521.11–23)

In other words, only Newton makes the decisive step from mechanical or mathematical moving forces to dynamical moving forces:

> The *laws* of motion were sufficiently established through Kepler's three analogies. They were altogether mechanical. Huygens knew also the compos-

32. This last conception is absolutely essential, for without it we have no license for applying the third law of motion (the equality of action and reaction) directly to the interaction between two spatially separated heavenly bodies. See notes 27 and 29 above. It is worth noting, moreover, that Huygens himself had no quarrel with Newton's derivation of the inverse-square law; where he balked was precisely at Newton's introduction of a universal attraction at a distance: see, e.g., Westfall [119], pp. 187–188, and Stein [104], pp. 177–180.

> ite ⟨yet derivative⟩ motion through the forces that flee and continually strive towards the center *(vis centrifuga et centripeta);* but as close as both [were] (for Galileo had long before already provided the law of gravity of falling bodies at heights associated with an ⟨approximately⟩ equal moment of their fall)—yet all that was erected was mere empiricism of the doctine of motion and always a universal and properly so-called principle was lacking: i.e., a concept of reason from which one could infer a priori, as from a cause to an effect, a law of force-determination; and this explanation was given by Newton, in that he called the moving force *attraction*—whereby it was emphasized that this cause is effected from the body itself immediately, and not through communication of motion, to another body, and thus not mechanically but rather purely dynamically. (528.15–529.5; and compare the closely related passages at 492.27–493.3, 513.3–25, 515.11–517.15, 518.10–20, 519.13–21, 520.9–5, 522.1–14, 523.7–18, 531.3–8, 538.26–539.4)

It is Newton, then, who thereby makes the transition from mathematical foundations of natural science to philosophical foundations: "Hereby the principles of natural science *(scientiae naturalis s. naturae scientia)* were established, ⟨as they must be,⟩ as belonging to philosophy—in which the mathematical is taken, not as immediately ⟨(direct)⟩ belonging to the system as a constituent, but rather only as means (indirect) and instrument for measurement" (517.3–7); "⟨Through philosophy—hence not through mathematics—Newton has made the most important conquest⟩" (513.24–25).

This last idea now enables us to come full circle, for the indicated philosophical foundation for the Newtonian "conquest" is provided by Kant himself in 1786. As we observed, the Newtonian argument requires, in addition to the purely kinematical analysis of Book I, the Newtonian conception of force as defined by the laws of motion, on the one hand, and the conception of a universal attraction acting immediately at a distance, on the other. The *Metaphysical Foundations* of 1786 is noteworthy for its philosophical elaboration and defense of both of these conceptions.

Thus the laws of motion—in the form of Kant's three laws of mechanics—are derived a priori from the three analogies of experience in the Mechanics of the *Metaphysical Foundations.* It is clear, moreover, that the conception of moving force developed there is exactly the Newtonian conception. The proof of the law of inertia (Prop. 3) emphasizes that any deviation from uniform rectilinear motion must have an *external* cause (4, 543.22–33), and the appended Observation stresses that the very possibility of proper natural science rests on the idea of the *lifelessness* of matter (and thus on the conception of force as external cause of change of motion)—which idea is explicitly opposed to *hylozoism* (544.2–30: apparently in opposition to the Leibnizean notion of "living force" or *vis viva*). Furthermore, the Observation to the first Definition of the Mechanics argues that mechanical moving forces—in the consideration of which all interacting bodies are viewed as necessarily in motion—presuppose dynamical moving forces that are active also in a state of rest:

> It is clear, however, that the movable would have no moving force *through its motion,* if it did not possess original-moving forces, through which it is active before all proper motion in any place where it is to be found . . . . (536.15–18)

> Therefore all mechanical laws presuppose dynamical [laws], and a matter, as moved, can have no moving force except by means of its repulsion or attraction, on which and with which it immediately acts in its motion and thereby communicates its proper motion to another. (536.23–537.4)

This conception of the relationship between mechanical and dynamical moving forces is in complete agreement with that of the *Transition* project.[33]

The conception of universal attraction acting immediately at a distance is given an a priori foundation in the Dynamics of the *Metaphysical Foundations.* Proposition 5 introduces the original force of attraction as a fundamental force "belonging to the essence of matter" (508.12–509.12). Propositions 7 and 8 then establish the two crucial properties of immediacy (direct action-at-a-distance) and universality:

> Proposition 7: *The attraction essential to all matter* is an immediate action of one matter on others through empty space. (512.17–19)

> Proposition 8: The original force of attraction, on which the very possibility of matter as such rests, extends in the universe from each part of matter to every other immediately to infinity. (516.22–26)

For Kant, therefore, in sharp contrast to Newton's own procedure at this point, these two crucial properties certainly do not here have a merely inductive or hypothetical status.[34] In any case, however, Kant's procedure

33. For example, from early 1799: "⟨Mechanical forces are those which are determined to motion as instruments of another matter; *dynamical* [forces] are those through which it is immediately determined. The former require the latter and are only possible through them as locomotive forces . . . ⟩" (21, 206.1–4); "⟨*Dynamical* [forces] at rest, *mechanical* from actual motion . . . ⟩" (206.22). Indeed, Kant is clearly in possession of this Newtonian, truly dynamical conception of force from at least 1756—when, in the Preface to the *Physical Monadology,* he contrasts "metaphysics" (the Leibnizean-Wolffian natural philosophy) with "geometry" (the Newtonian natural philosophy): "[Geometry] shows most exactly that attraction or universal gravitation is hardly to be explained by mechanical causes but rather from forces inherent in bodies that are active at rest and at a distance, [metaphysics] reckons this among the empty playthings of the imagination" (1, 475.27–476.2). For further discussion see the Introduction above.

34. Again, the basic idea of Kant's argument here is, I think, that these two properties

here and in the Mechanics is entirely consonant with the Newtonian conception of force: whereas the Mechanics provides the general framework for reasoning about any forces whatsoever (the laws of motion), the Dynamics introduces two particular such forces (the original forces of attraction and repulsion). Moreover, this structure or relationship mirrors—and is in turn mirrored by—the more general structure found in the first *Critique:* whereas the analogies of experience (to which the Mechanics corresponds) provide the most general framework for reasoning about any causal relations whatsoever (the concepts of substance, causality, and community), the anticipations of perception (to which the Dynamics corresponds) provide a priori constraints on the content (which can be given only empirically) of such relations—which content includes, paradigmatically, particular moving forces (A207/B252).[35]

Hence, when the *Metaphysical Foundations* derives moving forces from motion, as in Proposition 2 of the Phenomenology, this in no way contradicts the dynamical conception of force articulated in the *Transition* project. For what is inferred from circular motion here is itself a truly dynamical force in precisely the Newtonian sense: an external force—such as the attraction exerted by a central body—acting as an external cause via the law of inertia (Kant's second law of mechanics) so as to change the orbiting body's state of motion (in this case, its direction). There is no question here, in other words, of viewing the force in question as a mere effect of circular motion.[36] We are thus presented with an epistemic derivation of moving forces from motion, not a causal derivation, and this procedure is perfectly consistent with the manner in which the relationship between Kepler, Huygens, and Newton is conceived in the *Transition* project:

---

cannot be merely inductive because they (like the laws of motion) are necessary for *defining* the notion of true (as opposed to apparent) motion in the first place: without these two properties the Newtonian construction of a privileged center of mass frame necessarily fails, and hence the statement of the law of universal graviation is deprived not only of truth but also of *objective meaningfulness.* See, once again, note 29 above.

35. Although the Dynamics provides a priori constraints on the fundamental forces, it is also clear that neither the real possibility of these forces nor the particular laws according to which they act can be known a priori: see 524.18–525.25, 534.15–18.

36. Kant, in his later observations on Proposition 2, makes clear that what is at issue here are "active dynamical influences given through experience (gravity, or a stretched string)" (562.12–13); and it is noteworthy, furthermore, that he characterizes *centrifugal* force as due only to "actual motion—and in fact without any dynamical repulsive cause (as one can see from the example chosen by Newton at *Princ.Ph.N.* pag. 10 Edit. 1714*)" (562.4–8: the reference, to the second edition of 1713, is to the two rotating globes connected by a stretched string discussed at the end of the Scholium to the Definitions—[82], p. 53; [83], p. 12).

> ⟨The question is whether one could have inferred from the Keplerian analogies to the universal attraction of all matter or must have only contrived a hypothesis thereto.
>
> Mathematics appears to have here acted on philosophy as an instrument, as it were, through a postulate.
>
> One would have thought that, on the basis of the completely presented appearances, which constituted a system (although only as observation and experience), it could not fail to come about that the cause for the law that such appearances suggests would appear, and in fact according to the principle of identity. However, it is a *synthetic,* not an analytic, procedure to infer from the consequence to the ground that is the unique possible cause thereof.⟩ (22, 523.7–18)

Kant does not, of course, reject the inference from Kepler's laws to universal gravitation; his concern is rather to emphasize that this inference cannot proceed purely mathematically—it requires a prior, philosophically grounded, dynamics. Given such a dynamics, however, mathematics can serve as an indispensable instrument of philosophy in effecting precisely this Newtonian inference:

> Attraction and repulsion are the acts of the agitating forces of matter which contain a priori a principle of the possibility of experience and the step to physics, and it belongs to the metaphysical foundations of natural science and thus to philosophy to use mathematical principles as an instrument for the sake of philosophy in regard to the relations of given forces of matter, and [to proceed] from the Keplerian formulas (the three analogies) to the moving forces that act in accordance with them [and] to ground the system of universal gravitation by the original attraction, or motion in light and sound for optics and acoustics by repulsion, and so also with other relations of force to ground physics. (518.10–20)

It is clear, therefore, that the distinction Kant draws in the *Opus* between mathematical and dynamical moving forces, together with the closely related distinction between mathematical and philosophical foundations of natural science, cannot be read as a rejection and/or refutation of the fundamental principles of the *Metaphysical Foundations.* On the contrary, concerning this issue at any rate, the two works exhibit a remarkable agreement.

What, then, is lacking in the *Metaphysical Foundations?* Why should it be necessary to go beyond this work to the new project of the *Transition?* The answer, I think, is actually quite straightforward: the *Metaphysical Foundations* is correct as far as it goes (but see note 13 above), the problem is that it simply does not go far enough. Kant articulates the issue most clearly in the course of the outline-drafts from 1799–1800 on which we have recently been focusing:

⟨The transition to physics cannot lie in the metaphysical foundations (attraction and repulsion, etc.), for they yield no particular determinate properties specified [anzugebende] by experience, and one can contrive nothing specific of which one could know whether it is also in nature—or whether the existence of such may be provable—but rather can devise [such] only empirically or hypothetically to explain phenomena for a certain purpose.⟩ (282.12–18)

Whereas the *Metaphysical Foundations* deals with the universal forces of matter in general (the original forces of attraction and repulsion), it says nothing at all about any additional, more specific forces of matter—which, therefore, as far as the *Metaphysical Foundations* is concerned, are left solely to empirical physics. As far as the *Metaphysical Foundations* is concerned, any additional, more specific forces are thus left entirely without an a priori foundation, and the task of the *Transition* is to fill precisely this lacuna.

Earlier, in 1798, Kant writes:

This concept [of the *Transition*] is not that of matter in general (the *movable* in space) but rather of the *moving forces* of matter according to particular laws of motion (of experience), whose specific variety, however, can be cognized a priori as active causes through possible relations in space (of attraction and repulsion) as terms of the classification of motion. (21, 286.1–6)

The transition from the metaphysical foundations of natural science to physics must not consist wholly in a priori concepts of matter in general, for then it would be mere metaphysics ⟨(e.g., where merely attraction and repulsion in general are in question)⟩, but it must also not consist wholly of empirical representations, for then they would belong to physics (e.g., observations of chemistry)—[it belongs] rather, to a priori principles of the possibility of experience and thus to natural investigation [Naturforschung] . . . . (362.28–363.5)

⟨The particular moving forces of matter can be cognized only through experience and have a tendency to physics. In order to bring the latter about, however, principles of natural investigation [Naturforschung] (not logical [principles] applying to the subject in regard to method but elementary concepts that concern the object) must precede. To the transition belongs the anticipation of certain empirical representations that are required for the possibility of physics as a system.⟩ (530.18–24)

The problem of the *Transition* is therefore to establish something a priori, not solely concerning the two fundamental forces of attraction and repulsion as universal properties of matter in general (for the *Metaphysical Foundations* has already taken care of this problem), but rather concerning the rest of the moving forces that may be found in nature. The problem, in particular, is that these latter forces, precisely because they are specific and do not characterize matter in general, also do not appear susceptible to an a priori treatment.

What does Kant have in mind by such specific forces of matter? Some revealing examples are given in the earliest parts of the *Opus:*

> Chemical attraction is partial or biased, the world-attraction is impartial and is exerted merely in proportion to the quantity of matter. The former cannot be viewed as universal property of matter nor be cognized a priori according to its law.
>
> Both are general, but in regard to the last mentioned [*sic*] there is no *neutrality*—[rather] partiality, not only a limitation through space. (21, 444.16–22)

> The world-attraction is *impartial* (gravitation), extends to all distances, and is *generic.* Chemical attraction (in contact or striving thereto in *solidification*) is specific and has partiality. Local-attraction. (382.25–28)

> Attraction as surface-force is *cohesion* (attraction in contact). Attraction that proceeds also to the most distant parts is penetrating force. The latter, if it extends to all matter, as such and in general, is *gravitational-attraction;* but if it extends only to a few matters with free penetration of others (e.g., magnetism) then the attraction is *partial* and is to be called *elective-attraction* [*Wahlanziehung*] (yet still penetrating). ⟨The impartiality of the former attraction rests on a—matter in general.⟩ (390.24–391.2)[37]

The specific moving forces of matter therefore include those responsible for chemical phenomena (for example, chemical affinities), cohesion (of liquids and solids), solidification (that is, crystallization), and magnetism. They appear, in other words, to be just those forces comprising what is standardly called "experimental physics" in the eighteenth century.[38]

The way in which Kant contrasts the universal or generic force of gravitational attraction with the specific or partial attractions of chemistry

37. There is no doubt that gravitational attraction, in virtue of precisely its universality (impartiality) can be cognized a priori and thus belongs to the *Metaphysical Foundations:* "The doctrine of the laws of moving forces of matter in so far as they can be cognized a priori is called metaphysics,—so far, however, as they can only be derived from experience, physics. But that doctrine which considers only the a priori principles of the application of the former rational [doctrine] to the empirical can constitute the transition of natural philosophy from the metaphysics of corporeal nature to physics. Thus, for example, the doctrine of an attraction at a distance in general and its magnitude in inverse ratio to the square of the distance—as one can think these concepts a priori—[belongs] to the metaphysical foundations of natural science" (310.24–311.6).

38. Such "experimental physics" is contrasted with the "mathematical physics" exemplified by rational mechanics, optics, astronomy, and the Newtonian theory of universal gravitation: see, e.g., Heilbron [47], chap. I, §1; Hankins [42], chap. III; Kuhn, "Mathematical versus Experimental Traditions in the Development of Physical Science," in [62].

suggests, in particular, the influence of Herman Boerhaave's *Elementa chemiae* of 1732. For Boerhaave distinguishes the attractive forces responsible for chemical affinities from the attractive force of universal gravitation in exactly the same terms; moreover, he characterizes chemistry as precisely the science of the specific properties of specific types of bodies (as opposed to the universal properties common to all bodies). It is for this reason, Boerhaave argues, that chemistry cannot begin with definitions and general theory, but rather must be derived solely from particular experiments concerning particular types of bodies and their interaction—which latter cannot be foreseen a priori from universally applicable principles.[39]

The point, I think, can be put more generally. The problem of the *Transition* project is not an idiosyncratic one, arising solely from internal problems in the critical philosophy; it derives, rather, from one of the most important foundational problems facing eighteenth-century science. Newton had discovered the most general laws governing any forces whatsoever (the laws of motion); he had also established, by a most remarkable "deduction from the phenomena," the laws of one particular fundamental force: the universal attraction of gravity. Moreover, Newton had thereby decisively overthrown the mechanical natural philosophy, with its merely imaginary or "feigned" explanations of natural phenomena by means of contrived figures and motions of elementary corpuscles. Yet how was this brilliantly successful Newtonian paradigm to be extended beyond astronomy and terrestrial mechanics? How, in particular, were the laws of other types of attractions, such as those responsible for cohesion or chemical affinities, possibly to be discovered?[40]

Now chemical writers of the early to mid-eighteenth century, especially Boerhaave and the school of Georg Stahl (the developer of the theory of phlogiston), were virtually unanimous in their enthusiastic endorsement of the Newtonian paradigm. The extraordinarily complex phenomena studied by chemistry could not be comprehended within the mechanical natural philosophy: that is, solely in terms of motion and its communication through impact and pressure. Specific (Newtonian) forces of attraction and repulsion—which, unlike the universal attraction of gravity, depend on the particular types of matter under consideration—are un-

39. For Boerhaave see Metzger [79], part III. *Elementa chemiae* was perhaps the most well known chemical treatise of the eighteenth century; and it was certainly known by Kant, who refers to it in §I of the *New Exposition* of 1755 (1, 390.7), in the essay on *Negative Magnitudes* of 1763 (2, 186.16), in the *Dreams of a Spirit-seer* of 1766 (2, 330.33, 331.31), and in *On the First Grounds of the Distinction of Regions in Space* of 1768 (2, 377.7).

40. For this general problem in eighteenth-century Newtonianism, see Heilbron [47], chap. I, §§2–6, and Metzger [79], part I. In the succeeding discussion of specifically chemical writers I follow Metzger.

avoidably required. Mere mechanical moving forces could thus be at most "instruments" of more fundamental dynamical (Newtonian) forces. Similarly, these chemical writers unanimously endorsed the Newtonian ideal of "experimental philosophy": explanations of natural phenomena were not to be contrived a priori in terms of hypotheses concerning particular corpuscular configurations, but were rather to be painstakingly derived from the phenomena through careful observations and experiments—where the model for such derivation was of course Newton's argument for universal gravitation.[41]

Yet there was a fundamental problem here. In the case of the universal attraction of gravity, our theory of force and the phenomena it is intended to explain fit together perfectly: we observe the planetary (relative) motions, subject them to a mathematical analysis of centripetal acceleration, use the laws of motion to infer therefrom the existence of and mathematical law governing a dynamical force, and use these same laws of motion to arrive at the reciprocity and finally the universality of this force. In the case of chemical phenomena, however, it appeared that we must forever despair of even the possibility of this kind of tight fit between theory and experiment. For how are we to infer from our experiments concerning chemical reactions—careful observations of solutions, precipitations, crystallizations, and so on—to a characterization of the attractions and/or repulsions that are presumed to underlie such reactions? The problem, of course, is that the forces presumed to be operative here are exclusively microscopic, and thus the motions that are their immediate effects are entirely inaccessible to us. Here, unlike the case of gravitational attraction, there appeared to be no way to make theory and observation fruitfully meet one another, and chemistry appeared to be therefore in danger of remaining always a merely experimental science.[42]

It is for precisely this reason, moreover, that Kant himself despairs of

41. The view of mechanical moving forces as merely "instrumental," which, as we have seen, is central to the *Transition* project, is explicitly defended by both Boerhaave and Stahl. This same kind of conception concerning the relationship between the Newtonian natural philosophy and the mechanical natural philosophy is of course also defended (with reference to broadly chemical phenomena) in the General Observation to Dynamics of the *Metaphysical Foundations*.

42. In other words, the Newtonian natural philosophy appeared ultimately to be in no better position here than the mechanical natural philosophy: in the end, both seemed to be limited to contrived, merely hypothetical, explanations. Again, for a discussion of this Newtonian dilemma in chemistry, see Metzger [79], part I. One instructive example here is that of Buffon, who attempted to make do with *only* the universal force of gravity: variations in the law of attraction specific to particular types of matter, he taught, are due solely to variations in the *figures* of the elementary corpuscles constituting the various types of matter. Since there figures are wholly inaccessible, however, we have clearly made no progress beyond the mechanical natural philosophy whatsoever.

the possibility of a truly scientific chemistry—and consequently also of a philosophical foundation for this discipline—in the Preface to the *Metaphysical Foundations:*

> So long, therefore, as there is still no concept to be found for the chemical actions of matters on one another that can be constructed—i.e., no law of approach or withdrawal of the parts [thereof] can be specified according to which, perhaps in proportion to their densities or the like, their motions together with the consequences thereof can be made intuitive and presented a priori in space (a requirement that will only with great difficulty ever be fulfilled)—then chemistry can be no more than systematic art or experimental doctrine but never a proper science, because its principles are merely empirical and permit no a priori presentation in intuition. Consequently the principles of chemistry cannot in the least be made comprehensible according to their possibility, because they are unsuitable for the application of mathematics. (470.36–471.10)

Thus, since we do not know how to mathematize the required microscopic forces, we do not know how to give chemistry a scientific foundation. Approximately ten years later, however, when well embarked on the *Transition* project, Kant has apparently found new reasons for optimism: there is a way to comprehend a priori the specific or particular moving forces of matter after all; there is a way to effect a priori a transition from the metaphysical foundations of natural science to (empirical or experimental) physics. But what is the source of this newfound optimism?

II · The *Transition* Project and Reflective Judgement

The problem of the *Transition* project, as we have seen, is to build a bridge between the a priori doctrine of the universal properties of matter in general (the *Metaphysical Foundations*) and the empirical or experimental physics of the specfic properties and interactions of particular types of matter (for example, chemistry). Such a bridge is absolutely necessary if experimental physics is ever to amount to more than a mere empirical aggregate:

> The concept of a *natural science (philosophia naturalis)* is the systematic representation of the laws of motion of outer objects in space and time, in so far as they can be cognized a priori and therefore as necessary; for the empirical knowledge thereof, which concerns only the contingent cognition of these outer appearances acquired solely through experience, is not philosophy but only an aggregate of perceptions—whose completeness as a system is nevertheless an object for philosophy.
>
> Now the highest division of natural science according to its content can be no other than that into the *metaphysical foundations* thereof, which is

> grounded wholly on concepts of relations of motion and rest of outer objects, and *physics,* which systematically orders the content of empirical knowledge—and which, as was said, cannot, with its elements, in fact securely count on *completeness,* although it has the task of working towards this.
>
> Nevertheless, there can be a relation of the one mode of cognition to the other which is situated neither wholly on a priori principles, nor on empirical [principles], but rather merely on the transition of one to the other: how, namely, it is possible for us to seek out the elements of a doctrine of nature to be based on experience and to arrange them with the completeness required for systematic classification—[how it is possible] to order [them]—and to attain a *physics* which constitutes a comparatively complete whole, and which contains neither metaphysics of nature nor physics alone but rather merely the transition of the first to the second and the step that connects both banks. (21, 402.12–403.9)

The *Transition* project, then, is to connect the a priori with that which at first appears to be merely empirical, so as to show how the latter can attain a systematic and hence scientific status after all.

It has often been observed that such a description of the task of the *Transition* project bears a striking resemblance to the way in which Kant poses the problem of reflective judgement in 1790 (First Introduction to the *Critique of Judgement*):

> Such a concept is now that of an experience *as system according to empirical laws.* For although [experience] constitutes a system according to *transcendental* laws, which contain the conditions of the possibility of experience in general, it is still possible for there to be *such an infinite manifold* of empirical laws and *such a great heterogeneity of forms* of nature, which would belong to particular experience, that the concept of a system according to these (empirical) laws must be entirely foreign to the understanding—and neither the possibility, nor still less the necessity, of such a whole can be conceived. Nevertheless, however, a particular experience that is thoroughly coherent according to constant principles also requires this systematic coherence of empirical laws, in order that it be possible for the faculty of judgement to subsume the particular under the universal, always proceeding still within the empirical, up to the highest empirical laws and the natural forms in accordance with them, and therefore to consider the *aggregate* of particular experiences as a *system* thereof. For without this presupposition no thoroughgoing law-governed coherence, i.e., empirical unity [of particular experiences] can occur. (20, 203.3–21)

The analogy is striking indeed: as the universal principles of the *Metaphysical Foundations* stand to the particular phenomena of empirical physics in the *Transition* project, so the universal transcendental laws of the understanding stand to the empirical laws of "particular experience" for the faculty of reflective judgement.[43]

In the first *Critique* the problems later assigned to the faculty of reflective judgement are discussed under the rubric of the regulative use of reason, and the examples Kant presents there are especially noteworthy:

> If we survey our cognition of the understanding in its whole extent, then we find that what is entirely peculiar in that which reason prescribes to it and attempts to bring about is the *systematic* [element] of cognition: i.e., its coherence by means of a principle. This unity of reason always presupposes an idea, namely, that of the form of a whole of knowledge, which precedes the determinate cognition of the parts and contains the condition for determining a priori the place of each part as well as its relation to the rest. This idea therefore postulates complete unity of cognition of the understanding, whereby this becomes not merely a contingent aggregate, but a coherent system according to necessary laws. One cannot properly say that this idea is a concept of the object but rather of the thoroughgoing unity of such concepts, in so far as it serves as a rule for the understanding. These concepts of reason are not derived from nature; rather, we question nature according to these ideas and consider our cognition as defective so long as it is not adequate to them. It is admitted that *pure earth, pure water, pure air,* etc. are only to be found with great difficulty. Nevertheless, the concepts thereof are still necessary (which, thus, as concerning complete purity, have their origin solely in reason) in order to determine the share appropriate to each of these natural causes in the appearance. One therefore reduces all matters to the earths (as it were mere weight), salts and inflammable beings (as force), and finally to water and air as vehicles (as it were machines through which the former act), in order to explain the chemical interactions of matters among one another according to the idea of a mechanism. (A645–646/B673–674)

A second chemical example is found a few pages later:

> Much was already accomplished when the chemists could reduce all salts to two main species, acids and alkalis, and they even attempt to view this distinction too as merely a variety or diverse manifestation of one and the same basic material. One has attempted to reduce the various kinds of earths (the material of stones and even of metals) step by step to three and finally to two; but, not yet satisfied thereby, [chemists] cannot banish the thought that suggests that behind these varieties there is nevertheless a single species—or even, indeed, a common principle for these and the salts. (A652–653/B680–681)

In view of the close relationship, sketched above, between the project of

43. This connection between the project of the *Transition* and the *Critique of Judgement* has been pursued and developed especially by Lehmann: see, e.g., "Zur Problemanalyse von Kants Nachlaßwerk," in [66].

the *Transition* and the problem of providing a priori principles that could underlie the science of chemistry, Kant's use of such examples here is of course particularly intriguing.

Indeed, it is tempting to view Kant's discovery of the principle of reflective judgement as the key to the *Transition* project: this principle, that is, may be precisely the new a priori element that now—and only now—makes it possible to go beyond the *Metaphysical Foundations*. For the *Metaphysical Foundations* has recourse only to the transcendental principles of the understanding, which it proceeds to apply to the universal (although empirical) concept of matter in general. In this procedure we are essentially involved with the schematism of the universal concepts of the understanding and therefore with *determinative* judgement:

> With respect to the universal concepts of nature, under which an empirical concept in general (without particular empirical determinations) is first possible, reflection already has its instructions in the concept of a nature in general, and the faculty of judgement needs no particular principle of reflection, but *schematizes* these concepts a priori and applies these schemata to each empirical synthesis, without which absolutely no empirical judgement would be possible. The faculty of judgement is here in its reflection at the same time determinative, and the transcendental schematism [of judgement] serves here at the same time as the rule under which given empirical intuitions are subsumed. (20, 212.7–16)[44]

With the new principle of *reflective* judgement, however, we are able to go beyond mere application of the transcendental principles—beyond mere subsumption of the empirical under laws already present in the understanding. We may reasonably hope, therefore, that we can thereby extend our a priori anticipation of empirical nature further than anything envisioned in the *Metaphysical Foundations*.[45]

44. From the *First Introduction*—in the published Introduction the distinction between reflective and determinative judgement is explained as follows: "The faculty of judgement in general is the faculty of thinking the particular as contained under the universal. If the universal (the rule, the principle, the law) is given, then the faculty of judgement which subsumes the particular under it (even if, as transcendental faculty of judgement, it specifies a priori the conditions according to which alone [the particular] can be subsumed under this universal) is *determinative*. But if only the particular is given, for which the universal is to be found, then the faculty of judgement is merely *reflective*. The determinative judgement under transcendental laws, given by the understanding, is only subsumptive; the law is prescribed to it a priori, and it therefore has no need to think a law for itself in order to be able to subordinate the particular in nature to the universal . . . " (5, 179.19–31).

45. Compare Förster [31], pp. 545–546: "That any science proper has to exhibit, besides apodictic certainty, also the form of a system, Kant knew for a long time. He also knew that, like apodictic certainty, systematic unity cannot be gained empirically. What he did not know, at the time of writing the [*Metaphysical Foundations*], was how such systematic form might be anticipated a priori . . . . This situation was bound to change with the

There is a second, closely related source for new optimism here. Since the procedure of the *Metaphysical Foundations* is intimately involved with the schematism of the transcendental principles of the understanding—where this schematism, moreover, is directed towards the objects of *outer* sense—the a priori content thereby generated is necessarily connected with mathematics: namely, with the pure doctrine of motion. The metaphysics of corporeal nature is necessarily the foundation of a *mathematical* physics:

> However, in order that the application of mathematics to the doctrine of body, through which alone it can become natural science, be possible, principles for the *construction* of the concepts that belong to the possibility of matter in general must be stated first. Therefore, a complete analysis of the concept of matter in general must be taken as the basis, and this is a task for pure philosophy. For this purpose the latter requires no particular experiences, but only that which is met with in the isolated (although intrinsically empirical) concept itself, in relation to the pure intuitions in space and time, according to laws that already essentially follow upon the concept of nature in general; hence it is a genuine *metaphysics of corporeal nature.* (4, 472.1–12)[46]

Moreover, as we have seen, it is for precisely this reason that the *Metaphysical Foundations* is unable to envision an a priori foundation for chemistry: we are unable to specify mathematical microscopic force laws (analogous to the law of gravitational attraction) which would connect chemistry appropriately with the pure doctrine of motion.

Yet we are now in possession of an a priori principle (of reflective judgement) which does not proceed by such (mathematical) schematization of the transcendental principles of the understanding:

---

discovery of reflective judgement as an autonomous faculty. . . . This principle [of reflective judgement] itself, it may be worth pointing out, does not belong to the Transition, and certainly is not one of its parts. What it does, rather, is *prepare the ground* on which a transition from the metaphysical foundations of natural science to physics may first of all be carried out. For only if a principle of formal purposiveness can be set alongside Kant's general theory of matter, does something like an a priori 'Transition,' which goes beyond the [*Metaphysical Foundations*], become even possible." Note that Förster does not commit himself here to the idea that the principle of reflective judgement is the key to the *Transition* project. He does appear to suggest, however, that Kant's discovery of this principle is the crucial new event which first makes it possible to envision an extension of the *Metaphysical Foundations.*

46. Compare 470.13–15, 476.7–477.13, 478.21–31. Note that the quoted passage does not *identify* metaphysics of corporeal nature with mathematics. Indeed, in connection with 469.12–470.12, the point of this paragraph is precisely *to distinguish* the two: the task of metaphysics is not to do mathematics but rather to explain a priori the possibility of its application (see A733/B761, where Kant explicitly distinguishes "principles of the possibility" of mathematics from mathematics itself).

> The reflective judgement therefore operates with given appearances, in order to bring them under empirical concepts of determinate natural things, not schematically but rather *technically,* not as it were merely mechanically, as an instrument under the guidance of the understanding and the senses, but rather *artistically* [*künstlich*], according to the universal but yet indeterminate principle of a purposive ordering of nature in a system—as it were to favor our faculty of judgement, in the suitability of particular laws (about which the understanding says nothing) to the possibility of experience as a system, without which presupposition we cannot hope of finding our way in the labyrinth of the manifold of possible particular laws. (20, 213.23–214.8)

Once again, therefore, we might reasonably hope that this new type of a priori principle, unlike those found in the *Metaphysical Foundations,* could generate a priori content suitable for grounding even a non-mathematical science.

Now the development of Kant's thinking about the *Transition* project in the *Opus* appears to parallel the above line of thought rather closely, and this parallel emerges, in fact, in the course of Kant's reflections concerning mathematical versus philosophical foundations of natural science discussed in §I above. Recall that the issue is initially sharply posed in loose-leaf 6 of the fourth fascicle: "⟨*NB:* Of the mathematical foundations of physics. Whether this too belongs to the transition⟩" (21, 477.21–22). This remark is a marginal comment appended to an earlier passage on the same page:

> The moving forces of matter that can only be known through experience (hence not belonging to the metaph. found.) belong nevertheless to a priori concepts (thus also to metaphysics) in that which concerns their mutual relations to one another ⟨in a whole of matter in general⟩, if I understand under the moving forces only the motion itself: which is then, mathematically considered according to its direction and degree, *attraction* and *repulsion* either of the parts of matter with respect to one another or of one matter in relation to another outside of it—density and levity, and the like—which one thinks a priori at will and then can search in nature for which examples thereof are to be found, and thus one designates logical places for concepts (topics) of which one can determine a priori which appearances are appropriately fitted into one or the other. (475.17–29)

The idea here seems to be that the possible moving forces of matter can be classified or enumerated a priori (although not, of course, thereby known to exist as actual) in virtue of a mathematical consideration of motion. The totality of possible moving forces form an a priori system simply in virtue of the circumstance that all moving forces, considered merely as possible motions, necessarily relate to one another in space.

Earlier, in 1797–98, this idea of a mathematical system of the moving forces is stated even more explicitly:

Matter with its moving forces under empirical laws of motion is the object of physics.—However, because motion as alteration of relations of position in space in general is also subject to a priori principles, the moving forces of matter are also best organized according to the previously drawn classification of their functions by metaphysics in accordance with the categories, in order that they bring, according to the formal [element] of composition, even empirical principles into a whole of a system. (21, 527.18–25)

⟨The mathematical foundations of natural science are the laws of the moving forces which proceed from motion. The physical [foundations] are those from which motion arises. We here have to do with the first.—Physics towards which the meta. has a natural tendency is the complex of the laws of moving forces in a system whose form must precede a priori, and which is not fragmentarily aggregated but must rather contain the moving forces united by an idea in a whole.⟩ (528.11–18)

It appears, then, that the idea of a mathematical foundation of natural science (which is therefore closely linked to the pure doctrine of motion already philosophically grounded in the *Metaphysical Foundations*) rejected in loose-leafs 3/4 and 5 of the fourth fascicle is Kant's own earlier attempt to explain how an a priori system of moving forces is possible. Whereas Kant had earlier suggested that a unified system of moving forces is possible a priori in virtue of the mathematical unity of motion as alteration in space, he now sees that a fundamentally different type of unity is required.

At the same time, moreover, in loose-leaf 3/4 Kant begins to characterize the *Transition* project in a strikingly new way: namely, as a doctrine of *natural investigation* [Naturforschung]. This characterization is introduced—as far as I have been able to ascertain, for the first time—in the following important passage:

The metaphysical foundations have a tendency towards physics as a system of the moving forces of matter. Such a system cannot proceed from mere experiences—for that would yield only aggregates which lack the completeness of a whole—and can also not occur solely a priori—for that would be metaphysical foundations, which, however, contained no moving forces. Therefore, the transition from the meta. to physics, from the a priori concept of the movable in space (i.e., from the concept of a matter in general) to the system of moving forces, can [proceed] only by means of that which is common to both—by means of the moving forces precisely in so far as they act not on matter but rather united or opposed among one another, and thus form a system of the universal doctrine of forces *(physiologia generalis)* which stands between metaphysics and physics internally and combines the former with the latter in a system—in so far as it contains for itself a system of the application of a priori concepts to experience, i.e., natural investigation. The transition is properly a doctrine of natural investigation. (21, 478.11–26)

The idea here, I think, is that the desired system of moving forces cannot proceed purely a priori (and therefore not purely mathematically), because the task of the *Transition* project is precisely to unite the a priori with the properly empirical. And this task—"the application of a priori concepts to experience"—can only be accomplished by methodological principles (not merely by mathematical principles) belonging ultimately to the faculty of judgement.[47]

The link between natural investigation and the faculty of judgement appears in §II of the First Introduction to the *Critique of Judgement:*

> This in itself (according to all concepts of the understanding) contingent law-governedness, which the faculty of judgement (only on its own behalf) presumes for nature and presupposes in it, is a formal purposiveness of nature, which we indeed *assume* in it, but whereby it grounds neither a theoretical cognition of nature nor a practical principle of freedom; nevertheless, however, a principle for the judging and investigation of nature [Nachforschung der Natur] is still given, in order to seek the universal laws for particular experiences, according to which we have to establish them to bring about that systematic connection that is necessary for a coherent experience, and which we have a priori ground for assuming. (20, 204.1–11)[48]

47. The passage from 21, 478.11–26, is crucial to Tuschling's conception of "phoronomy-critique" (see note 20 above), for it appears explicitly to assert that the *Metaphysical Foundations*—contrary to its own intentions—"contained no moving forces." Yet it is clear, I think, that Kant means to exclude from the *Metaphysical Foundations* "the moving forces of matter that can only be known through experience" (475.17–18), but *not* the original forces of attraction and repulsion: see, e.g., 310.24–311.6, 362.28–363.5; 22, 282.12–18, 518.10–20. The deeper point implicit in 478.11–26 is that even the *Metaphysical Foundations* itself cannot proceed "solely a priori"—for the *real possibility* of *any* forces can only be given empirically. See 483.24–29: "Empirical concepts, e.g., gravity, whose moving forces can be thought a priori (e.g., attraction and repulsion), although their existence can only be given through experience, belong to this topic of the transition.—This class of moving forces can belong to physiology—namely, pure [physiology], etc."; and 310.24–311.15. Compare also Kant's remarks about fundamental forces and his dynamical concept of matter in the General Observation to Dynamics of the *Metaphysical Foundations:* "in the case of fundamental forces their possibility can never be comprehended" so that "if the material itself is transformed into fundamental forces . . . all means escape us for *constructing* this concept of matter and presenting as possible in intuition what we think universally" (4, 524.39–525.13). Indeed, this is why Kant's dynamical concept of matter is an *empirical* rather than an a priori concept (see A220–224/B267–272: in the case of empirical concepts "their possibility must either be known a posteriori and empirically, or it cannot be known at all"—A222/B269–270). Thus, a *Metaphysical Foundations* that really proceeded "solely a priori" in fact "would . . . contain no moving forces."

48. Compare 203.3–21 cited above. The notion of natural investigation [Naturforschung] occurs in the first *Critique* at A694/B722—linked to the regulative use of reason and teleology—and similarly in the *Critique of Judgement* at 5, 441.34 (§85).

Natural investigation is therefore concerned with the *search* for empirical laws and thus with regulative principles—as Kant explains in 1798:

> ⟨NB: The transition must certainly not ingress into physics (chemistry, etc.). It merely anticipates the moving forces which can be thought a priori according to their form and classifies the empirically-universal only in such a way as to regulate the searching-out of experience [Aufsuching der Erfahrung] for the sake of a system of natural investigation (regulative principle).⟩ (22, 263.1–6)

Accordingly, Kant soon makes the link between natural investigation and the faculty of judgement completely explicit:

> The transition from the metaphysical foundations of natural science to physics must not consist wholly in a priori concepts of matter in general, for then it would be mere metaphysics ⟨⟨e.g., where merely attraction and repulsion in general are in question⟩⟩, but it must also not consist wholly of empirical representations, for then this would belong to physics (e.g., observations of chemistry)—[it belongs] rather, to a priori principles of the possibility of experience and thus to natural investigation: i.e., to the subjective principle of the schematism of the faculty of judgement to classify the empirically given moving forces as such according to a priori principles and thus to make the step from an aggregate of [these forces] to a system, as compilation, to physics as a system thereof. (21, 362.28–363.9)

It is thus clear beyond the shadow of a doubt that the faculty of judgement is now an essential ingredient of the *Transition* project.[49]

Does it follow, however, that the principle of reflective judgement contains the key to the *Transition* project—that this principle now makes it possible for Kant to do what he could not do in 1786? Does this principle, in other words, now enable us to give an a priori foundation for even non-mathematical sciences such as chemistry in particular and empirical or experimental physics generally? I think not, for it is extremely difficult to envision how, on the basis of the principle of reflective judgement alone, such an a priori foundation could possibly proceed.

To begin with, what *is* the principle of reflective judgement? Kant gives one formulation in §V of the First Introduction:

> The peculiar principle of the faculty of judgement is thus: *Nature specifies its universal laws to empirical* [laws], *according to the form of a logical system, on behalf of the faculty of judgement.* (20, 216.1–3)

In the same section Kant emphasizes that this principle is completely indeterminate:

49. The importance of the passage from 21, 362–363, is therefore rightly emphasized by Lehmann: [66], p. 100.

> The reflective judgement therefore operates with given appearances, in order to bring them under empirical concepts of determinate natural things, . . . according to the universal, but at the same time indeterminate principle of a purposive ordering of nature in a system. . . . Therefore the faculty of judgement itself a priori makes the *Technic of Nature* the principle of its reflection, yet without being able to explain this or even to determine it more exactly, or being able to have an objective ground of determination of the universal concepts of nature thereto (from a cognition of things in themselves), but only in order to be able to reflect according to its own subjective laws—according to its own need—although at the same time in agreement with laws of nature in general. (213.23–214.14)

Although the principle of reflective judgement postulates a systematic unity of empirical laws under the a priori laws of the understanding, it says nothing whatsoever concerning the content of such a unified system of empirical laws—that is, concerning what the laws belonging to this system will be. All that we can say a priori about this postulated unity is that our cognitive faculties are adequate for (eventually) discovering it. In this way the principle of reflective judgement prescribes a law only to itself but not to nature.[50]

Accordingly, reflective judgement yields no constitutive principles specifying the structure of the unified system of empirical laws it postulates, but only regulative principles for investigating empirical nature so as (eventually) to discover such a unified system. We thereby generate only what Kant calls "maxims of the faculty of judgement, which lie at the basis of the investigation of nature [Nachforschung der Natur] a priori" (5, 182.11–13), and, as examples of such maxims, we have:

> "Nature takes the shortest way *(lex parsimoniae);* yet it makes no leaps, neither in the sequence of its alterations nor in the coordination of specifically different forms *(lex continui in natura);* its great manifoldness in empirical laws is yet unity under few principles *(principia praeter necessitatem non sunt multiplicanda)*"; and so on. (182.19–25; see also 20, 210.20–26)

In other words, what the principle of reflective judgement actually generates here are merely the heuristic or methodological principles presented in

50. See §IV of the published Introduction: "Now this principle [of reflective judgement] can be no other than: that, since the universal laws of nature have their ground in our understanding, which prescribes them to nature (although only according to the universal concept of it as nature), the particular empirical laws, in relation to that which is still left undetermined by the former, must be considered in accordance with such a unity as if, as it were, an understanding (although not ours) had given them on behalf of our cognitive faculty in order to make a system of experience according to particular laws of nature possible. Not as if, in this way, such an understanding must be assumed as actual (for it is only the reflective judgement for which this idea serves as a principle—for reflecting, not for determining); rather, this faculty thereby gives a law only to itself and not to nature." (5, 180.18–30)

the first *Critique* as products of the regulative use of reason at A652–663/B680–691.[51]

The faculty of reflective judgement, through its regulative maxims, thus operates purely methodologically in the progress of science. Starting from given empirical data, we are directed to search always for more and more general and unified empirical laws under which to subsume such data. The postulated end-point of this search procedure is an ideal *complete science* in which all the empirical data are subsumed under a maximally general and unified system of empirical laws. Such an ideal complete science is never actually reached, however, but is only to be approached asymptotically. Moreover, as far as the faculty of reflective judgement is concerned, the ideal of a complete science is also entirely indeterminate: we can say nothing at all about its content, but only that the empirical laws to be found therein—whatever they may turn out to be—will constitute such a maximally unified system. Further to anticipate the content of the ideal complete science would be to proceed constitutively rather than regulatively and thus to prescribe laws a priori to nature—which latter is wholly beyond the power of the faculty of reflective judgement.[52]

It is extremely difficult to conceive, therefore, how the principle of reflective judgement could generate scientific content of the kind actually required in the *Transition* project: that it could help us to anticipate a priori the totality of possible empirically given moving forces, for example. In the *Critique of Judgement* itself this principle is not of course used to generate any scientific content whatsoever. On the contrary, it acquires content solely through its application to aesthetic judgement, on the one hand, and physico-teleology—and thus, in the end, physico-theology—on the other. The purely methodological application to empirical natural science is clearly secondary in this work, and, in any case, it in no way goes

51. Compare especially A663/B691: "What is remarkable about these principles, and also what alone concerns us, is this: that they appear to be transcendental, and, although they yet contain mere ideas for the observance of the empirical employment of reason—which the latter can follow only as it were asymptotically, i.e., approaching, without ever attaining them—they nevertheless, as synthetic a priori propositions, have objective but indeterminate validity, and serve as rules for possible experience, and thus can be actually employed with good success in the elaboration thereof, as heuristic principles . . . . "

52. See §IV of the published Introduction: "The faculty of reflective judgement, which has the obligation of ascending from the particular in nature to the universal, therefore requires a principle, which it cannot borrow from experience, because it is to ground precisely the unity of all empirical principles under equally empirical but higher principles, and thus the possibility of the systematic subordination of such under one another. Such a transcendental principle can therefore be given as law only by the faculty of reflective judgement to itself—not taken from outside (for it would otherwise be determinative judgement) or prescribed to nature: because reflection concerning the laws of nature adjusts itself in accordance with nature, and the latter does not adjust itself to that condition according to which we endeavor to attain a concept of it which is wholly contingent in relation to nature" (5, 180.5–17).

beyond the methodological regulative maxims already enumerated in the first *Critique.*

Indeed, how could such purely regulative methodological maxims possibly help us with the real scientific problem faced by the *Transition* project? That problem, as we have seen, concerns the lack of tight fit between theory and observation in experimental sciences such as chemistry. In sharp contrast to the paradigmatic case of Newtonian gravitation theory, there appears to be no way rigorously to connect the hypothetical microscopic forces thought ultimately to underlie the phenomena in question here with the macroscopic experimental data (concerning observed reactions, solutions, and precipitations, for example) that are accessible to us. It appears, in other words, that here our higher-level theoretical conceptions will always remain mere hypothetical contrivances, and thus that the sciences under consideration will always remain *merely* empirical or experimental. Yet the regulative maxims of reflective judgement themselves concern what Kant, in the first *Critique,* calls the hypothetical employment of reason—which, as such, has neither constitutive nor properly probative force:

> The hypothetical employment of reason based upon ideas, as problematic concepts, is properly not *constitutive,* namely not such that one can thereby, if one wants to judge in all rigor, infer the truth of the universal rule assumed as hypothesis; for how is one to know all possible consequences, which, in that they follow from the assumed principle, *prove* its universality? Rather, it is only regulative, in order thereby, so far as it is possible, to bring unity to the particular cognitions and thereby *to approximate* the rule to universality. (A647/B675)

Such a merely problematic employment of reason cannot itself lead to a more than hypothetical status for the theoretical conceptions introduced into the empirical or experimental sciences.

Finally, we should remind ourselves that, as just noted, Kant is already in possession of the relevant methodological maxims at the time of the first *Critique.* Moreover, as also observed above, Kant even has explicitly in mind there the application of these maxims to chemical theory in particular. During the very same period, however, Kant, in the *Metaphysical Foundations,* explicitly denies that chemistry is a science properly so-called and clearly has no conception whatever of what will later become the *Transition* project.[53] It follows, therefore, that the regulative maxims of

53. Compare paragraph four of the Preface to the *Metaphysical Foundations:* "in accordance with demands of reason, every doctrine of nature must finally lead to natural science and terminate there, because such necessity of laws is inseparably joined to the concept of nature and therefore must certainly be comprehended. Hence, the most complete explanation of given appearances from chemical principles still always leaves behind a certain dissatisfac-

reflective judgement cannot, by themselves, constitute the key to the *Transition* project. To be sure, at the time of the *Metaphysical Foundations* and the first *Critique* Kant has not yet considered reflective judgement as an autonomous faculty and has not yet formulated the principle of reflective judgement as an autonomous a priori principle. Yet he does consider the regulative use of reason and the relevant methodological maxims; and since, as we have seen, it is also clear that the principle of reflective judgement itself generates no new scientific content whatsoever (generating content only in connection with aesthetics and physico-teleology), it is entirely obscure how this new autonomous status can possibly account for the *Transition* project.

Hence, the principle of reflective judgement cannot, I think, explain Kant's new optimism concerning the prospects for an a priori foundation for the empirical or experimental sciences. This principle can, however, help us to understand why there is now a new *problem* for the critical philosophy: more precisely, when considered in connection with the *Metaphysical Foundations* of 1786, it can help us to understand why there is now in fact the danger of a most significant "gap" in the critical system.

Recall that the *Metaphysical Foundations* proceeds in the opposite direction from reflective judgement. Whereas reflective judgement proceeds from the bottom up, as it were, from the particular empirical facts towards the transcendental principles of the understanding; the *Metaphysical Foundations* proceeds from the top down, from the transcendental principles of the understanding to the metaphysical principles of the doctrine of body which further specify the transcendental principles through the application thereof to the empirical concept of matter.[54] As such, the *Metaphysical Foundations*, unlike reflective judgement, is constitutive, and, I would suggest, what the *Metaphysical Foundations* is constitutive

---

tion, because one can cite no a priori grounds for such principles which, as contingent laws, have been learned merely from experience." (4, 469.4–11)

54. Compare §V of the published Introduction to the *Critique of Judgement:* "A transcendental principle is that through which is represented a priori the universal condition under which alone things can be objects of our cognition in general. On the other hand, a principle is called metaphysical if it represents a priori the condition under which alone objects, whose concept must be empirically given, can be further determined a priori. Thus, the principle of the cognition of bodies as substances and as alterable substances is transcendental, if it is thereby asserted that their alterations must have a cause; it is metaphysical, however, if it is thereby asserted that their alterations must have an *external* cause: because in the first case bodies may be thought only through ontological predicates (pure concepts of the understanding), e.g., as substance, in order to cognize the proposition a priori; but in the second case the empirical concept of a body (as a movable thing in space) must be laid at the basis of this proposition—however, as soon as this is done, that the later predicate (motion only through external causes) belongs to body can be comprehended completely a priori" (5, 181.15–31).

of is precisely the empirical concept of matter. What the *Metaphysical Foundations* does, that is, is to schematize the empirical concept of matter by providing spatio-temporal conditions for the application of this concept to objects of experience: specifically, by means of the pure and applied doctrine of motion, the *Metaphysical Foundations* explains how the concept of matter as *the movable in space* can be determinately applied to experience.[55]

Moreover, the *Metaphysical Foundations* thereby provides an a priori foundation for one particular natural science: the Newtonian theory of universal gravitation. This latter, unlike the *Metaphysical Foundations* itself, is certainly an empirical science; for the law of universal gravitation is inferred from the observed "phenomena" described by Kepler's laws of planetary motion. Yet the point of the *Metaphysical Foundations* is to provide an a priori grounding for the law of universal gravitation by proving a priori the Newtonian laws of motion, on the one hand, and the two crucial properties of immediacy and universality of gravitational attraction, on the other. It is therefore possible to view Newton's argument for universal gravitation as a "deduction from the phenomena" in a very strong sense and thus to attain a more than merely inductive or hypothetical status for that law. In the case of this particular natural science, then, we actually see what it means for the understanding to prescribe laws a priori to nature, and, accordingly, we here understand exactly *how* nature can specify its universal or transcendental laws to empirical laws.[56]

In this way the *Metaphysical Foundations* provides an a priori foundation for the most general empirical concept (the empirical concept of matter) and the most general empirical law (the law of universal gravitation), which characterize and govern all matter as such—regardless of the specific differences of various distinct types of matter. Reflective judgement, by contrast, proceeds from the most specific empirical concepts and laws, and attempts always to unify and consolidate these under more and more general empirical concepts and laws. Reflective judgement thereby remains

55. Compare paragraph fifteen of the Preface to the *Metaphysical Foundations* (4, 476.7–477.13). See also the important passage at A664/B692, where Kant explains the precise sense in which the dynamical principles of pure understanding are *constitutive:* "Nevertheless these dynamical laws in question are certainly constitutive in relation to *experience,* in that they make the *concepts,* without which no experience can occur, possible a priori. On the other hand, principles of pure reason can never be constitutive in relation to empirical *concepts,* because no corresponding schema of sensibility can be given to them, and they can therefore have no object *in concreto.*" Compare also A301/B357: "That everything which happens has a cause can absolutely not be inferred from the concept of that which happens in general; on the contrary, this principle is what alone first shows one how a determinate empirical concept of that which happens can be obtained."

56. See Chapter 4 above, and also [36].

always merely regulative, for it merely aims at and searches for—without ever actually reaching—the asymptotic ideal of a maximally unified complete science. Moreover, this latter, from the point of view of reflective judgement, remains entirely indeterminate; for, as we have seen, reflective judgement cannot itself in any way specify the content of the ideal complete science. We know only that the ideal system of maximally unified empirical concepts and laws will, in some indeterminate and unexplained fashion, ultimately stand under the transcendental principles of the understanding.

But now a serious problem emerges—and, indeed, the possibility of precisely a "gap" in the critical system. For, from the point of view of the *Metaphysical Foundations,* the ideal of a complete science is *not* entirely indeterminate: we know something quite determinate and specific about the most general empirical concept and the most general empirical law. We know, that is, that Newtonian physics—with its quite specific concepts of mass, force, and interaction and its quite specific law of universal gravitation—is necessarily a part (and, indeed, the most general part) of the ideal complete science. To be sure, the Newtonian theory of gravitation by no means exhausts the content of the ideal complete science, which, as truly complete, can only be approached asymptotically. Nevertheless, from the point of view of the *Metaphysical Foundations,* we know in advance that however this completion may proceed universal gravitation must always remain a necessary part of it. And now the problem or possible "gap" is just this: What reason is there to suppose that there is any connection whatsoever between the constitutive procedure of the *Metaphysical Foundations* and the regulative procedure of reflective judgement? Why, in following the regulative maxims of reflective judgement, should we necessarily proceed in the direction of the *Metaphysical Foundations* and the theory of universal gravitation? Why, on the contrary, should we not, in proceeding from the less general to the more general, diverge arbitrarily far, as it were, from what the *Metaphysical Foundations,* proceeding in the opposite direction, has already established as most general? For all we know so far, in following the purely methodological recommendations of reflective judgement, we might converge asymptotically towards a complete science that has no connection at all with specifically Newtonian physics.

The problem can be stated in somewhat more precise terms as follows. Reflective judgement, for Kant, guides the process of empirical classification or concept formation—the procedure of systematizing lower-level empirical concepts in terms of higher-level empirical concepts:

> Now there belongs hereto, if one proceeds empirically and ascends from the particular to the universal, a *classification* of the manifold, i.e., a comparison

> of several classes, of which each stands under a determinate concept, among one another; and, if the former is complete according to a common characteristic, its subsumption under higher classes (species) proceeds until one reaches the concept that contains the principle of the entire classification (and constitutes the highest species). (20, 214.25–32)

For this purpose reflective judgement postulates or assumes that nothing will interrupt or interfere with the procedure of empirical systematization—the asymptotic progression towards the highest species:

> Now it is clear that reflective judgement, according to its nature, cannot undertake to *classify* the whole of nature in accordance with its empirical variety if it does not presuppose that nature *specifies* even its transcendental laws according to some principle. This principle can be no other than that of the suitability to the faculty of reflective judgement itself, to find sufficient affinity in the immeasurable manifoldness of things in accordance with empirical laws in order to be able to bring them under empirical concepts (classes) and these under more universal laws (higher classes) and thus to achieve an empirical system of nature. (215.14–24)

And clearly, so long as we have no independent information concerning the content of such higher species—so long, that is, as they are left entirely indeterminate—nothing can stand in the way of our proceeding according to the maxims of reflective judgement. The problem, however, is that the *Metaphysical Foundations* has itself already specified the very highest species of empirical classification: namely, the empirical concept of matter. What assurance do we have that the procedure of empirical systematization, as guided by the regulative maxims of reflective judgement, will aim asymptotically at this particular, already given, empirical concept? In other words, we now have two entirely independent ways of approaching the highest species of empirical classification, and we have no principle whatsoever for coordinating them. This difficulty, I suggest, indeed opens up the possibility of a most significant "gap" in the critical system.

Were it not for the *Metaphysical Foundations*, on the other hand, this particular difficulty could not arise. We would then have only one route to the highest species of empirical classification and the most general empirical laws, namely, that marked out by reflective judgement itself. For the transcendental laws of the understanding are themselves completely undetermined with respect to all *empirical* laws:

> To be sure, empirical laws as such can in no way derive their origin from pure understanding—no more than the immeasurable manifold of appearances can be adequately comprehended from the pure form of sensible intuition. (A127)

> Pure understanding is not, however, in a position, through mere categories, to prescribe to appearances any a priori laws other than those which are

> involved in a *nature in general,* that is, in the law-governedness of all appearances in space and time. Particular laws, because they concern empirically determined appearances, can *not be completely derived* therefrom, although they one and all stand under them. (B165)[57]

Were it not for the *Metaphysical Foundations,* in other words, we would have only the empirically undetermined conception of a nature in general articulated by the transcendental principles of the understanding and the entirely indeterminate conception of the systematic unity of nature provided by the regulative maxims of reflective judgement.

Yet it would not be appropriate to attempt to resolve our problem simply by dropping the *Metaphysical Foundations*—together with its more determinate conception of corporeal nature articulated in the metaphysical doctrine of body—from the critical system. For, in the first place, although empirical laws are not of course derived from the transcendental laws of the understanding, it is equally important to the critical system that empirical laws somehow "stand under" the latter, that they are in some sense "special determinations" of the a priori laws of nature in general:

> Yet all empirical laws are only special determinations of the pure laws of the understanding, under which and in accordance with the norm of which they first become possible, and the appearances take on a lawful form—just as all appearances, notwithstanding the diversity of their empirical form, nonetheless also must always be in accordance with the conditions of the pure form of sensibility. (A127–128)[58]

57. Indeed, it is for precisely this reason that there is a need for reflective judgement. See §IV of the *First Introduction:* "We have seen in the *Critique of Pure Reason* that the whole of nature as the totality of all objects of experience constitutes a system according to transcendental laws, namely such as the understanding itself supplies a priori (namely for appearances in so far as they, combined in one consciousness, are to constitute experience). . . . But it does not follow therefrom that nature is also a system *comprehensible* by the human faculty of cognition according to *empirical laws,* and that the thoroughgoing systematic coherence of its appearances in an experience—and thus the latter itself as system—is possible for mankind. For the manifoldness and diversity of empirical laws could be so great that, although it would in fact be possible in part to connect perceptions according to incidentally discovered particular laws, it would never be possible to bring these empirical laws themselves to unity of affinity under a common principle—if, namely, as it is still in itself possible (at least as far as the understanding can constitute a priori), the manifoldness and diversity of these laws, and similarly the natural forms in accordance with them, were infinitely great and presented us here with a crude chaotic aggregate without the least trace of a system, although we equally had to presuppose such [a system] according to transcendental laws" (20, 208.22–209.19).

58. Compare A158/B198: "Even laws of nature, when they are considered as principles of the empirical employment of the understanding, at the same time carry with themselves an expression of necessity and thus at least the suggestion of a determination from grounds

And it is only the *Metaphysical Foundations,* I suggest, that first makes it clear what the nature of this crucially important relationship between transcendental laws and empirical laws actually is. For the *Metaphysical Foundations* further specifies the transcendental laws of the understanding by means of the empirical concept of matter to obtain the metaphysical doctrine of body, and then makes it clear how the latter, in turn, serves as an a priori ground or foundation for the empirical law of universal gravitation. By the same token, it is only the *Metaphysical Foundations* that first makes it clear how the dynamical principles of pure understanding are constitutive with respect to (generate schemata for) empirical concepts—in this case, the empirical concept of matter. Without the *Metaphysical Foundations,* therefore, we would have no conception whatsoever of *how* the principles of pure understanding are actually constitutive with respect to experience.[59]

In the second place, however, although the transcendental principles of the understanding are indeed completely undetermined with respect to all empirical laws, the transcendental concept of a nature in general is itself not *entirely* indeterminate. Constitutive of this latter concept, for example, are quite definite conceptions of substance (as necessarily conserved in its total quantity) and causality (as necessarily effected in a strictly deterministic fashion). And there appears to be no more guarantee that the regulative procedures of reflective judgement must always be in harmony with an ideal complete science embodying the general conceptions embodied in the transcendental principles of the understanding than there is analogously in the case of the more specific principles articulated in the *Metaphysical Foundations.* So dropping the *Metaphysical Foundations* from

---

that hold a priori and antecedently to all experience. Yet all laws of nature without distinction stand under higher principles of the understanding, in that they merely apply these to particular cases of appearance. These principles alone therefore give the concept that contains the condition, and as it were the exponent, of a rule in general; but experience gives the case that stands under the rule."

59. It is no wonder, then, that in the Preface to the *Metaphysical Foundations* Kant explicitly asserts that it alone provides "sense and meaning" for the transcendental principles of the first *Critique:* "It is also indeed very remarkable (but cannot be expounded in detail here) that general metaphysics, in all instances where it requires examples (intuitions) in order to provide meaning for its pure concepts of the understanding, must always derive them from the general doctrine of body—and thus from the form and the principles of outer intuition; and, if these are not exhibited completely, it gropes uncertainly and hesitantly among mere sense-less concepts. . . . And so a separated metaphysics of corporeal nature does excellent and indispensable service for general metaphysics, in that the former furnishes examples (instances *in concreto*) in which to realize the concepts and propositions of the latter (properly transcendental philosophy), that is, to provide a mere form of thought with sense and meaning" (4, 478.3–20). Compare the General Note to the System of the Principles at B291–294.

the critical system would at best *postpone* the problem of a possible "gap" or mismatch between the constitutive and regulative domains.[60]

In any case, Kant himself never entertains the possibility of dropping the *Metaphysical Foundations* from the critical system. Instead, he embarks on a fundamentally new project: the transition from the metaphysical foundations of natural science to physics. Moreover, it is important to observe that this project proceeds *from* the metaphysical foundations *to* physics: that is, the *Transition* project is a continuation of the "top down" procedure already initiated in the *Metaphysical Foundations,* not an elaboration of the "bottom up" procedure characteristic of reflective judgement. Kant makes this explicit in loose-leaf 6: "It is here not [an] ascent from experience to the universal but the transition is [a] descent" (21, 476.11–12).[61] Similarly, Kant speaks frequently of the "tendency" of the metaphysical foundations of natural science towards physics:

> ⟨. . . the metaphysical foundations of natural science have a natural tendency towards physics in themselves—namely, *to step over* from the former to the latter: i.e., to bring about an empirical system of moving forces, for that is the goal of this philosophy as doctrine of nature.—But this stepping over from one territory into another makes it necessary that both regions are in contact with one another according to the law of continuity, without which the transition would be a leap and there would be no intermediate concept which unified both according to a universal principle.
>
> The transition from the metaphysical foundations of natural science therefore constitutes for itself a particular system—namely that of the moving forces of matter—which is connected above with the metaphysics of nature and below with physics in prospect (the rational principle with the empirical) [so as to] build a bridge over a cleft *(hiatus),* without which this tendency would be of no consequence.⟩ (21, 617.16–30)

The *Transition* project therefore aims at a continuous connection between metaphysics and physics, between the rational and the empirical, between the progression from "above" and that from "below."

It is striking, furthermore, that the passages in which Kant speaks of the "tendency" of the metaphysical foundations towards physics overlap considerably with those in which he speaks of a "gap" (or equivalently a "cleft") whose filling constitutes the task of the *Transition* project. Con-

60. For a more detailed discussion of the determinate content of the transcendental principles of the understanding, considered in relation to the more specific principles of pure natural science articulated in the *Metaphysical Foundations,* see [36]—which also attempts to provide a more detailed interpretation of the precise sense in which empirical laws "stand under" the transcendental principles.

61. Compare 21, 643.29–30, where Kant speaks of "The transition from the metaphysical foundations of natural science to physics, not conversely from [the] empirical to cognition a priori . . . ."

sider, for example, the very first passage where the "gap" is mentioned (from loose-leaf 5 in 1798):

> But this tendency in the transition from the meta. to physics cannot happen immediately and through a leap, for the concepts which must carry us over from a system of one particular kind to that of another must ⟨on the one hand⟩ carry with them a priori principles, but, on the other, also empirical [principles] which, because they contain comparative universality, can also be used as the most general [principles] for the system of physics.—There is thus between the metaphysical foundations of nat. sci. and physics still a gap to be filled, whose filling can be called a transition from the one to the other. (21, 482.19–27)[62]

And the outline-drafts entitled Farrago 1–4 (from 1798–99), where the "gap" is mentioned several times, are also those where the "tendency" of the *Metaphysical Foundations* towards physics is mentioned most frequently.[63] This circumstance seems to me to confirm the above interpretation of the "gap" as arising precisely from the "top down" constitutive procedure of the *Metaphysical Foundations*—when this procedure is juxtaposed with that of reflective judgement. The filling of the "gap" therefore calls for a further elaboration of this "top down" procedure, whereby constitutive a priori principles are extended even further into the domain of the empirical in order ultimately to meet or connect with the "bottom up," merely regulative, procedure of reflective judgement. In this way "a gap in the system of pure natural science *(philosophia naturalis pura)* is now filled, and the circle of all that belongs to a priori cognition is closed" (21, 640.4–6).[64]

Indeed, during this same period Kant begins to characterize the *Transition* project as the articulation of principles that are, at the same time, both regulative and constitutive. Thus, immediately before Farrago 1–4 Kant writes (in 1798):

62. The importance of this passage has been rightly emphasized by Förster [31], p. 549.

63. See the compilation of "tendency" passages in Tuschling [113], p. 92. Compare also, e.g., 21, 161–165, 360–361, 366.28–367.12, 504.22–506.10; 22, 166–167, 178.22–29, 182.22–24.

64. Why then does Kant only first mention the "gap" in 1798–99? It appears from our above discussion of loose-leafs 6, 3/4, and 5 that Kant had first envisioned an extension of the *Metaphysical Foundations* that was to have proceeded purely mathematically—on the basis of the purely mathematical unity of all possible moving forces in space. He then came to see, however, that a fundamentally different type of unity—that of reflective judgement (natural investigation) is required. It is therefore at this point, and only at this point, that the idea of a possible extension of the constitutive procedure of the *Metaphysical Foundations* is self-consciously and deliberately juxtaposed with the regulative procedure of reflective judgement. (I am indebted to Eckart Förster for prompting me to make this interpretive suggestion explicit—although he himself would no doubt strongly disagree with it.)

⟨The relation to a doctrinal system [of the moving forces] lies a priori in the concept of physics as that towards which the metaphysical foundations has the tendency, and this concept of the system is also the regulative principle of their unification into a whole.

However, the constitutive principle of the system of the empirically given moving forces of matter is this transition—where this is not sought for fragmentarily (for this yields no whole according to principles) but is rather considered as contained in a system.⟩ (22, 178.26–33)[65]

And immediately after Farrago 1–4 we have (in 1799):

This transition is not mere propaedeutic, for that is an unsteady concept and concerns only the subjective [element] of cognition. It is a not merely regulative but also constitutive formal principle, subsisting a priori, of natural science for a system. (22, 240.25–28)

Regulative principles which are, at the same time, constitutive. (241.19)

Hence, to carry out the *Transition* project—and thus to fill the "gap" in the critical system—what we now require is a kind of *intersection* between the constitutive domain (of the *Metaphysical Foundations*) and the regulative domain (of reflective judgement), for only so can there be a continuous connection between the two, formerly entirely independent domains.

It seems to me, moreover, that the apparently paradoxical idea of an intersection between the constitutive and regulative domains—the demand for principles that are, at the same time, both constitutive and regulative—is nonetheless an unavoidable problem for the critical philosophy. Perhaps the most basic thought of this philosophy is that empirical judgements, as such, are somehow made possible by a priori principles; indeed, this represents the true vocation and entire point of a priori principles:

. . . and the categories are, in the end, of no other than a possible empirical use, in that they serve merely thereto, through grounds of an a priori necessary unity (due to the necessary unification of all consciousness in an original apperception), to subordinate appearances to universal rules of synthesis, and to thereby make them suitable for a thoroughgoing connection in one experience.

But all our cognitions lie in the whole of all possible experience, and in the universal relation to the latter consists the transcendental truth that precedes all empirical [truth] and makes it possible. (A146/B185)

65. As Förster has emphasized to me, we find the following passage several pages later: "⟨The idea of the a priori cognizable system of the empirically given moving forces of matter as the filling of a gap through the regulative principle of synthetic cognition⟩" (182.22–24). Read in conjunction with the first passage, however, this does not seem to me to contradict the idea that the *Transition* is *both* regulative and constitutive.

It follows, then, that the a priori categories of the understanding must somehow provide a rational ground for all empirical truth: not merely for abstract principles such as the law of causality, but also for each and every particular empirical law—and, in the end, for each and every particular empirical judgement.

Yet, as Kant himself repeatedly emphasizes, the categories and principles of the understanding do not themselves make it evident how such a necessary grounding of the domain of the properly empirical is to proceed: they do not make it evident how the transcendental concept of a nature in general is to be determinately applied to the specific, empirically given nature with which we are in fact confronted. Indeed, the only strictly transcendental principles we have for coming to terms with nature as empirically given are those of reflective judgement, and these, as merely regulative, necessarily fail to make it clear how the domain of the properly empirical is to be determinately connected with the a priori constitutive principles of the understanding as well. Nevertheless, since properly empirical truth can, in the end, have no other rational ground than that of the transcendental principles of the understanding, it is a necessary demand of reason that the two domains should ultimately be brought into thoroughgoing connection:

> For the law of reason, to seek [the systematic unity of nature], is necessary, because without it we would have absolutely no reason, but without this no thoroughly coherent employment of the understanding, and in the absence of this no sufficient mark of empirical truth—and we must therefore, with respect to the latter, presuppose throughout the systematic unity of nature as objectively valid and necessary. (A651/B679)

Kant is therefore committed, it seems to me, to the idea that even the regulative use of reason concerning the empirically given nature with which we are in fact confronted (the demand for the systematic unity of nature) is ultimately grounded in the constitutive requirements of the understanding. Even the regulative use of reason, that is, is itself ultimately grounded in the conditions of the possibility of experience.

Now, as we have seen, it is really only the *Metaphysical Foundations* that first makes it clear how the domain of the properly empirical can be determinately grounded in the transcendental concept of a nature in general. The *Metaphysical Foundations* accomplishes this for the particular case of the Newtonian theory of gravity—by applying the transcendental principles of the understanding to the empirical concept of matter and thereby grounding the empirical law of universal gravitation. Yet this achievement of the *Metaphysical Foundations,* as impressive as it is, in no way suffices to elucidate how experience in general is grounded in the a priori, for the Newtonian theory of gravity comprehends only a very small fragment of the totality of empirically given phenomena of nature. In particular, the phenomena studied by the emerging new sciences of heat, light, electricity and magnetism, and chemistry remain entirely out of account; for, as far as one can see, such phenomena remain entirely unconnected with universal gravitation. In the end, therefore, the seemingly paradoxical task of the *Transition* project is and must be an unavoidable problem for Kant. For otherwise we have no determinate conception whatever of how experience—that is, experience as a whole—is possible.

It is no wonder, therefore, that Kant soon comes to characterize the *Transition* project as essentially involving the conditions of the possibility of experience. Thus, for example, in 1798 he writes:

> It is therefore not to be avoided in a system of natural science that therein a leap *(saltus)* does not take the lead—if consideration is not taken of an *intermediate concept* [*Mittelbegriff*] (not the logical [concept] in a syllogism which merely concerns the form of inference but rather the real [concept] which reason presents an object), which is connected, on the one hand, with an a priori concept of the object and, on the other, with the condition of the possibility of the experience ⟨in which this concept can be realized⟩; for only so does such a concept serve for the transition from the metaphysical foundations of natural science to physics which is then no leap. (21, 285.22–31)[66]

And it is no wonder, finally, that when Kant then attempts, in the outline-drafts entitled Transition 1–14, to carry out the *Transition* project by means of an a priori proof of the existence of the aether (which, as we have seen, is thought to underlie precisely the phenomena involving the "specific variety of matter" studied by the emerging new sciences) he bases this aether-deduction entirely on the conditions of the possibility of experience.

III · The Chemical Revolution

We saw above that Kant's delineation of reflective judgement as an autonomous faculty in 1790 does not, by itself, appear to be a sufficient source for the new optimism concerning the prospects for a philosophical foundation of the empirical or experimental sciences (the theory of heat, chemistry, and so on) manifest in the *Transition* project. The juxtaposition of the regulative procedure of reflective judgement with the constitutive procedure of the *Metaphysical Foundations* creates a problem for (a "gap" in) the critical system, but the principle of reflective judgement alone yields

66. Compare 21, 330.19–22: "⟨The transition from an a priori knowledge of nature to physics contains the principles according to which one is to determine in advance the manifold of laws of nature through reason in order to establish experience⟩," and 21, 476.23–25, 477.28–478.5.

no intimation of how this problem is to be solved. I now want to suggest that Kant's new optimism is based more on developments taking place in the empirical or experimental sciences themselves than on any independently motivated philosophical considerations. Briefly, between the period of the first *Critique* and the *Metaphysical Foundations* and that of the *Opus* Kant is increasingly aware of developments in the theory of heat and in chemistry through which these disciplines are in fact becoming genuine sciences (and no longer mere empirical "aggregates") after all. And, since they are now becoming true sciences in fact, a philosophical foundation therefore must be possible. The scientific developments in question fall under the rubric of the so-called chemical revolution associated with the name of Antoine Lavoisier.

In the period of the *Metaphysical Foundations* and the first *Critique* Kant's chemistry is the traditional phlogistic chemistry developed especially by Georg Stahl. This is made perfectly explicit in the Preface to the second (1787) edition of the *Critique,* where Kant gives the following example of "natural science, in so far it is grounded on *empirical* principles":

> . . . in more recent times *Stahl* transformed metals into calxes and the latter back into metal, in that he extracted something from them and then restored it . . . . (Bxii–xiii)[67]

Moreover, the two chemical examples discussed in the Appendix to the Transcendental Dialectic at A645–646/B673–674 and A652–653/B680–681 respectively (cited at the beginning of §II above) are from Stahlian (phlogistic) chemistry as well.

Thus in the first passage Kant reinterprets the ancient doctrine of the four elements—earth, water, air, and fire—as concepts of reason supplying an explanatory framework for chemistry:

> One therefore reduces all matters to the earths (as it were mere weight), salts and inflammable beings (as force), and finally to water and air as vehicles (as it were machines through which the former act), in order to explain the chemical interactions of matters among one another according to the idea of a mechanism. (A646/B674)

The use of "inflammable beings [brennliche Wesen]" here refers to Stahl's phlogiston or inflammable principle, and the idea of water and air as

67. Kemp Smith renders "Kalk" as "oxides" here, which is most unfortunate. It is of course Lavoisier's theory that the calcification of metals is a process of oxidation by which metals *absorb* oxygen from the air to become calxes—which latter, in turn, can be reduced back to the metallic state by the *extraction* of oxygen. It is Stahl's theory, by contrast, that calcification is a phlogistic process by which phlogiston is *extracted* from the metal—reduction then takes place via the *absorption* of phlogiston.

"vehicles" is traditional Stahlian doctrine.[68] According to Stahl's theory of combustion air in fact does play an essential, albeit non-chemical or purely mechanical role: it is a "vehicle" or "instrument" for carrying off the phlogiston to be extracted from the inflammable body—without which, however, the body cannot burn.[69] Similarly, water is an essential mechanical "vehicle" by which "salts" (that is, acids and alkalis) manifest their volatility (see Metzger [79], especially pp. 165–166; *Danziger Physik:* 29, 161.31–32).

In the second passage Kant explicitly uses the traditional terminology of "salts" to embrace acids and alkalis (whereas it was Lavoisier who first used "salts" in the modern sense to designate the so-called neutral salts or compounds of an acid plus a base), and he further allies himself with Stahlian chemistry in the remainder of the passage:

> One has attempted to reduce the various kinds of earths (the material of stones and even of metals) step by step to three and finally to two; but, not yet satisfied thereby, [chemists] cannot banish the thought that suggests that behind these varieties there is nevertheless a single species—or even, indeed, a common principle for these and the salts. (A653/B681)

According to Stahlian doctrine a metal is a composite of an earth plus phlogiston, which latter is then extracted on calcification (whereas according to Lavoisier a metal is a simple body to which oxygen is added on calcification). Similarly, according to Stahl a "salt" (for example, an acid) is a composite of an earth plus water.[70]

It is clear, then, that the chemistry to which Kant denies the status of

68. Compare side II of Reflexion 45 (approximately 1775–1777) and Adickes's comments thereto (14, 371–396). Reflexion 45 speaks of "salts and inflammable beings, *hypomochlion* of chemical actions" (371.4–5) and then asserts "Both, salts and phlogiston, constitute the chemical potencies" (396.5–6). According to the *Danziger Physik* of 1785 (cited by Adickes at 14, 385): "One has inflammable beings wherein only a part is inflammable. We here take only the proper phlogiston or inflammable into consideration, which constitutes only a small part of even spirit of wine, in that the other part is water. This is therefore proper chemical potency" (29, 161.27–31).

69. Hence, when air is fully "saturated" with phlogiston it cannot support combustion—thus deoxygenated air (nitrogen gas) was first called "phlogisticated air," and oxygen gas was first called "dephlogisticated air."

70. Again, see Metzger [79]. According to Stahl's predecessor J. J. Becher, mineral bodies (including metals, of course) consist of three earths: vitreous earth or calx, combustible or sulpherous earth, and fluid or mercurial earth. Becher held that the second or combustible earth "burnt off" in the calcification of metals, and it is this earth that was then transformed into Stahl's inflammable principle or phlogiston. For Stahl's theory of acids or "salts" see [79], part II, chap. IX: vitriolic acid (sulphuric acid) is a composite of vitreous earth plus water, acid of nitre (nitric acid) is a composite of vitriolic acid plus phlogiston, marine acid (hydrochloric acid) is a composite of acid of nitre plus mercurial earth—hence vitriolic acid is a "universal acid" or common principle for all the acids.

a true or "proper" science in the *Metaphysical Foundations* is essentially that of Stahl. There is no doubt, however, that by the time Kant is well under way on the *Transition* project he has enthusiastically adopted the chemistry of Lavoisier. Thus in 1797, in the Preface to the First Part of *The Metaphysics of Morals,* Kant writes:

> So the *moralist* correctly says: there is only one virtue and doctrine thereof, i.e., one unique system that combines all duties of virtue through a single principle; the *chemist:* that there is only one chemistry (that according to *Lavoisier*); the *teacher of medicine:* that there is only a single principle for a system of disease classification (according to *Brown*)—without, however, because the *new system* excludes the others, thereby belittling the merits of the older moralists, chemists, and teachers of medicine . . . . (6, 207.11–17)

In 1798, towards the end of the *Anthropology From a Pragmatic Point of View,* Kant asks:

> What amount of knowledge, what discovery of new methods would now lie already in store, if an Archimedes, a Newton, or a Lavoisier had, with their industry and talent, been favored by nature with a lifetime lasting through a century of undiminished vitality? (7, 326.1–5)

And the juxtaposition of Lavoisier with Archimedes and Newton here is of course particularly suggestive. In any case, however, Kant has evidently experienced the chemical revolution for himself between the critical period and 1797–98.

We shall attempt below to document Kant's awareness of the developments constituting the chemical revolution more precisely, and we shall also, of course, discuss the impact of these developments on the *Transition* project of the *Opus.* But it may first prove useful to sketch the main lines of these developments as they appeared to those who experienced them. For this purpose we can distinguish three principal phases or stages of the chemical revolution: the development of pneumatic chemistry by Joseph Black, Carl Wilhelm Scheele, Joseph Priestley, and Henry Cavendish; the advances in the science of heat by Joseph Black and Johan Carl Wilcke; and Lavoisier's construction of a new systematic chemistry, which effectively integrates the first two phases with the traditional analytic chemistry, minerology, and metallurgy of Boerhaave, Becher, and Stahl via Lavoisier's oxygen theory of combustion and acidity.[71]

71. Here I follow Guerlac. Compare [40], p. xvii: "In the person of Lavoiser two largely separate and distinct chemical traditions seem for the first time to have been merged. At his hands, the pharmaceutical, mineral, and analytical chemistry of the continent was fruitfully combined with the results of the British 'pneumatic' chemists who discovered and characterized the more familiar permanent gases." To be sure, Guerlac here mentions only the first and third phases distinguished above. However, in [41] Guerlac makes it admirably clear how important the second phase was to Lavoisier himself. It was equally important, I may

Two main discoveries were made by the pneumatic chemists: first, that there are a large number of gases or substances in the aeriform or "permanently elastic" state and that common atmospheric air is not an element but rather a mixture of such gases; second, that "air"—including the new "factitious airs" just mentioned—can be "fixed" or chemically combined with other bodies, wherein it no longer manifests its most obvious sensible properties such as elasticity. This latter discovery undermines the traditional view (held by both Stahl and Boerhaave, for example) of "air" as incapable of true chemical interaction—as functioning merely "instrumentally" or as a "vehicle" in chemical reactions—and is obviously a most essential ingredient in Lavoisier's oxygen theory of combustion.

It appears that the existence of gases other than common atmospheric air was recognized in the work of John Mayow in the late 1660s and early 1670s (Partington [95], vol. 3, p. 109—Mayow's work is discussed extensively in vol. 2, chap. XVI). However, the development of modern pneumatic chemistry perhaps most properly begins with Stephen Hales's *Vegetable Staticks* (1727). Hales collected the "air" released in the distillation of a number of substances, which he accordingly viewed as existing in a prior "fixed" or inelastic state chemically combined in the substance in question. And it appears that Hales's work was actually the most important influence on Lavoisier in this regard (Guerlac [40], chap. 1). Yet Hales did not himself recognize the qualitative variety of gases or "airs," and he considered common air as an element (Partington [95], vol. 3, pp. 112–123). The first to recognize explicitly that "fixed" or combined "air" is not necessarily common atmospheric air was Joseph Black, who in 1752–1756 collected carbon dioxide gas evolved from various substances and demonstrated that its properties are quite distinct from common air (Partington [95], vol. 3, pp. 136–140). The terminology "fixed air" was then customarily used for this particular gas (CO_2). The discovery of other such chemically evolved "airs" quickly followed: most notably, "mephitic air" or "phlogisticated air" (nitrogen gas) and "inflammable air" (hydrogen gas) by Cavendish in 1764–1766, and "fire air," "eminently respirable air," or "dephlogisticated air" (oxgen gas) by Scheele and Priestley independently in 1770–1774. That atmospheric air is a mixture principally of nitrogen and oxygen was stated by Torbern Bergmann and Scheele in 1775 and definitively established by Lavoisier in 1776.[72]

add, to those who first accepted Lavoisier's theory (including Kant, as we shall see below). That Guerlac does not emphasize the second phase in the above-cited passage reflects the significant differences, discussed below, between how we understand the chemical revolution today and how it was understood by those who experienced it.

72. See Partington [95], vol. 3, chaps. V–VIII, and, for Lavoisier's work on the composition of nitric acid, pp. 410–416. This latter work of Lavoisier also contains the first statement

During approximately the same period major advances were being made in the quantitative science of heat. In particular, the notions of temperature and heat were first clearly distinguished via the discovery and quantitative investigation of latent and specific heats. Once again, the first to recognize and precisely investigate these phenomena was Joseph Black, in the years 1757–1764. Since Black's work was only officially published posthumously in 1803, however, many on the Continent gave priority to Johan Wilcke, who published a paper concerning latent heat independently of Black in 1772 and a paper on specific heats in 1781. In any case, it was these advances that were systematized and perfected by Lavoisier and Laplace in their *Memoir on Heat* of 1783.[73]

Black developed the notion of specific heat, which he called "capacity and attraction for heat," in self-conscious opposition to Boerhaave's view that heat is distributed equally in equal volumes in a system of bodies in thermal equilibrium. This is plainly to confuse temperature (intensity of heat) with quantity of heat, for neither equal volumes nor equal masses of different substances at the same temperature manifest equal rates of cooling, for example. Black refined and quantified the notion of "capacity for heat" by mixing equal weights of different substances at different temperatures: the ratio of their heat capacities is inversely proportional to the changes of temperature of each in the resulting mixture.[74]

Perhaps even more significant, however, was Black's articulation of the concept of latent heat. He first arrived at this concept by closely considering the temperature changes observed in melting ice, noting that the temperature of the resulting water remains at the freezing point until all the ice is melted. This suggests that, contrary to common sense, a small amount of heat added to solid matter at the melting point is quite insufficient either to effect its fusion (liquifaction) or to raise its temperature. On the contrary, a significant amount of heat is required simply to effect fusion or the transition to the liquid state—which heat then lies *latent* in the liquid state with no sensible effect on the temperature. By careful and precise experiments, Black was able to assign definite numbers to the latent heat

---

of his anti-phlogistic theory of acidity (he finds that nitric acid contains oxygen and argues that it does not contain phlogiston—as Stahl's theory would have it: see note 70 above).

73. See McKie and Heathcote [75] and Guerlac [41]. An anonymous version of Black's Lectures was published in London in 1770. German translations of Wilcke's papers by A. G. Kästner appeared in 1776 and 1782 respectively. Adair Crawford's *Experiments and Observations on Animal Heat* (1779), written after attending the lectures of Black's colleague William Irvine at Glasgow, was instrumental in stimulating interest in the new developments on the Continent—for Lavoisier and Laplace in particular.

74. Thus, for example, Black mixed a pound of gold at 150° F with a pound of water at 50°; the resulting mixture was at 55°; therefore, the heat capacity of gold is to that of water as 5:95 or 1:19. See [75], pp. 12–15.

of fusion of ice in terms of the amount of heat required to increase the temperature of water a given number of degrees.[75] Moreover, Black was able to extend his results to the fusion of other solid substances, and also to the transition from the liquid to the vaporous state. In the latter case as well a substantial amount of heat, a latent heat of vaporization, is required simply to effect the required state transition—which heat again lies latent in the vaporous state with no sensible effect on the temperature (McKie and Heathcote [75], pp. 20–27).

Black's work therefore results in a general doctrine of latent heat:[76] the three states of aggregation or phases—solid, liquid, and vaporous—depend directly on a quantity of heat that is latent in or somehow combined with the substance in question. This heat is not manifest as temperature and serves solely as the cause of liquifaction or vaporization respectively. The doctrine is stated explicitly by Black in his posthumously published *Lectures:*

> I consider fluidity as depending, immediately and inseparably, on a certain quantity of the matter of heat, which is combined with the fluid body, in a particular manner, so as not to be communicated to a thermometer, or to other bodies, but capable of being extricated again by other methods, and of re-assuming the form of moveable or communicable heat. ([7], vol. 1, p. 144)

Similarly, in vaporization:

> . . . a particle of water, in the instant of its becoming a particle of vapour, attracts and unites with itself one or more atoms of this cause of heat . . . and retains them as parts or ingredients of its vaporous form . . . when a particle of vapour again becomes water, these atoms of heat are set at liberty by the fixed laws of chemical affinity. ([7], vol. 1, p. 165)

As these statements make evident, the general doctine of latent heat suggests, almost irresistibly, a conception of heat as a particular material substance or element capable of *chemical* combination—via chemical affinity—with other substances. By the same token, the notion of specific heat or heat capacity is naturally understood in terms of the differing chemical affinities for the matter of heat exhibited by different substances.

75. Thus, for example, Black compared the rates of warming of a given mass of ice with the same mass of water—both at as near as possible to 32° F. From the times in which it took the two masses to attain a temperature of 40° he estimated that the amount of heat required simply to melt the ice would be sufficient to raise the temperature of the water 140°: these 140° have been absorbed by or are latent in the fusion. Further experiments based on mixing water and ice confirmed the approximate validity of this number. See [75], pp. 15–19.

76. This is certainly not true of Wilcke's work, which is confined to the study of melting snow and makes no mention of either the fusion of other solid substances or vaporization: [75], pp. 78–94.

Indeed, it was this general doctrine of latent heat that persuaded the preponderance of enlightened scientific opinion of the time of the correctness of the material or caloric theory of heat. The opposition between a mechanical or kinetic conception, according to which heat consists simply in motions or vibrations of the parts of ordinary matter, and a material conception, according to which heat is a particular (extraordinarily subtle and elastic) type of matter in its own right, was of course well known. Yet Black's discovery that heat could apparently combine chemically with other substances so as to effect the transition from the solid to the liquid and thence to the gaseous or vaporous state was extremely hard to reconcile with the mechanical theory.[77] The material or caloric theory, on the other hand, could very naturally comprehend the idea of a chemical combination of heat with other substances and thus appeared to hold the key to the three states of aggregation. In particular, the material theory appeared to hold the key to the newly discovered gaseous state, which could now be viewed as a *solution* of a given type of matter in the extraordinarily subtle and elastic caloric fluid.[78]

The caloric theory of heat and of the three states of aggregation, and especially the caloric theory of the newly discovered gaseous state, played a fundamental role in Lavoisier's new systematic chemistry. Lavoisier's conception of the gaseous state was first published in a paper of 1777 ("On the Combination of the Matter of Fire with Evaporable Fluids and the Formation of Elastic Aeriform Fluids") where he argues that a gas or "elastic fluid" is a solution of a volatile body in the matter of fire: more precisely, a gas is literally a compound of a particular individuating substance or "base" plus caloric.[79] The caloric theory of the three states of

77. Similarly, Black's discovery of specific heats was also difficult to comprehend on the mechanical theory: if heat is a form of motion it should propagate according to the laws of motion and thus heat capacity should be proportional to density—which is frequently not the case. This problem was raised by Black himself, who, although he remained officially agnostic on the question, appeared definitely to lean towards the material or caloric theory: see [75], pp. 27–30.

78. The caloric theory of heat, and, in particular, the "caloric theory of gases"—which explains the elasticity or pressure of a gas statically, in terms of the repulsive forces exerted by the caloric fluid in which the matter of the gas is dissolved—was really not definitively overthrown, despite Count Rumford's celebrated work on "the mechanical equivalent of heat" in 1798, until the articulation of the principle of the conservation of energy around 1850: see Fox [30]. Thus, for example, we now understand latent heats without relying on the idea of a chemical combination of heat with the substance in question: latent heat of vaporization represents the energy required to overcome the intermolecular (van der Waals) forces holding the molecules of a liquid in close proximity; similarly, latent heat of fusion represents the energy required to overcome the (bond-forming) forces responsible for the close-packed crystalline molecular structures characteristic of solids.

79. See Guerlac [41], pp. 206–209, and Partington [95], vol. 3, pp. 420–421. As Partington points out, Lavoisier's conception is an extension of Black's doctrine of latent heats:

aggregation—the idea that solidity, liquidity, and the gaseous state are indeed three *states* (rather than three elements: earth, water, and air), which any substance may assume depending on the amount of heat with which it is combined—is then stated by Lavoisier in a publication of 1780. Moreover, the caloric theory of the states of aggregation, which Guerlac has called Lavoisier's "covering theory,"[80] is introduced in the first chapter of Lavoisier's *Elements of Chemistry* (1789), entitled "Of the Combinations of Caloric, and the Formation of Elastic Aeriform Fluids or Gases":

> The same [as of water] may be affirmed of all bodies in nature: They are either solid or liquid, or in the state of elastic aeriform vapour, according to the proportion which takes place between the attractive force inherent in their particles, and the repulsive power of the heat acting upon these; or, what amounts to the same thing, in proportion to the degree of heat to which they are exposed.
>
> It is difficult to comprehend these phenomena, without admitting them as the effects of a real and material substance, or very subtile fluid, which, insinuating itself between the particles of bodies, separates them from each other; and, even allowing the existence of this fluid to be hypothetical, we shall see in the sequel, that it explains the phenomena of nature in a very satisfactory manner. ([63], pp. 18–19; [64], p. 4)[81]

---

whereas Black states that ice + latent heat = water and water + latent heat = steam, Lavoisier argues, for example, that oxygen gas = oxygen base + caloric—moreover, Black himself explicitly recognized this relationship between Lavoisier's theory and his own doctrine ([95], vol. 3, p. 154).

80. See [41], pp. 200–216. It appears that Lavoisier held this "covering theory" as early as 1772, when he formulated a draft of a "System of the Elements" which argues that water, air, and fire are all capable of existing in either a "free state"—in which they manifest their characteristic sensible properties of fluidity, elasticity, and warmth respectively—or a "fixed state"—in which they are chemically combined with other substances and no longer manifest their characteristic sensible properties ("fixed" water is combined with salts as "water of crystallization"). See also [40], pp. 90–101 and Appendix IV (pp. 215–218)—where the "System of the Elements" is reproduced.

81. As Howard Stein has emphasized to me, Lavoisier adopts an officially agnostic attitude towards the caloric theory on the very next page: "We are not obliged to suppose this to be a real substance; it being sufficient, as will more clearly appear in the sequel to this work, that it be considered as the repulsive cause, whatever that may be, which separates the particles of matter from each other, so that we are still at liberty to investigate its effects in an abstract and mathematical manner"—and this of course echoes the official agnosticism adopted in Lavoisier's and Laplace's *Memoir on Heat* (1783). Yet this official agnosticism is quite inconsistent with Lavoisier's *chemical use* of the theory of heat. Thus, for example, Lavoisier defines "combined caloric" (latent heat) as "that which is fixed in bodies by affinity or elective attraction, so as to form part of the substance of the body, even part of its solidity" ([63], p. 28; [64], p. 19) and uses the idea of chemical affinity as an essential part of his theory of combustion: "This experiment proves, in a most convincing manner, that, at a certain degree of temperature, oxygen possesses a stronger elective attraction, or affinity, for phosphorous than for caloric; that, in consequence of this, the phosphorous attracts the

This pivotal appearance of the caloric theory of the states of aggregation in the *Elements* mirrors its pivotal role in Lavoisier's theory.

Thus, although we tend to forget this today, Lavoisier's theory of combustion and calcination in fact had two distinct yet interrelated aims: first, to explain the *chemical* changes in the inflammable body or metal and in the surrounding air (the absorption by the former of oxygen from the latter); second, and equally important, to explain the obvious *physical* processes taking place (the production of heat and light). These latter physical processes were explained precisely by the changing chemical affinities involving caloric and other substances underlying combustion and calcination.[82] This emerges most clearly, perhaps, in Lavoisier's Notes to R. Kirwan's *Essay on Phlogiston* (1784):

> We do not therefore affirm, that vital air [oxygen gas] combines with metals to form metallic calces, because this manner of enunciating would not be sufficiently accurate: but we say, when a metal is heated to a certain temperature, and when its particles are separated from each other to a certain distance by heat, and their attraction to each other is sufficiently diminished, it becomes capable of decomposing vital air, from which it seizes the base, namely *oxygen,* and sets the other principle, namely the *caloric,* at liberty. This explanation of what passes during the calcination, is not an hypothesis, but the result of facts. ([58], p. 13)

Similarly, in combustion: "There is likewise a total absorption of vital air, or rather of the *oxygen* which forms its base. . . . In these operations, the caloric and the light, which maintained the *oxygen* in a state of expansion, are set at liberty . . . " ([58], p. 14). In general, then: "the property of burning is nothing else but the property which certain substances possess of decomposing vital air by the great affinity they have for the oxygenous principle" ([58], pp. 15–16). Hence:

> An inflammable body is nothing else but a body which has the property of decomposing vital air, and taking the base from the caloric and light, that is to say the oxygen which was united to them. When this decomposition of the air is rapid, and as it were instantaneous, there is an appearance of flame, heat, and light; when, on the contrary, the decomposition is very slow, and quietly made, the heat and light are scarcely perceptible.
>
> If therefore we attach to the word Inflammability the idea of disengagement of caloric and light, such as takes place in the phenomena of combustion and calcination, we must conclude, that vital air or oxygenous gas

base of oxygen gas from the caloric, which, being set free, spreads itself over the surrounding bodies" ([63], p. 52; [64], p. 57). Only a real material substance can obey the laws of chemical affinity.

82. That light (flame) as well as heat is produced is accommodated by defining caloric, rather vaguely, as "the matter of heat and light" or "the matter of fire or of light": see Partington [95], vol. 3, pp. 462–465.

> is the inflammable body most eminently, since it is principally and almost entirely from this substance that the caloric and light are disengaged. ([58], pp. 20–21)

The caloric theory of the gaseous state is therefore an absolutely central part of Lavoisier's argument: in particular, of his argument that oxygen gas rather than phlogiston is the principle of inflammability.

This fact tends to be forgotten today because we of course do not now accept the caloric theory of heat and of the gaseous state. For us, Lavoisier's achievement is exhausted by the purely chemical part of his theory: by the precise and ingenious experiments through which he demonstrated that the increase in weight of the inflammable body or metal is exactly matched by the loss in weight of oxygen from the surrounding air. Here is where Lavoisier showed himself to be a master in the use of the balance and the principle of conservation of matter, especially as applied to the newly discovered gaseous products and agents involved in chemical reactions.[83] And here is where Lavoisier did in fact demonstrate to virtually everyone's satisfaction that oxygen is absorbed by the inflammable body or metal in combustion and calcination respectively.[84] Yet this demonstration, as impressive as it was, was also quite insufficient to overthrow

83. See Guerlac [40], p. xviii: "Methodologically, the key to the [Chemical] Revolution was Lavoisier's systematic application of his special 'regent,' the balance, not merely to solids and liquids, but also to gases. While the British chemists of the eighteenth century . . . came gradually to perceive that gases make up a third class of substances as important to the chemist as solids and liquids, their work was often more physical than strictly chemical. It was Lavoisier who most convincingly and systematically demonstrated . . . that this newly discovered group of substances must be regularly accounted for in strict chemical bookkeeping if the constitution of familiar substances and the nature of familiar reactions were to be correctly understood. Perhaps it is not too much to say that the Chemical Revolution—to hazard a metaphor—supplanted a two-dimensional by a three-dimensional quantitative chemistry."

84. Lavoisier thereby laid the foundations for his oxygen theory of acidity. Thus, for example, in burning sulphur he obtained sulphurous acid (actually sulphurous anhydride, SO_2, which yields sulphurous acid when combined with water) and in burning phosphorous he obtained phosphoric acid (actually phosphoric anhydride, P_2O_5). Lavoisier asserted accordingly that sulphurous acid (together with sulphuric acid) and phosphoric acid are oxides of sulphur and phosphorous respectively—which latter are simple substances or elements, contrary to the Stahlian theory of acidity. After he was able to show that nitric acid too contains oxygen, Lavoisier concluded that all acids consist of an "acidifiable base" or "radical" plus oxygen. There are some acids, it is true, for which the decomposition into a radical and oxygen had not yet been discovered—marine acid or muriatic acid (hydrochloric acid) in particular—but Lavoisier was confident that such a decomposition would soon be discovered and was content, in his new chemical nomenclature, to refer to the "unknown base of muriatic acid" as simply the "muriatic radical." This acid (HCl), of course, turned out to be the exception that proves the rule. For Lavoisier's experiments and resulting theory of acidity, see Partington [95], vol. 3, pp. 384–416, 421–425. For the theory of acidity in particular, see Metzger [80], pp. 22–33.

the phlogiston theory—for both Lavoisier and his contemporaries. Many respectable chemists could and did cheerfully accept the fact of oxygen's absorption while continuing to insist that phlogiston simultaneously escapes in combustion and calcination—and, indeed, that phlogiston, not oxygen, is the principle of inflammability.

Such a view was in fact maintained, at least for a time, by almost all the leading pneumatic chemists: Scheele, Priestley, Cavendish, and Kirwan.[85] For these chemists it was obvious that something—some "fire" or inflammable principle—must be escaping from the inflammable body during combustion, and they tended to identify this principle (which, following Stahl, they continued to call "phlogiston") with the newly discovered "inflammable air" (hydrogen gas).[86] Moreover, this version of the phlogiston theory was confirmed by the evolution of "inflammable air" from metals acted upon by dilute acids, which appeared to show that the Stahlian conception of a metal as a composite of a calx plus phlogiston was essentially correct.[87] In any case, however, Lavoisier's chemical demonstration of the absorption of oxygen did not, by itself, touch the physical question of the origin of the obvious heat and light evolved in the reactions under consideration. On the contrary, here the phlogiston theory, which very naturally located the source of this light and heat in the given inflammable body or metal, appeared to be as strong as ever.

It was therefore incumbent upon Lavoisier to produce an alternative account of the physical phenomena of combustion and calcination—the obvious heat and light thereby evolved. And he found such an alternative account, as we know, in the doctrine of latent heat and the caloric theory of the states of aggregation: the heat and light evolved in combustion and calcination do not come from the inflammable body or metal, but rather from oxygen gas—which relinquishes its fixed or latent caloric while the "oxygenous principle" or oxygen base unites with the given substance. The importance of this physical side of the question is underscored by the circumstance that, although Lavoisier's experiments on the absorption of oxygen date from 1772–1776, he waited until he had mastered the advances in the science of heat resulting in the doctrine of latent heat and the

85. Only Joseph Black, who recognized Lavoisier's theory of combustion as an extension of his own doctrine of latent heat, quickly adopted Lavoisier's new chemistry—from at least 1784 (see Partington [95], vol. 3, pp. 488–494).

86. For Scheele, "inflammable air" = phlogiston = "the simple inflammable principle," and "fire air" (oxygen gas) + phlogiston = heat. For Cavendish, after his discovery of the synthesis of water (1784), hydrogen = water + phlogiston, and oxygen = water − phlogiston. See Partington [95], vol. 3, chaps. VI–VIII.

87. Thus, for example, Cavendish explained a reaction of the form metal + acid = "salt" + "inflammable air" as (calx + phlogiston) + acid = (calx + acid) + phlogiston: Partington [95], vol. 3, p. 444.

caloric theory before publishing his celebrated anti-phlogistic manifesto, "Reflections on Phlogiston," in 1786. Moreover, a substantial part of this paper, published after the *Memoir on Heat* of 1783, is devoted to Lavoisier's "covering theory" (the caloric theory) and to the consequent account of the physical side of the problem.[88] It is clear, then, that one of the most important factors contributing to the initial acceptance of Lavoisier's anti-phlogistic chemistry was the elegant way in which it united the chemical theory of combustion and calcination with the new physical discoveries in the science of heat. The unification of chemistry with this new branch of experimental physics, together with Lavoisier's theory of the composition of water and his experiments on the decomposition of water by metals, was decisive in securing the increasing acceptance of Lavoisier's system from 1787 on.[89]

The factors responsible for the developing acceptance of Lavoisier's system, especially in Germany, are clearly exhibited in J. S. T. Gehler's *Physikalisches Wörterbuch* [37], first appearing in 1787–1796—which, as we shall see, was quite familiar to Kant. Gehler was a lucid, informed, and judicious expositer of the latest scientific discoveries; and it is particularly illuminating to trace his evolving appreciation for Lavoisier's theory in the articles "Phlogiston" (1790: [37], vol. III, pp. 461–475), "Heat [Wärme]" (1791: [37], vol. IV, pp. 532–569), and "Anti-phlogistic System" (1796: [37], vol. V, pp. 31–49).

88. See Partington [95], vol. 3, pp. 462–466; and especially Guerlac [41], where Lavoisier's evolving knowledge of the advances in the science of heat, together with the influence of this knowledge on his chemical theory, is carefully documented. Guerlac sums up the main point as follows ([41], p. 256): "the deeper understanding of heat phenomena, stimulated by the work of the British school, and by the experiments and discussions with Laplace, put the capstone on Lavoisier's antiphlogistic theory of combustion. Just as the experiments of 1777 with Laplace encouraged Lavoisier to come out openly against the phlogiston theory, though with considerable caution, so the calorimetric experiments of 1782–1784 and their theoretical comprehension contributed important, if not decisive, elements of his famous manifesto, his 'Réflexions sur le phlogistique' of 1786."

89. For the initial reception of Lavoisier's system see Partington [95], vol. 3, pp. 488–494. For the importance of the theory of the composition of water see Guerlac [41], pp. 261–266, and Partington [95], vol. 3, pp. 443–450, 457–460. This latter theory permitted Lavoisier to explain the "inflammable air" evolved from metals acted upon by dilute acids—which, as noted above, had furnished substantial confirmation for the phlogiston theory. Given the composition of water from oxygen and hydrogen, Lavoisier could now assert that the evolved hydrogen gas came not from the metal but from the water in which the acid was diluted: a reaction of the form metal + acid = "salt" + "inflammable air" could now be explained, for the case of zinc dissolved in dilute sulphuric acid, say, as zinc + (hydrogen + oxygen) = zinc oxide + hydrogen; zinc oxide + sulphuric acid = sulphate of zinc. As Partington points out (p. 444), whereas Lavoisier viewed sulphuric acid as an oxide (sulphuric acid anhydride, SO_3), at present it is viewed as an oxide plus water ($H_2O + SO_3 = H_2SO_4$), so that the hydrogen really comes from the acid.

The article on phlogiston (also called the "inflammable being [brennbares Wesen]") expounds Stahl's theory as the correct account of combustion and calcination. Phlogiston is a purely hypothetical entity, however, so opinions vary as to its precise nature: Stahl himself views it as fire combined with a type of earth; other view it as fire itself; still others, such as Scheele, view it as the newly discovered "inflammable air." In any case, there is no doubt that the discoveries of the last fifteen years concerning the different types of gases have produced advances in the theory of combustion and phlogiston. In particular, Lavoisier has convincingly shown that the increase in weight of the products of combustion and calcination is due to an absorption of the newly discovered "dephlogisticated air." Lavoisier himself views "dephlogisticated air" as a combination of oxygen base plus caloric ("Feuerstoff"), and, on this basis, he has even attempted to explain the phenomena in question without phlogiston:

> That which, according to Stahlian conceptions is extraction of the inflammable [being], is here considered as combination with the principle [Grundstoff] of acids; that which one customarily views as uniting with phlogiston is here called release of the acidifying principle; and so on. . . . Thus the explanations of this system are precisely the reverse of the usual ones. ([37], vol. III, pp. 468–469)

One cannot deny that Lavoisier's explanations are simple and natural, and one must admit that he has shown that "dephlogisticated air" is in fact absorbed in combustion. It is also true, however, that the idea that "dephlogisticated air" = oxygen base plus caloric is just as hypothetical as Stahl's phlogistic conception.[90] And there is no reason, finally, that there cannot be a *substitution* of substances in combustion: as oxygen is absorbed phlogiston is released.

The article on heat, appearing one year later, is most instructive. Gehler vigorously defends the caloric theory of heat and of the states of aggregation, and he bases this defense squarely on the new discoveries concerning latent and specific heats due to Wilcke and Black. For these discoveries show precisely that heat obeys the chemical laws of combination or affinity rather than the mechanical laws of motion. One must distinguish, in particular, between latent or bound heat, which is combined with another substance in the same way that any solvent is combined with its solute, and free heat, which acts sensibly on the thermometer, expands on all

90. This point is perfectly correct, of course, and Lavoisier accordingly attempted to gloss over his own use of hypothetical principles. Indeed, Lavoisier's idea that oxygen gas is a combination of oxygen base plus caloric (with the latter disengaged in combustion) is uncomfortably close to Stahl's idea that phlogiston is a combination of an earth plus fire (with the latter disengaged in combustion). See Partington [95], vol. 3, p. 463, for example; and Metzger [80], pp. 36–44.

sides in all spaces, and penetrates all bodies and vessels. In its bound form heat is the cause of the transition from solidity or rigidity to liquidity, and also of the transition to the vaporous state and to the permanently elastic or gaseous state. It follows that rigidity and fluidity are not essential properties of any substance and that heat is the sole proper cause of fluidity; similarly, vapor too is a mere state, also due solely to heat. These facts concerning the chemical combinations of heat are so well established that scarcely any physicist now doubts the existence of caloric as a real material substance—which, accordingly, cannot be viewed as merely hypothetical.[91] Finally, Lavoisier has shown that gases in general are solutions in caloric and has used this result in his new theory of combustion: for Lavoisier, combustion consists in the decomposition of certain substances by means of caloric, which substances experience an absorption of pure air accompanied by light and a very copious release of bound heat according to a complicated network of elective affinities. Only the future will tell, however, whether this account is correct, and the entire theory of heat and fire now appears to be approaching a decisive crisis.

In the article on the anti-phlogistic system, appearing five years later in a supplemental volume, the transformation is complete. Gehler reports that the anti-phlogistic system has become gradually better known in Germany since 1789 and has now achieved a definitive victory. (In the meantime, Lavoisier's *Elements* has been translated by S. F. Hermbstädt in 1792, and C. Girtanner has published a version of the new chemical nomenclature in 1791 and a textbook—*Anfangsgründe der antiphlogistischen Chemie*—in 1792.[92]) No physicist will now deny that among all the hypothetical systems Lavoisier's has pride of place, due to the outstanding simplicity and ease of its explanations. The entire system proceeds from the action of caloric, which penetrates the smallest parts of bodies through its elasticity and transforms them into a state of "drop-forming fluidity" (liquidity) or, when the elasticity is sufficient to overcome the pressure of the atmosphere, into a state of "elastic fluidity" or gas.[93] The air of the atmosphere consists of two gases in the ratio 27:72, the first of which consists of the basis oxygen plus caloric. Phosphorous, sulphur, and carbon all have the property of separating oxygen from its bound caloric at a sufficiently high temperature, which caloric is then

91. For more on the importance of the discovery of specific and latent heats in the establishment of the caloric theory, especially in Germany, see Adickes's notes to Reflexion 54: 14, 448–456.

92. For the reception of Lavoisier's system in Germany, see Partington [95], vol. 3, pp. 492–494 and chap. XII.

93. In basing the entire system on the caloric theory, Gehler is of course simply following the lead of Lavoisier himself in the first chapter of the *Elements*—as does Girtanner in the second chapter of his textbook.

released and manifests itself as heat and light—and in this consists combustion. The products of combustion are acids, all of which therefore consist of a basis plus oxygen. Water too is a composite and consists of hydrogen plus oxygen in the ratio 15:85.

In general, then, Gehler continues, what is most distinctive about the new system is not so much the rejection of phlogiston but rather a radically new conception of the elements that reverses the old ordering of the simple and the complex: air and water are no longer elements but complexes; carbon, sulphur, phosphorous, the basis of marine acid (muriatic radical), and the metals are now simple; acids are now relatively complex (oxides); gases are complexes of a base plus caloric; and so on. And it is precisely this reordering of the elements that immediately solves a number of important problems: in particular, the absorption of oxygen naturally explains the increase in weight in the products of combustion, and the release of caloric naturally explains the heat evolved in the process of combustion. It is true that opponents have attacked the new theory for employing arbitrary assumptions and circular explanations; since 1782, however, experiments by Lavoisier and others on the composition of water and on latent heat (the calorimetric work of Lavoisier and Laplace) have together filled the gaps in the system and grounded its calculations. Nevertheless, it remains a hypothesis, and we cannot here expect the kind of certainty we have concerning the Copernican system.

Gehler's physical dictionary therefore exemplifies the three stages in the developing acceptance of Lavoisier's system emphasized above. In the first stage the phlogiston theory is still accepted while, at the same time, the importance of the newly discovered gases or "elastic fluids" is also explicitly recognized. In particular, it is explicitly recognized that Lavoisier has shown that oxygen gas is absorbed in combustion and calcination. In the second stage the recent discoveries concerning specific and latent heats are assimilated, and these are conceived as supporting the caloric theory of heat via the idea of a chemical combination or affinity of heat with other material substances. Moreover, Lavoisier's conception of the gaseous state is seen to follow naturally from the caloric theory, and, on this basis, considerably more force is granted to his anti-phlogistic theory of combustion. In the third stage the discovery of the synthesis and decomposition of water, together with the systematization and perfection of the theory of latent and specific heats by Lavoisier and Laplace, leads to the final triumph of the anti-phlogistic system.

If Gehler is to be relied upon in this matter, therefore, it appears that the assimilation of Lavoisier's doctrine in Germany closely followed Lavoisier's own theoretical development. In any case, however, it is now time to return to Kant. What can be said more specifically about Kant's own evolving knowledge of the developments we have been reviewing?

Kant's published writings unfortunately provide us with very little information in this regard. We know that in the critical period Kant still accepted the chemistry of Stahl, whereas by 1797, in the Preface to the First Part of *The Metaphysics of Morals,* he has officially embraced the chemistry of Lavoisier. Moreover, neither the first *Critique* nor the *Metaphysical Foundations* manifests explicit recognition of any of the developments we have been considering. Thus, for example, the passage at A645–646/B673–674 discussed above considers only "*pure earth, pure water, pure air,* etc."—and there is no indication whatever that neither water nor air is a simple substance or element. And, when Kant touches on chemical topics in the General Observation to Dynamics of the *Metaphysical Foundations* (4, 526–532), there is no hint of the new developments in pneumatic chemistry resulting in the discovery of a variety of "elastic fluids" or gases. In #2 (526.12–529.25) the only states of matter considered are solidity and fluidity, and the notion of fluidity here clearly includes only liquids. Similarly, #3 (529.26–530.7) considers only a single example of "expansive elasticity": namely, *air* (in the singular). Nevertheless, there is an intriguing remark about air and the "matter of heat":

> Thus air possesses a derived elasticity by means of the matter of heat, which is intimately united with [the air], and whose elasticity is perhaps original. On the other hand, the basic matter [Grundstoff] of the fluid that we call air must nevertheless still possess its own elasticity, which is called original. (530.1–5)

But this remark can hardly be construed as an expression of either the caloric theory of gases or the general doctrine of latent heat.[94]

On the other hand, Kant touches on both the new developments in pneumatic chemistry and the advances in the science of heat in *On the Volcanos on the Moon* (1785). In particular, Kant attempts to explain the origin of heat in the solar system by means of the idea that condensing vapors generate heat:

> Thus *Crawford's* discoveries supply a hint to make comprehensible the generation of as large a degree of heat as one wishes simultaneously with the formation of the heavenly bodies. For, if the element of heat by itself is uniformly distributed everywhere in the universe, but is only attached to

94. No "elastic fluid" other than air is mentioned; there is no suggestion that solids and liquids can be transformed into gases by the addition of heat; finally, there is no commitment to a *chemical* combination of air with the matter of heat: on the contrary, Kant's more detailed discussion at 522.24–38 suggests a *mechanical* action by which the matter of heat acts on "air-particles" in accordance with "the law of communication of motion through vibration of elastic matters" (which, Kant thinks, can be used to explain the Boyle-Mariotte Law)—there is no question here, that is, of a genuine *solution* of air in the matter of heat (compare #4 at 530.8–532.9).

> various matters in the proportion in which they variously attract it; if, as he proves, vaporous matters contain—and indeed require for a vaporous expansion—heat that they can hold as soon as they pass into the state of denser masses, i.e., unite into heavenly spheres: then these spheres must contain an excess of heat-matter over the natural equilibrium with the heat-matter in the space where they are found—i.e., their relative heat with respect to that of the universe is increased. (Thus, vitriolic acid air, when it comes in contact with ice, immediately loses its vaporous state, and thereby augments the heat in such an amount that the ice melts in a moment.) (8, 74.27–75.11)

In this passage Kant explicitly recognizes the existence of gases other than common air, as well as an important phenomenon connected with the discovery of latent heats: namely, heat of condensation. Yet there is still no clear indication of the general doctrine of latent heat, or even of the specific idea that a definite quantity of heat is required simply to bring about a state transition.[95]

A much clearer statement of the doctrine of latent heats is found in §58 of the *Critique of Judgement,* where Kant discusses the formation of solid bodies via crystallization from a liquid:

> The commonest example of this type of formation is freezing water; in which straight ice-rays are first generated which join together in angles of 60 degrees, while others attach themselves in the same way to each vertex, until all has become ice—in such a way that during this time the water between the ice crystals does not become continuously more viscous, but is as completely fluid as it would be at a much greater heat, and is yet perfectly ice cold. The matter that disengages itself, which is suddenly released at the moment of solidification, is a considerable quantity of caloric [Wärmestoff], whose absence, since it was merely required for the state of fluidity [zum Flüssigsein], leaves this now present ice not in the least colder than the water that shortly before was fluid. (5, 348.24–35)

Moreover, on the next page, Kant speaks of "an atmosphere, which is a mixture of different types of air [ein Gemisch verscheidener Luftarten]" (349.28–29). By 1790, then, it appears that Kant has assimilated the first

95. Moreover, Kant appears to be confused about the relation between latent and free heat, for he suggests that condensing vapors *retain* their heat of condensation in the resulting solid state. Crawford, in his 1779 book, deliberately refrains from discussing the theory of latent heats—confining himself to specific heats—and simply states that cold is produced by the evaporation of water, heat by the condensation of vapor. Indeed, Crawford himself rejected the doctrine of latent heats in the form of a *chemical* combination of heat with matter, and instead adopted Irvine's theory according to which the different states of aggregation have different *specific* heats (see Partington [95], vol. 3, pp. 156–157). For this reason Crawford is sharply criticized by Gehler in the above-cited article on heat: [37], vol. IV, pp. 564–566. (Kant, however, in the quoted passage, appears clearly to accept the chemical action of heat.)

two stages of the chemical revolution: the new developments in pneumatic chemistry and the advances in the science of heat.[96]

A somewhat more fine-grained picture emerges from a consideration of the evolution of Kant's lectures on theoretical physics. We know that Kant lectured in 1776, 1779, 1781, and 1783 using as his textbook J. C. T. Erxleben's *Anfangsgründe der Naturlehre,* which appeared in 1772 (second edition 1777)[97]—and the 1776 lectures are reported on in the *Berliner Physik*. Erxleben does not expound any of the new developments in chemistry (pneumatic chemistry and the theory of heat) in either the first or second edition—and, accordingly, neither does the *Berliner Physik.* In 1785, however, Kant chose a new textbook for his lectures: W. J. G. Karsten's *Anleitung zur gemeinnützlichen Kenntniß der Natur* (1783)—see Arnoldt [3], pp. 599–600. Moreover, we are fortunate enough to have a report on the 1785 lectures, the *Danziger Physik* (29, 95–169—Karsten's textbook is conveniently reprinted in full on pp. 171–590).

The *Danziger Physik,* following Karsten, does expound some of the new chemical developments: the developments in pneumatic chemistry in particular. Chapter X concerns the elements and begins with a discussion of "pure fire, water, air, etc." (161.4–5), which, just as in the first *Critique* at A645–646/B673–674, are classifications originating in reason "which precede experience and according to which we then order our experiences" (161.8–9). Now, however, it is asserted that "Air is no element,

96. Reflexion 64 (14, 482: from the late 1780s) is an extract from F. X. Baader's *Vom Wärmestoff, seiner Vertheilung, Bindung und Entbindung, vorzüglich beim Brennen der Körper* (1786), concerning the "heat-binding force" responsible for specific heats: see Adickes's notes at 14, 482–483 (Gehler lists Baader's book as a source for the caloric theory of heat in his article on heat in his physical dictionary). This confirms the idea that Kant was well on the way to assimilating the advances in the science of heat by the late 1780s. Adickes appears to think that Kant may have even embraced the caloric theory of latent heat before 1779—see his comments on R. 54: 448–456. This seems to me to be very unlikely, however. Reflexion 54 itself says nothing whatever concerning latent heat (although it does mention the "element of fire [Feuerelement]"); and Adickes's main ground for suggesting such an early indication of the doctrine of latent heat here is that in R. 45 (1776) Kant already paraphrases Kästner's 1776 translation of Wilcke's 1772 paper at 391.2–3: "With boiling water one can melt only 1⅓ as much snow, and the mixture is ice cold." Again, however, this is simply a phenomenon explained by the theory of latent heat, not that theory itself (moreover, it is actually quite difficult to extract a theoretical notion of latent heat from Wilcke's paper: see [75], pp. 78–94). In any case, however, it is very unlikely that Kant has made the connection between the idea of latent heat and the existence of an "element of fire" in 1776: Adickes takes R. 45 as notes for Kant's physics lectures of 1776 (392.9–10), but in the *Berliner Physik* itself (a report on these lectures) there is no mention either of the idea of latent heat or of the "matter of heat"—on the contrary, this work appears clearly to embrace a *mechanical* conception of heat (29, 82.33–35, 83.19–20).

97. See Arnoldt [3], pp. 572–590. A copy of the 1772 edition of Erxleben, "with numerous written observations," was found in Kant's library: see Warda [118], p. 28.

since it seems to be a certain form [Gestalt] into which everything can be transformed" (162.1–3), for "*One can bring all matters into a vaporous form [Dunstform] by means of the burning glass and perhaps even the earth was such at the beginning*" (162.10–12). "Phlogisticated air" (163.8) and "fixed air" (163.28) are mentioned specifically, and, in general: "The various types of air that one extracts from chalk, fermenting beer, etc. one calls gas" (163.24–26). Thus we here have a significant advance on the view of the elements expressed in the first *Critique*.

The account in the *Danziger Physik* breaks off with these remarks on chapter X of Karsten. Karsten himself goes much further, however: he considers the newly discovered types of air in much greater detail in chapter XVI and some of the new opinions concerning the nature of fire in chapter XXVI. In particular, §§464–472 present Crawford's theory of specific heats, and §§491–495 consider some of the new views on phlogiston—including a very sympathetic presentation of Lavoisier's anti-phlogistic hypothesis.[98] Earlier, in chapter XII (§§194–197), Karsten presents the concept of latent heat in terms of a "fire-matter" obeying the laws of chemical affinity and elective attraction. Nevertheless, he appears still not to have fully grasped the caloric theory of gases, for he upholds the traditional view of vaporization as a dissolution in *air* (325.29–30). This is quite hard to reconcile with the modern view of atmospheric air as a mixture of "elastic fluids" (355.12–19), of course, so Karsten also admits that in some cases of vaporization "the dissolving action is to be ascribed more to the *fiery basic principle of heat* [*feurigen Grundstoff der Wärme*] then to the air" (351.9–11).[99]

There is a very interesting—and I think very significant—divergence between Karsten and the *Danziger Physik* concerning the respective roles of mathematics, physics, and chemistry. Karsten argues (173–194) that physics or natural science has been mistakenly identified with applied

98. Karsten holds that Lavoisier's work on the absorption of oxygen is decisive but not at all incompatible with Stahl's theory: there can be a *substitution* of matters whereby the oxygen principle is absorbed while phlogiston is released. In this connection he refers explicitly to the theories of Kirwan and Crawford: 29, 518–519. (According to Crawford, heat and phlogiston are opposing principles: when one enters the other is removed. In combustion, then, heat or fire is released from the air with which it is bound, while the inflammable body loses phlogiston which is then gained by the air.)

99. The *Danziger Physik* itself does not mention the doctrine of latent heat. It does, however, state the material theory of heat (118.23–119.7), and even asserts the intertransformability of the three states of aggregation through heat: "All bodies become solid, if one extracts sufficient heat from them; fluid, if the heat is permitted to rise very high; and, if the heat is still increased, they transform themselves into elastic vapors" (121.1–4). This is not yet the doctrine of latent heat, of course, for there is no suggestion that a certain fixed amount of heat is required simply to effect the state transition, with no corresponding increase (or decrease) in temperature.

mathematics. However, since physics is the science of the basic materials found in nature, their properties, and their actions upon one another, it actually makes more sense to identify physics with chemistry than with mathematics—for applied mathematics or the doctrine of magnitude yields only the measure or quantity of an effect and not its properties. Chemistry, on the other hand, is quite naturally a part of physics, for chemistry is simply the closer study of what can be worked up from the basic materials found in nature. Thus the customary classification of the universal doctrine of nature into physics, chemistry, and natural history is quite artificial: chemistry is really nothing but applied physics, and, moreover, chemistry includes minerology which is part of natural history.

The *Danziger Physik,* on the contrary, maintains the traditional division of the universal doctrine of nature into "mathematical physics, chemistry, and natural description" (94.25–26). Physics is based on a priori principles (97.25) and, in particular, on mathematics (97.7–8). Chemistry is based on a posteriori principles (97.26), for:

> Mathematics certainly does not suffice to explain the outcome of chemistry or one has not yet been able to explain a single chemical experiment mathematically: therefore one left chemistry out of the doctrine of nature, because it has no a priori principles, in that here neither figure nor magnitude comes into consideration, but rather merely the most intimate action that one matter has on another. (97.12–18)

Natural description, finally, is based on no principles at all, but merely records the historically given variety of matters through observation (97.26–28). And here one can scarcely even begin applying mathematics (97.22–33). Nevertheless, natural description "presupposes chemical knowledge, just as much as chemistry in turn presupposes mathematical knowledge of nature" (97.23–24). A difficult question thus arises:

> If mathematics is necessary for natural knowledge, then: how much mathematics is necessary, or how much mathematics is one to bring in? Karsten, an otherwise scrupulous physicist, has too little mathematics, and, in general, this question is difficult to decide. (99.28–31)

The divergence between Karsten and the *Danziger Physik* is therefore evident.

According to the *Danziger Physik,* however, there has recently been a significant change in chemistry:

> Chemistry has raised itself to greater perfection in recent times; it also rightfully deserves the claim to the entire doctrine of nature: for only the fewest appearances of nature can be explained mathematically—only the smallest part of the occurrences of nature can be mathematically demonstrated. Thus, e.g., it can, to be sure, be explained according to mathematical propositions:

> when snow falls to the earth; but why vapors transform into drops or are able to dissolve—here mathematics yields no elucidation, but this must be explained from universal empirical laws of chemistry, and philosophy always belongs to chemistry, for it is a matter for the philosopher: to discover the universal laws of the action of matter through experience and to derive everything therefrom systematically. (97.30–98.7)

It is hard to resist the conclusion that this situation—Kant's developing awareness (in 1785) of the new chemical developments and of the general importance of chemistry, coupled with his continuing insistence on the foundational role of mathematics in natural science—constitutes the immediate background for the discussion of chemistry and mathematics in the Preface to the *Metaphysical Foundations*.[100]

In the *Danziger Physik* an intriguing passage then follows:

> A few scholars have attempted to combine chemistry with physics—such as Erxleben, who has begun a chapter of chemistry, but this is also quite insufficient: for he does nothing more than speak of solutions and precipitations. Lichtenberg has already carried it forward somewhat more extensively. (98.10–14)[101]

And this is especially intriguing because Kant once again adopted a new textbook the very next time he lectured on theoretical physics, in 1787–88: G. C. Lichtenberg's extensively annotated third edition (1784) of Erxleben's *Anfangsgründe der Naturlehre* [24] (see Arnoldt [3], pp. 615–616). Moreover, there can be no doubt that the quoted passage is referring to precisely this work.[102]

Unfortunately, we do not yet have available a report on Kant's lectures from 1787–88. Nevertheless, a consideration of the content of Lichtenberg-Erxleben—especially of the extensive additions and corrections made by Lichtenberg—may perhaps shed light on the development

100. Compare also 100.21–102.9 of the *Danziger Physik*, which touches on many key themes of the *Metaphysical Foundations*.

101. Note that the complaint lodged against Erxleben here applies equally to Kant's own discussion of chemistry in #4 of the General Observation to Dynamics in the *Metaphysical Foundations* (4, 530.8–532.9), which considers only the abstract and general features of chemical solution.

102. Compare Lichtenberg's remarks in the Preface to his edition: "Various [matters] which properly belong in chemistry have been brought forward—not only some new discoveries that could not be passed over here, but those occasioned by the author himself. He speaks of *solution, precipitation, acidification, crystallization, etc.* which makes necessary explanations that cannot be wholly left to oral delivery if one wants [the matter] to be understandable to the majority of the audience. In precisely the same way, the report on the different types of air, which cannot be dispensed with for a better understanding of some following chapters, makes necessary an acquaintance with various chemical means of dissolution" ([24], pp. xxvi–xxvii).

of Kant's thinking here. For present purposes the most important of Lichtenberg's additions are the following: an appendix to chapter 6 (on fluids) outlining the basic concepts and classifications of chemistry, an insertion in chapter 7 (on air) considering the newly discovered gases or "factitious airs," and an addition of eighteen new sections (§§494b–s) to chapter 9 (on heat) presenting the recent findings concerning latent heats, specific heats, and the theory of combustion.

Much of the new material added by Lichtenberg overlaps with Karsten. Thus the system of chemistry outlined in the appendix to chapter 6 is essentially traditional Stahlian chemistry (it clearly expounds the basic classification of substances into salts, earths, inflammable matters, and metals as well as the phlogistic account of combustion and calcination), while, just as in Karsten, the new discoveries in pneumatic chemistry are simply grafted onto this framework (there is speculation that phlogiston may be identical to "inflammable air," for example). What is most striking, however, is the much more sophisticated and systematic grasp of the caloric theory of heat. The new theory is presented in §§494b–s as a "Brief Outline of Crawford's Theory of Fire."[103] Lichtenberg begins by embracing the material theory of heat and the idea that heat or the matter of fire combines with other bodies according to the rules of chemical affinity. He distinguishes free fire or heat from bound fire or heat and, on this basis, explains the concepts of specific and latent heat. Black's estimate of approximately 130° F is given for the latent heat of fusion of ice, Watt's estimate of approximately 800° for the latent heat of vaporization of steam. That Lichtenberg—unlike Karsten[104]—explicitly considers latent heat of vaporization is significant; for Lichtenberg—again unlike Karsten—makes a point of correcting the traditional view (upheld by Erxleben) of vapors as solutions in *air:* on the contrary, argues Lichtenberg, air has nothing to do with the question, and both vapors and "permanently elastic fluids" are best viewed as solutions in the "matter of fire" ([24], pp. 378ff.). In other words, Lichtenberg has made the decisive step of endorsing the caloric theory of gases.[105]

103. Lichtenberg, following J. H. Magellan's *Essai sur la Nouvelle Théorie du Feu Élémentaire, et de la Chaleur des Corps* (1780), mistakenly attributes the new discoveries to Crawford—although Crawford himself is merely following Irvine's lecures on Black's work and, moreover, does not in fact develop the theory of *latent* heat at all in his 1779 book: see [75], pp. 38–44. In any case, it is interesting to note that Lichtenberg's discussion here is taken as a source for the caloric theory of heat by both Gehler (in the article on heat in his physical dictionary) and Girtanner (in chapter 2 of his textbook on anti-phlogistic chemistry).

104. Karsten, following Wilcke, discusses only latent heat of fusion and ignores latent heat of vaporization: 29, 316–319.

105. Perhaps because of this, Lichtenberg is also quite open-minded about Lavoisier's

Lichtenberg's reasons for adding so much properly chemical material to a textbook on physics are especially instructive. When first introducing chemical subjects in chapter 6 he explains that discoveries since the last (1777) edition in the theories of air and fire make necessary the consideration of more chemical information—which, as such, one could previously ignore in a customary physics text. The insertion on gases in chapter 7 amplifies on this: Erxleben had left the study of "factitious airs" to chemistry alone, but now such study is unavoidably required by the physicist; for it makes clear for the first time the true nature of our atmosphere, and, moreover, the new "factitious airs" are intimately connected with newly discovered facts concerning the nature of fire, which, in turn, shed much light on the nature of bodies in general—for example, solid bodies can easily be transformed into "permanently elastic" bodies and vice versa. It appears, then, that Lichtenberg is here advocating a more intimate relationship between physics and chemistry precisely because of the connections, emphasized above, between the new developments in pneumatic chemistry and the caloric theory of heat—connections effected by the caloric theory of gases and of the states of aggregation.

Now if, as appears not unlikely, Kant has assimilated this new material in Lichtenberg-Erxleben in 1787–88, then Kant also has new grounds for optimism concerning the unification of physics and chemistry in 1787–88. In other words, Kant himself may have also seen, following Lichtenberg, that the caloric theory of heat and of the states of aggregation dramatically enhances the prospects for a "physicalization" of chemistry that could ground it as a genuine science.[106] And, as emphasized above, while it in

views. Although he appears to prefer Crawford's theory of combustion (which occupies a kind of intermediate position between Lavoisier and Stahl: see note 98 above), he also grants that Lavoisier's anti-phlogistic conception is a genuine option (§494s). Moreover, he appears to be quite well disposed towards Lavoisier's theory of the composition of water ([24], pp. 170, 215). Nevertheless, Lichtenberg delayed his official acceptance of Lavoisier's system until 1794, when he proclaimed the triumph of the new theory in his Preface to the sixth edition of Erxleben. Even here, however, he remained suspicious of what he considered to be the artificial simplicity of "French chemistry": in particular, he thought that when *electricity* was properly brought into consideration significant changes in Lavoisier's system would be required. See, for example, Gehler's article on the anti-phlogistic system: [37], vol. V, pp. 45ff. In this respect Lichtenberg proved to be prophetic indeed: for Davy's electrochemical work on chlorine and fluorine destroyed Lavoisier's theory of acidity, and Berzelius's general electrochemical theory made the caloric theory of heat chemically superfluous—fire and heat of combustion and chemical reaction were now viewed as electrical in origin: see Partington [95], vol. 4, pp. 51–59 and pp. 168–174 respectively.

106. In this connection it is interesting to note that the second (1787) edition of the transcendental deduction gives as examples the application of the category of quantity to "the empirical intuition of a house" and the application of the category of causality to a state transition from fluidity to solidity (B162–163)—where the latter replaces the famous

no way forces an official acceptance of Lavoisier's chemistry, to see this is nonetheless to be extremely close to such an acceptance. Yet Kant did not accept Lavoisier's system in 1787, of course, for, as we have seen, he makes a special point of citing Stahl's theory of the calcination of metals in the Preface to the second edition of the first *Critique* at Bxii–xiii.[107] Moreover, it is quite difficult to specify more exactly when, between 1787 and 1797, Kant actually came to embrace Lavoisier.

The first explicit mention of Lavoisier with which I am acquainted is a fragment from 1789–90 (R. 66):

> According to Lavoisier, when something (according to Stahl) is dephlogisticated, something is added (pure air); when it is phlogisticated, something (pure air) is removed . . . . (14, 489.8–10)[108]

And this fragment is striking for the way in which it appears to introduce a reconsideration of Bxii–xiii. It is of course impossible to determine from Reflexion 66 alone, however, whether Kant means to accept Lavoisier's view, to dispute it, or merely to consider it. As far as I know, the first more definite indication occurs in a report on Kant's lectures on metaphysics in 1792–93 *(Metaphysik Dohna):*

> Is water an element? No; for it can still be decomposed, it consists of respirable air [Lebensluft] and inflammable air [brennbarer luft], and we call something elementary that contains no species. (28, 664.14–16)[109]

But from this report we cannot determine with certainty what Kant himself thought at the time, and, in any case, to assert the compositeness of water is not necessarily explicitly to embrace Lavoisier (recall that Lichtenberg does the former but not the latter in his 1784 edition of Erxleben).

However, in a letter to S. T. Soemmerring of August 10, 1795, Kant himself asserts the compositeness of water in a way that is unmistakably suggestive of Lavoisier's particular doctrine:

---

example of the ship being driven downstream by the current from the Second Analogy (A191–193/B236–238): a *mechanical* example of causality via the "communication of motion" has been replaced by a *physico-chemical* example.

107. This is in no way incompatible with Kant's having fully assimilated Karsten and Lichtenberg-Erxleben by 1787. For both Karsten and Lichtenberg expound Stahl's theory of metals as composites of metallic calx plus phlogiston: see Karsten, §156; Lichtenberg-Erxleben [24], pp. 157–164.

108. Adickes conjectures (493.20–33) that R. 66 may have been written in 1790 following the passage from Gehler's article on phlogiston cited above.

109. The importance of this passage is rightly stressed by Tuschling [113], p. 43 (n. 16). More generally, I entirely agree with Tuschling about the importance of new developments in science (since the *Metaphysical Foundations*)—and, in particular, of the new developments in chemistry—for the *Transition* project: [113], pp. 39–46.

> Pure common water, until recently still held to be a chemical element, is now separated through pneumatic experiments into two different types of air. Each of these types of air, outside of its basis, still contains caloric—which can perhaps in turn be decomposed by nature into light-material and other matter, just as light can be further decomposed into various colors, etc. (12, 33.31–34.2)

This clear juxtaposition of the composition of water with Lavoisier's caloric theory of gases makes it highly probable that Kant has explicitly adopted Lavoisier's system by 1795 at the latest.[110]

Thus, although our evidence is rather sparse, it appears from what we do know that Kant's assimilation of the developments constituting the chemical revolution between the critical period and that of the *Opus* follows the same pattern we have seen unfolding several times above. It appears that by 1785 he has become aware of the new discoveries in pneumatic chemistry and, in particular, of the composition of the atmosphere; between 1785 and 1790 he has assimilated the developments in the science of heat, including the doctrine of latent heat and the caloric theory of the states of aggregation; between 1790 and 1795 he has completed the conversion to Lavoisier's system of chemistry (and, in particular, has become aware of the composition of water). Moreover, it appears not unlikely that Kant came, following Lichtenberg, to view the unification of the caloric theory with the developments in pneumatic chemistry effected by the caloric theory of gases as a possible basis for a new, more physical chemistry together with a new, more chemically sophisticated physics—a new kind of *physical chemistry*, which was then actually con-

110. Adickes, on the basis of Reflexionen 72–73 (1793–94)—which consider phlogistic alternatives to Lavoisier's theory of the composition of water (reminiscent of Cavendish's theory mentioned in note 86 above)—maintains that Kant embraced Lavoisier *only* in 1795 (14, 490–495, 503–516), and even goes so far as to assert that from 1793 until 1795 Kant "sharply attacked" Lavoisier's conception: see [1], vol. I, p. 63. I follow Tuschling (see note 109 above) in thinking that Adickes has become overly zealous here. For, as Adickes himself points out, R. 72–73 concern views of J. H. Voigt (14, 506.35–509.42, 513.25–514.9), and they are perfectly consistent, as far as I can see, with the idea that Kant is simply recording or considering Voigt's views—just as he is recording or considering Gren's views in R. 75, 77, and 78. Indeed, R. 72 begins "Supposing that . . . " (510.3); R. 73 begins "Perhaps . . . " (513.3). In any case, in addition to the above report from the *Metaphysik Dohna* (which Adickes does not mention), a further relevant consideration is the circumstance that Girtanner had apparently sent Kant a copy of his textbook of anti-phlogistic chemistry in 1792: a letter from J. B. Erhard in January 1793 reports that Girtanner is anxiously waiting for Kant's reaction (11, 408.30–31), and a copy of the textbook was found in Kant's library (Warda [118], p. 34). It is possible, then, that Kant accepted Lavoisier in 1792–93 upon reading Girtanner (who argues, in chap. 13, that the discovery of the composition of water is one of the most important results of anti-phlogistic chemistry). All we really know for certain, however, is that between 1792 and 1795 Kant is following the debate over Lavoisier's system with ever-increasing interest.

structed by Lavoisier. Could it not be this growing awareness of the new physical chemistry which, more than any other factor, fuels the new optimism about the empirical or experimental sciences manifest in Kant's *Transition* project?[111]

IV · The Aether-Deduction

The *Opus* itself exhibits a thoroughgoing involvement with the developments reviewed above. Very early on, for example, in loose-leaf 26/32 (before 1790), Kant states the caloric theory of gases: "*Vapor,* is the state of a matter which is dissolved in fire and therefore becomes elastic" (21, 417.17–19; see also, e.g., 21, 464.22–24, 381.8–9, 384.6–9; 22, 213.26–27, 214.27–215.3; 21, 521.5–7, 480.11–12, 481.17–23). During the same early period, in loose-leaf 43/47, the problem of solidity is closely connected with the concept of latent heat: "A basic principle to explain solidity is the loss of absolute heat in becoming ice and the equality of relative heat with freezing water" (21, 424.15–17; see also, e.g., 21, 382.29–383.10, 395.10–12; 22, 213.6–17; 21, 248.21–23, 523.9–10, 477.24–26, 479.11–17, 479.28–480.4, 481.8–14). And the following fragment on the science of chemistry from loose-leaf 23 dates from several years later:

> ⟨What is chemistry? The science of the inner forces of matter.⟩
> Dissolution (chemical) is the separation of two matters penetrating one another through attraction.—It is either quantitative when the matter is divided into homogeneous or qualitative when it is divided into its inhomogeneous (specifically different) matters. (a) Water into vapor (b) into two types of air. The latter is properly called decomposition[.] (21, 453.19–24)

Thus, as one would expect at this time, Kant illustrates the science of chemistry with Lavoisier's doctrine of the composition of water (see also, e.g., 21, 432.4–7, 401.2–4; 22, 251.26–28, 238.28–31, 508.26–509.1).

A striking—and I think most significant—passage occurs in loose-leaf 3/4 (1798), which, as we have seen several times already, is of pivotal importance in the development of Kant's thinking in the *Opus.* Kant begins by asking if the phenomenon of latent heat of fusion requires the notion of the binding of caloric:

111. Förster argues, on the basis of Kant's correspondence with Kiesewetter, that the idea of the *Transition* project goes back at least to 1790 (and thus eight years before the "gap" in the critical system is first mentioned): [31], pp. 536–537. The above suggestion about the source of the *Transition* project is perfectly consistent with this, I think, for it appears that Kant had assimilated the caloric theory, and had seen the prospects for a unification of physics and chemistry effected thereby, by the late 1780s. (Concerning the the timing of the *Transition* project and the first explicit mention of the "gap," compare note 64 above.)

> Is it really necessary [to assume] a binding of caloric in order to explain the unchanging temperature in the melting of ice? Answer: if one uses the word *binding* for deprivation of the customary manifestations and characteristics through application to other effects, then any heat that no longer acts on the thermometer can be called bound. (21, 479.28–480.4)

An extensive paraphrase from the article on heat in Gehler's physical dictionary discussed above then follows (480.9–481.24; see 22, 811–813, for a detailed comparison with Gehler's article).

In particular, Kant, closely following Gehler, recognizes the chemical combination of heat with other matter suggested by the doctrine of latent heat as a primary ground for the postulation of caloric:

> One now considers caloric as something which binds itself chemically with bodies according to its various affinities and thereby can lose the effectiveness which it had in the free state.—This matter, which is not wholly hypothetical, can not present itself to the eyes, be enclosed in vessels, and be subjected to immediate experiments.—Free heat disperses the parts of bodies, expands vessels, transforms bodies into elastic matters.—A space empty of heat is unthinkable. . . . This matter is a universal means of dissolution, which, like all *menstrua,* loses a part of its dissolving force by means of actual combination and with complete saturation exhibits this force not at all; but, after release, exerts this force once again: precisely as acids lose their corrosive force through combination with alkalis and manifest it again after separation therefrom—Elective affinity.—That caloric which makes matters fluid, vaporous and gaseous is always found in the bound state. (480.21–481.5)

Accordingly, Kant paraphrases the three principles articulated by Gehler as encapsulating the caloric theory of the states of aggregation: the transformations from the solid to the liquid state, from the liquid to the vaporous state, and from these to the gaseous state always involve a binding of caloric; the reverse transformations always involve a release of sensible heat (481.8–24).

It is no wonder, then, that Kant devotes a substantial portion of the earlier part of the *Opus*—namely, that part written before Transition 1–14 (1799)—to an elaboration of his own particular version of the caloric theory of the states of aggregation. This elaboration is based on a particular conception of caloric, which is clearly articulated in the following passage from the Octavio-outline (1796–97):

> In general, that there is such a matter [caloric] is mere hypothesis according to which one can explain many phenomena of expansion and contraction of the matter penetrated by it—e.g., of bound and latent and, on the other hand, of released and free heat—according to their quantity mechanically/ or chemically; in the first manner through friction and striking as modes of setting rigid solid matter and thereby caloric in the state of vibration and

> thus of radiation, in the second manner through affinities of fluid matters with others to decompose the latter and thereby to expel this [caloric] or to absorb it. (21, 383.5–14)

Kant's conception is therefore a kind of compromise between a mechanical theory of heat and a true caloric fluid theory: heat is not literally a fluid that flows from one body to another while maintaining a constant total quantity, it consists rather in vibrations communicated from one body to another—which vibrations, however, themselves subsist in a special matter or vehicle that can in turn combine chemically with other matters. Kant's caloric thus unites mechanical and chemical properties.

Moreover, this caloric is distributed everywhere in the universe:

> . . . it can only be a matter which is expansive into the infinity of space but is also, through this very infinity alone, attractive and thereby a self limiting quantum—i.e., it constitutes the aether as the basis of matter filling all the universe, whose inner motion, set into eternal vibrations by the first impact, constitutes a living force (not a dead force through pressure). (380.4–10)

As everywhere distributed and eternally vibrating, Kant's caloric or heat-aether can then naturally be identified with a light-aether:

> We wish to call the aether the *empyreal* expansum (fire-air) which acts in two modes (namely as light and also as heat) progressively and in oscillation, and both contains and penetrates all matter constituting the universe. (383.31–14)[112]

In any case, Kant's conception of the matter of heat as a universally distributed continuum in a state of perpeptual vibration most closely resembles the conception of Boerhaave.[113]

The three states of aggregation—which Kant typically designates as

112. Compare also, e.g., 381.9–28, 256.1–23, 503.15–25; 22, 214.2–22. Scheele, in his *Chemische Abhandlung von der Luft und dem Feuer* (1777), calls oxygen gas "empyreal air" or "fire-air"—but this is not what Kant has in mind here: "One could call the aether empyreal air (not in the sense of Scheele where it signifies a respirable type of air, but rather as an expansive matter which contains the ground of all types of air by its penetration)" (22, 214.27–215.3). Scheele's work on radiant heat also made the identification of heat-aether with light-aether—together with the related conception of heat as transmitted via vibrations in the heat(-light)-aether—more natural.

113. See Metzger [79], pp. 219–226. Boerhaave too attempted to reconcile the material and mechanical theories of heat by conceiving of a universally distributed matter of heat whose *vibrations* are communicated from one body to another; and, in this way, Boerhaave hoped to accommodate the phenomenon of friction (as Kant also does in the first quoted passage). Yet Boerhaave (who was of course unaware of the phenomena of latent and specific heats) did not envision a *chemical* combination of heat with other matters and instead maintained that heat was active only in the otherwise empty pores of bodies. He is sharply criticized for this merely *mechanical* theory by Gehler ([37], vol. IV, pp. 544–545).

solidity, attractive-fluidity, and *elastic-fluidity*—then depend on the amount of heat present in a body in the familiar fashion:

> All attractive-fluids appear eventually to transform ⟨wholly and completely⟩ into the elastic-fluid [state] with the addition of heat, which is transposed into a permanent state (that does not cease with the subtraction of heat).—On the other hand, the attractive-fluid [state] transforms eventually with the subtraction of this matter which penetrates all bodies into the state of solidity. (22, 247.1–7)

In particular, the gaseous state or state of elastic-fluidity is easily conceived as an excess of internal caloric over external caloric by which the expansive force of the penetrating internal caloric is greater than the cohesive force exerted by the external caloric; the matter consequently expands without limit: "An elastic-fluidity (type of air) free of all attraction in contact can be repulsive in all degrees and expansible up to imponderability through heat" (21, 283.3–7; see also, e.g., 22, 246.16–25; 21, 296.10–15).

Yet a problem arises for this mode of representation in accounting for *three* states of aggregation. If the state of aggregation depends on the ratio of internal repulsive or expansive force (exerted by the penetrating matter of heat) to attractive force of cohesion (which, for Kant, is due to the surrounding external caloric), then it seems that there should be only two states of aggregation: solid when the attractive forces predominate, gaseous when the repulsive forces predominate.[114] Moreover, this problem is in fact raised by Lavoisier in the very first chapter of his *Elements.* If only the attractive force of cohesion (which, for Lavoisier, is due to a microscopic attraction between the particles of bodies) and the repulsive force of heat existed, then, he argues, bodies would pass instantaneously from the solid to the gaseous state as soon as the latter force exceeded the former. Hence, there must be some third power operative here, and this, for Lavoisier, is the pressure exerted by the atmosphere.[115]

114. It will not suffice to define a third state, liquidity, by an exact equality or balance of attractive and repulsive forces. For liquids manifest some degree of cohesion and thus resist separation: Whence comes this resistance if there is zero net attractive force? A state defined by an *exact* balance of forces would be highly unstable as well.

115. See [63], pp. 18–21; [64], pp. 3–8: "the particles of all bodies may be considered as subjected to the action of two opposite powers, the one repulsive, the other attractive, between which they remain in equilibrium. So long as the attractive force remains stronger, the body must continue in a state of solidity; but if, on the contrary, heat has so far removed these particles from each other, as to place them beyond the sphere of attraction, they lose the adhesion they before had with each other, and the body ceases to be solid. . . . But, if these two powers only existed, bodies would become liquid at an indivisible degree of the thermometer, and would almost instantaneously pass from the solid state of aggregation to that of aeriform elasticity. . . . That this does not happen, must depend on the action of

Kant, on the other hand, attempts to account for the three states of aggregation while invoking no power other than the actions of caloric. A liquid, for Kant, is a matter that is cohesive (resists the separation of its parts) but not rigid (does not resist the displacement or rearrangement of its parts). Rigidity is an additional property over and above mere cohesion—manifest only in solid bodies possessing additional internal frictional forces over and above the attraction of cohesion. For Kant, then, solidity or rigidity is the problematic case.[116] Accordingly, much of the earlier part of the *Opus* is devoted to a singular and rather ingenious account of solidity, which account has two basic ideas: first, the parts of solid bodies, in addition to being cohesive, are also structured in a crystalline form, and it is this which provides the resistance to displacement or rearrangement; second, such structures are formed by a process of crystallization from an inhomogeneous liquid in which specifically different types of matter are separated *and bind different amounts of caloric according to their different specific heats*. These latter differing amounts of bound caloric then set the specifically different, now separated matters into different characteristic modes of vibration which, together with the action of the external caloric, give rise to a periodic structure or crystal.[117] In this way, Kant has combined the idea of a perpetually vibrating heat-matter with the caloric theory of the states of aggregation (based on the idea of latent heat) so as to produce an account that is substantially closer to our present conception than is that of Lavoisier.[118]

---

some third power. The pressure of the atmosphere prevents this separation. . . . Whence it appears that, without this atmospheric pressure, we should not have any permanent liquid, and should only be able to see bodies in that state of existence in the very instant of melting, as the smallest additional caloric would instantly separate their particles, and diffuse them through the surrounding medium." Gehler summarizes Lavoisier's theory in his article on the anti-phlogistic system in his physical dictionary ([37], vol. V, p. 32).

116. This account of the problem is taken over unchanged in the *Opus* from the *Metaphysical Foundations* (#2 in the General Observation to Dynamics: 4, 526.12–529.25). The account there concludes: "thus how solid bodies are possible is still very much an unsolved problem, although the customary doctrine of nature believes that it has just as easily been done with it" (529.23–25).

117. See, e.g., 21, 384.24–385.27, 390.24–394.10, 273.9–280.26; 22, 218.18–220.7; 21, 298.9–299.8; 22, 594.10–598.18. Kant believes that a caloric operating by means of vibrations or oscillations—and not simply through pressure—is necessary for explaining the simple cohesion of liquids also; for only so, he argues, can we explain the natural spherical shape taken up by "drop-forming" fluids: see, e.g., 21, 270.20–272.2.

118. For us, a liquid is an unstructured arrangement of molecules held together by intermolecular attractive forces (van der Waals forces); the molecules of a gas move freely (inertially) unaffected by such forces; the molecules of a solid are rigidly fixed in crystalline structures by chemical bonds. Of course we no longer hold the caloric theory and give quite a different interpretation to the concept of latent heat (see note 78 above); nor do we require that crystallization always arises from inhomogeneous ingredients.

However, I do not propose further to explore the details of Kant's elaboration of the caloric theory of the states of aggregation here. It is sufficient for present purposes to observe that, during the same earlier part of the *Opus,* Kant naturally comes to view the caloric theory as constituting the key to, or basis of, a unified system of the moving forces of matter. This becomes especially clear in the outline-drafts No. 1–No. 3 η from 1798:

> ⟨The inner moving forces of matter as a whole rest principally on caloric.
> The enclosability [Sperrbarkeit] or unenclosability of the heat-matter constitutes the difference of fluidity or solidity. The heat-matter is in itself unenclosable.
> Attractive-fluid, expansive-fluid matters both through heat-matter that can be bound and released, is partly coercible, partly incoercible⟩. (22, 264.4–10)[119]

> ⟨Unity of the active principle of the combination of all moving forces of matter. To know it from the parts to the whole and vice versa.
> The moving forces of matter are in the end based on the universally distributed caloric.⟩ (267.5–9)

> ⟨Heat-matter is the cause of fluidity and dissolution as well as solidity and crystallization: dry as well as wet.
> Heat-matter is incompressible and inexhaustible.—It is the force combining all moving forces, the universal means of combination and dissolution.⟩ (21, 530.11–15)[120]

The caloric theory of the states of aggregation can thus easily be viewed as a possible foundation for a unified system of physics, and this is especially true if one identifies the heat-matter in question, as Kant does, with a light-matter or light-aether.[121]

119. The unenclosability [Unsperrbarkeit] of caloric refers to the circumstance that, unlike other "aeriform fluids," caloric can penetrate any vessel (and thus, in Kant's terminology, is incoercible). The idea expressed in the passage seems to be that caloric can be *relatively* enclosed or coerced in its *bound* state.

120. Compare also, from A/B Transition (early 1799): "The moving forces of matter—according to their quality in so far as it is fluid or non-fluid (solid)—rest on an all-penetrating movable and moving matter that is required for one as much as the other (fluidity as well as solidity), ⟨namely on caloric⟩" (22, 232.22–26).

121. In this connection, it is important to observe that the aether theory of the *Opus* differs essentially from Kant's earlier speculations concerning the aether in the *Metaphysical Foundations* and in Reflexionen 44–54 from the 1770s (14, 287–449). In these earlier speculations the aether is conceived as the cause of cohesion through its external pressure, but this aether, identified as the vehicle of light (see, e.g., R. 45 at 349.1–350.4), is *not* identified with caloric or the matter of heat. Indeed, Kant does not seem clearly to endorse the material theory of heat until at least R. 54 (probably late 1770s). Moreover, while there is an attempt to understand the states of solidity and liquidity (there is no mention of the vaporous or gaseous state) in terms of an interplay between internal heat vibrations and

The idea of a unified system of the moving forces of matter based on the caloric theory has thus become undeniably attractive—but it also presents us with an extremely serious problem. Have we not become involved throughout with particular explanations of empirically given phenomena? Is not the caloric theory of the states of aggregation, however natural and attractive, still a mere empirical theory or hypothesis? Does not Kant's specific elaboration of this theory, especially his rather ingenious account of solidity and crystallization, have an even more hypothetical status? What, in general, does such merely empirical theorizing have to do with Kant's philosophical or transcendental enterprise? How do we thereby effect *a priori* a transition from the metaphysical foundations of natural science to physics?

The problem can be raised more specifically with reference to the paraphrase from Gehler in loose-leaf 3/4 with which we began. For it is there asserted that caloric "is not wholly hypothetical," while, at the same time, the primary ground for the postulation of caloric appears to be nothing more or less than its explanatory power in the theory of latent heat and the states of aggregation. Kant raises this problem himself in the course of the same outline-drafts No. 1–No. 3 η:

> On these grounds one cannot properly call the heat-matter an elastic-fluid, although it certainly makes all *others* elastic and fluid. It cannot be considered as a merely hypothetical matter and yet this matter cannot be presented separately.—But we cannot specify it here "as a universal means of dissolution *(menstrum),* which, like all *menstrua,* loses a part of its dissolving force by means of actual combination and with complete saturation exhibits this force not at all; but, after release, exerts this force once again (precisely as acids lose their corrosive force through combination with alkalis and manifest it again after separation therefrom), i.e., according to laws of elective affinity"—for we would thereby stray over into physics, and here we are still limited to the solution of the problem according to which principles we have to make the transition from the metaph. found. of N.S. to physics. (21, 297.8–24)[122]

---

external aether pressure (see, e.g., R. 46 at 418.1–428.3), no use is made of the concept of bound or latent heat. Accordingly, there is no attempt to explain the states of aggregation through the actions of caloric or heat-matter alone. In these earlier speculations, therefore, Kant neither deploys the caloric theory of the states of aggregation nor identifies the universally distributed aether with the matter of heat.

122. Compare also 22, 263.8–24 (quoted above) and, from Farrago 1–4 (1798–99), 21, 625.6–20: "In this didactic, i.e., doctrinal, system these moving forces are certainly objects of experience but their combination into a separate part of natural science rests on an a priori principle, and the first law thereof is that their institution does not stray over into physics as a system of empirical natural knowledge [Naturkunde], because it then exceeds its territory (of the elementary concepts of the moving forces).—Thus, in this doctrinal system nothing concerning solution and precipitation and the moving forces belonging

Thus, Kant here returns to the same passage from Gehler he had earlier paraphrased, and the problem is made perfectly explicit a few pages later:

> ⟨"A space empty of heat is unthinkable" (Gehler) *Why not*[?]⟩ (303.11–12)

The question, in other words, concerns precisely how the existence of a universally distributed caloric is to obtain a necessary and non-hypothetical status—for such a status cannot possibly proceed from mere physical or empirical theorizing.[123]

It is therefore incumbent upon Kant to inject some kind of transcendental content into the idea of caloric, to connect it not so much with explanations of particular empirical phenomena as with the conditions of the possibility of experience. And Kant attempts to do just this in the immediately succeeding outline-drafts, particularly in Elem.Syst. 1–7 (late 1798) and A.Elem.Syst. 1–6 (early 1799). The argument developed there proceeds from the concept of quantity of matter. Our application of this concept, Kant argues, depends on the procedure of weighing, and this procedure, in turn, requires a mechanical instrument: namely, the balance. Yet the balance cannot be conceived as a mere mathematical lever without thickness or breadth: on the contrary, it is a real physical body requiring physical (and not merely mathematical) rigidity. But physical rigidity, as we know, is only possible via the action of a universally distributed, perpetually vibrating, imponderable material—caloric:

> ⟨The capacity for weighing presupposes the coercibility of the matter of the balance, which resists the bending and breaking of the latter as well as the tearing of the cord from which the weight hangs. The mechanics of the moving forces is only thinkable under presupposition of the dynamical [moving forces], objective capacity for weighing [only under presupposition] of the preceding subjective [capacity]. A living force of the matter penetrating the body must be the cause of the dead [force] of pressure or traction, which effects an infinite series of contacts immediately subordinate to one another, and thereby effects moving surface-force of a mass, i.e., attractive cohesion. Coercibility, permeability, and perpetuity or attraction—thus the moving force of caloric is required merely for the balance as the instrument of weighing.⟩ (22, 138.28–139.6)

---

thereto can appear; this belongs in the field of chemistry as a part of physics and thus to the empirical system of natural science, and here, on the other hand, the transition from the metaph. found. of N.S. to physics is not yet occupied with this but rather remains with the combination of the elementary concepts that can be thought a priori in a system in order only to present completely the original moving forces of matter in themselves and in their relation to one another"; and 633.3–16.

123. That the *necessity* of caloric is in question here is confirmed several lines below: "according to modality is *necessity* and its empirical function: the permanence of the moving forces by means of caloric" (303.18–20).

> There must therefore be a respectively imponderable matter which makes possible the ⟨mechanical capacity for weighing⟩, without which we could have no experience of the quantity of matter and its moving force through weight. (158.10–13)

Hence, the cohesive and rigidifying action of caloric is not simply a particular empirical phenomenon among others; it is built into the conditions of application of one of our most fundamental concepts necessary for the experience of matter. (Compare also, e.g., 22, 196.19–198.18.)

Kant concludes that the assumption or postulation of caloric has an a priori and not merely hypothetical status:

> ⟨Therefore, already contained in the *concept* (hence a priori) of capacity for weighing *(ponderabilitas objectiva)* is the assumption and presupposition of a matter penetrating all bodies which has primitive moving force—without needing empirically to stray over into physics (through observation and experiment) or to contrive any hypothetical matter for the explanation of the phenomenon of weighing, which is rather postulated here.⟩ (22, 587.10–16)

The existence of caloric cannot be derived hypothetically from any physical experiment, for the very possibility of such experiment already presupposes its existence:

> In all this the theory is in no way based on experience and borrows nothing from physics, but is rather based ⟨merely⟩ on concepts of the possibility of certain acting causes according to laws of motion in so far as they a priori make experience possible and precede [experience] as necessary presupposition for the sake of experiments. (595.14–18)

More generally, all our physical experiments require machines or mechanical instruments such as the balance or lever, and that which is a necessary precondition of these mechanical instruments must itself be viewed as a transcendental condition of the possibility of physical experience.[124]

It is clear, however, that this attempt at a transcendental grounding for the postulation of caloric is quite insufficient. For the argument depends entirely on a very particular account of cohesion and rigidity: namely, the account developed in earlier outline-drafts and discussed above. Is not this

124. Compare 598.5–13: "Finally, that there must be an imponderable and at the same time incoercible ⟨motion of⟩ matter as a fluid—distributed everywhere in the universe, penetrating all bodies in substance, and continuing for itself from the beginning in all time through oscillation (via mutual attraction and repulsion), which receives its proof through no observation or experiment as hypothetical matter—because it extends over all experience of machinery—but rather can only proceed from the concept of the possibility of experience and constitutes the transition from the elementary system to the cosmic system"; and also 21, 192.21–30; 22, 606.27–608.21. That Kant places so much emphasis on the balance as a condition of the possibility of physical experiment is surely to be understood in terms of a commitment to the experimental method established by Lavoisier: see note 83 above.

account itself simply one empirical hypothesis among others? Why *must* cohesion and rigidity be explained by the penetration of a universally distributed heat-matter—and not, for example, by microscopic action-at-a-distance forces? The latter style of explanation was in fact favored by many thinkers of the period, such as Boscovich, Buffon, Priestley, and Cavendish; and Kant was undoubtedly aware of this "atomistic" program.[125] Until such alternative accounts of cohesion and rigidity are somehow ruled out—and, moreover, ruled out a priori—the claim that Kant's caloric theory "is in no way based on experience and borrows nothing from physics" must appear hollow indeed.

Further, the above attempt at a transcendental argument appears clearly deficient from the point of view of the critical philosophy as well. For the machinery of that philosophy is in no way effectively deployed there. In particular, there is no attempt (as in the *Metaphysical Foundations*) to apply or instantiate the transcendental principles of the understanding so as to generate additional a priori content via the schematization of an empirical concept, and, what is perhaps even more significant in the present context, no use whatever is made of the transcendental principle of reflective judgement (as we would certainly expect given Kant's developing awareness of the importance of this principle for the *Transition* project traced out in §II). The argument from the possibility of estimating quantity of matter by means of the balance cannot therefore qualify as a genuinely transcendental argument. It is no wonder, then, that Kant devotes the whole of the next series of outline-drafts, Transition 1–14 (mid-1799), to just such a genuinely transcendental aether-deduction.[126]

125. Note that this program, precisely through its explicit appeal to Newtonian action-at-a-distance forces, does not fall under the rubric of the "mathematical-mechanical" explanatory scheme rejected in the General Observation to Dynamics of the *Metaphysical Foundations*. Some further considerations are therefore necessary in order to reject this newer style of "atomism."

126. Tuschling [113], especially chap. VI, argues that the early part of the *Opus* up to Transition 1–14 is to be understood as the gradual evolution of a genuinely "transcendental" dynamics. For Tuschling, however, the relevant distinction is between "transcendental" and "metaphysical"—as this distinction is developed in paragraphs six and seven of the Preface to the *Metaphysical Foundations* (4, 469.26–470.35) and in §V of the published Introduction to the *Critique of Judgement* (5, 181.15–31: quoted above in note 54). Tuschling, that is, sees Kant as moving away from a conception of synthetic a priori physical principles generated via an *application* of transcendental laws to *empirically given* concepts (as in the *Metaphysical Foundations*) towards a conception on which physical principles and physical concepts (namely, the aether) are generated directly from the "transcendental subject." The basis for this reading is the idea of "phoronomy-critique" and a consequent rejection of the *Metaphysical Foundations* in the *Transition* project—an idea we have criticized in §I. From the present point of view, the relevant distinction is rather between "transcendental" and "empirical"—as this distinction is articulated at A56–57/B80–81. What matters here, in other words, is the distinction between transcendental philosophy, which explains and

The argument of Transition 1–14 does indeed deploy machinery from the critical philosophy—in a most surprising and bewildering fashion. The object is to show that caloric or the aether is a necessary condition of the possibility of experience as such (not merely of the use of the balance or physical experiment, say), but the notion of the possibility of experience is given a much stronger interpretation than it had in the critical period. This comes out very clearly when Kant introduces the problem:

> If, with respect to the existence of a certain matter of a peculiar quality, the question is raised: whether it is a priori provable (demonstrable) or only empirically provable *(probabilis)*, then we can expect only *subjective* conditions of the possibility of its cognition—i.e., those of the possibility of an *experience* of such an object. For existence is not a certain ⟨particular⟩ predicate of a thing, but rather the absolute positing of the thing with all of its predicates.—There is thence only One experience, and, if *experiences* are spoken of, this signifies only the *distributive* unity of manifold perceptions, not the *collective* unity of its object itself in its thoroughgoing determination; from which it then follows that, if we wish to judge a priori concerning objects of experience, we can demand and expect only principles of the agreement of the representation of the objects with the conditions of the *possibility* of the experience of them. (22, 549.6–550.3)

> If it can be proved that the unity of the *whole* of possible experience rests on the existence of such a matter [caloric] (with its stated properties), then the actuality of this matter is also proved—⟨to be sure⟩ not **through** experience ⟨but yet⟩ a priori, merely from conditions of the *possibility* of its [existence] for the sake of the possibility of experience. (550.18–23)

Thus the notion of the possibility of experience is here connected with the idea of complete or thoroughgoing determination (durchgängige Bestimmung), an idea which is developed in the first *Critique* in §2 of chapter III of Book II of the Transcendental Dialectic: the Transcendental Ideal.[127]

The principle of thoroughgoing determination states that every thing or object is characterized affirmatively or negatively by each one of the totality of all possible predicates (A571–572/B599–600). This principle is not

---

thereby grounds the possibility of both mathematical and physical scientific knowledge, and the content of that scientific knowledge itself. More precisely, the relevant distinction is between "the whole of pure philosophy including critique, . . . the investigation of all that can ever be known a priori as well as the presentation of that which constitutes a system of pure philosophical cognitions of this type—but is distinct from all empirical as well as mathematical employment of reason" (A841/B869), and actual scientific knowlege. (As A845–848/B873–876 makes clear, the *Metaphysical Foundations* itself—and therefore (special) metaphysics—certainly belongs to this "pure philosophy.") I am indebted to Graciela De Pierris for emphasizing to me the importance of the transcendental/empirical distinction.

127. Förster has very clearly brought out the connection between the aether-deduction and the transcendental ideal, together with the historical antecedents of the latter in Kant's pre-critical works, in [32].

merely formal or logical but has a transcendental presupposition: namely, the representation of the totality of all possible predicates or the totality of all possibility (A573/B601). Since, however, the totality of all possible predicates can never be instantiated in experience, the representation in question is a mere idea of reason:

> Thoroughgoing determination is therefore a concept which we can never present *in concreto* according to its totality, and thus is based on an idea which resides solely in reason—which latter prescribes to the understanding the rule of its complete employment. (A573/B601)

Although this idea is therefore wholly indeterminate with respect to the possible predicates which are to constitute it, it nevertheless leads to the concept of an individual (einzeln) object that is itself completely determined thereby (A573–574/B601–602).

This comes about when we distinguish, within the class of all possible predicates in the merely logical sense, those predicates truly expressing a being—as opposed to mere non-being or lack of being. These predicates alone express transcendental affirmation: that is, reality or thing-hood (A574–575/B602–603). And the resulting sub-totality of predicates or possibilities is

> . . . a transcendental substrate which as it were contains the whole supply of material from which all possible predicates of things can be taken. Thus this substrate is nothing other than the idea of an All of reality *(omnitudo realitatis)*. All true negations are consequently nothing but *limitations,* which they could not be called if the unlimited (the All) were not the basis. (A575–576/B603–604)

But now the All of reality *(omnitudo realitatis)* is itself completely determined; for, of each possible pair of contradictorily opposed predicates, one (and one only)—namely, that expressing transcendental affirmation—necessarily belongs to it. This transcendental idea is therefore the concept of an individual (einzeln) being or transcendental *ideal.* In fact, it is "the single [einzige] proper ideal of which human reason is capable; for only in this single case is an in itself universal concept of a thing completely determined through itself and cognized as the representation of an individual" (A576/B604).

As an idea of reason, however, the transcendental ideal does not have objective existence:

> For reason employs it only, as the *concept* of all reality, as the basis of the thoroughgoing determination of things in general, without requiring that all this reality is objectively given and itself constitutes a thing. The latter is a mere fiction, through which we comprehend and realize the manifold of our idea in an ideal, as a particular being . . . . (A580/B608)

On the contrary, the transcendental ideal is a mere idea or concept through which reason guides the empirical employment of the understanding.

Now, in this latter employment

> . . . an object of the senses can only be completely determined when it is compared with all predicates of appearance and therewith represented affirmatively or negatively. Because, however, that which constitutes the thing itself (in the appearance), namely the real, must be given (without which it can also not even be thought), while that wherein the real of all appearance is given is the one all-inclusive experience [die einzige allbefassende Erfahrung]; it follows that the matter for the possibility of all objects of the senses must be presupposed as given in a totality, on whose limitation alone can rest all possibility of empirical objects, their distinction from one another, and their complete determination. (A581–582/B609–610)

Yet precisely here we initiate a dialectical illusion:

> But that we hereby hypostatize this idea of the totality of all reality comes about as follows: because we dialectically transform the *distributive* unity of the empirical employment of the understanding into the *collective* unity of a whole of experience, and we think a single [einzeln] thing in this whole of appearance which contains all empirical reality in itself, which then, by the above-mentioned transcendental subreption, is confused with the concept of a thing which stands at the source of the possibility of all things and supplies the real condition of their complete determination. (A582–583/B610–611)

The crucial mistake, in other words, is to move from the notion of the real of appearance—which, by its very nature, can only be given progressively in the course of the advance of experience—to that of experience as a single given (finished and complete) totality.

As §40 of the *Prolegomena* puts the matter with particular clarity, this very same dialectical transformation necessarily leads to a transcendent use of reason:

> The empirical employment to which reason limits the understanding does not fulfill its own whole determination. Each single experience is only a part of the whole sphere of its domain, but the *absolute whole of all possible experience* [das *absolute Ganze all möglichen Erfahrung*] is itself no experience and therefore a necessary problem for reason—which requires wholly other concepts for its mere representation than the concepts of the understanding, whose employment is only immanent, i.e., extends to experience, so far as it can be given, whereas the concepts of reason extend to the completeness, i.e., the collective unity, of the whole of possible experience and thereby extend beyond any given experience and become *transcendent.* (4, 327.31–328.10)

Moreover, as Kant explains in the *Critique,* to take the collective unity of the whole of possible experience as an object is to misunderstand a regulative employment of reason as constitutive:

> Reason therefore properly has only the understanding and its purposive operation as object, and, as the latter unites the manifold in the object through concepts, so the former unites the manifold of concepts through ideas, in that it posits a certain collective unity as the goal of the activities of the understanding, which otherwise is occupied only with distributive unity.
>
> I accordingly assert: the transcendental ideas are never of constitutive employment, so that concepts of certain objects are thereby given, and when understood in this way they are merely sophistical (dialectical) concepts. On the other hand, they have an excellent and indispensably necessary regulative employment: namely, to direct the understanding towards a certain goal, at the prospect of which the lines of direction of all its rules converge in a point . . . . (A644/B672)

It is clear, therefore, that when concepts of the understanding are established as conditions of the possibility of experience, and thereby acquire constitutive force, what is at issue here is the *distributive* unity of experience—the conditions that make possible each and every given experience. The *collective* unity of experience—the unity of all experience as a single totality—is comprehended by no such constitutive conditions, but only by the merely regulative principles of reason.

As we saw above, however, the notion of the possibility of experience deployed in the aether-deduction of Transition 1–14 is precisely that of the collective unity of experience: "the *collective* unity of its object itself in its thoroughgoing determination" or "the unity of the *whole* of possible experience." The goal of the aether-deduction is to show that an everywhere distributed aether or caloric exists, because only so is the possibility of the *collective* unity of experience secured:

> The object of an all-inclusive [allbefassenden] experience contains in itself all subjectively-moving, and thus sense-affecting and perception-effecting, forces of matter, whose totality is called caloric, as the basis of this universal force-excitation which affects all (physical) bodies, and hereby also the subject itself, and from the synthetic consciousness [of these forces], which is not empirical, the formal conditions of these sense-moving forces may be developed in attraction and repulsion. (22, 553.21–28)

> Caloric is actual because its concept (with the attributes we ascribe to it) makes the totality of experience possible—not as hypothesis concerning perceived objects in order to *explain* their phenomena, but rather given immediately through reason in order to ground the possibility of experience itself. (554.12–17)

Here, unlike the parallel case of the transcendental ideal, we are indeed able to assert the objective existence of an object underlying the possibility of experience as a whole, and thus underlying the possibility of complete determination.

Paradoxically, however, it is through the exact correspondence between the attributes of the object of the aether-deduction and those of the trancendental ideal that we are now able to establish the objective existence of the former:

> There is objectively only One experience and all perceptions subsist in a given system (which is not contrived) of the absolute whole of experience: i.e., "there exists ⟨an absolute-whole as⟩ system of the moving forces of matter because the concept of such is objectively an empirical concept and thus an object so thought is *actual*" (here, but also only in this single case, it can be said *a posse ad esse valet consequentia*). This concept is *singular* [*einzig*] in its kind *(unicas),* and this because its object is also *individual* [*einzeln*] *(conceptus singularis);* for the *All* of matter signifies not a distributive but a collective universality of objects which belong to the absolute unity of all possible experience. (21, 592.5–15)

Yet it is of course entirely obscure how we are now able to assert, and indeed even to prove, the existence of such an "*All* of matter." Why does this concept, like the parallel case of the transcendental ideal, not signify merely a regulative idea employed simply to direct the empirical activities of the understanding, activities which themselves exemplify distributive but never collective universality?

What is motivating Kant becomes clear, I think, from the considerations of §II above. We there argued that the regulative procedure of reflective judgement (discussed in the first *Critique* under the rubric of the regulative use of reason), when juxtaposed with the constitutive procedure of the *Metaphysical Foundations,* creates a need for principles that are at the same time both regulative and constitutive. The regulative procedure of reflective judgement aims at the indeterminate and forever unrealizable idea of a complete empirical science—which, were it *(per impossibile)* ever actually to be realized, would in fact ground the collective unity of experience as a whole. The constitutive procedure of the *Metaphysical Foundations,* on the other hand, schematizes and articulates the highest concept of empirical classification—the empirical concept of matter—which then extends over all outer experience distributively. Thus, for example, all our particular experiences of matter (of any kind whatsoever) are subject to the Newtonian laws of motion and involve the two fundamental forces of attraction and repulsion.[128] We therefore know—and

128. The distributive character of the universality involved here may be illustrated by the circumstance that the Newtonian laws of motion provide a framework for reasoning about any and all forces—without, however, enabling us either to enumerate the totality of empirically given forces or to systematize such a totality.

know constitutively—that the ideal complete science must contain at least this much. Since, however, the idea of a complete science has thereby ceased to be wholly indeterminate, we need some assurance that the regulative procedure of reflective judgement will intersect or meet with the constitutive procedure of the *Metaphysical Foundations*—otherwise there is indeed a substantial "gap" in the critical system.

My suggestion, then, is that after the execution of the *Metaphysical Foundations* and the articulation of reflective judgement as an autonomous faculty, it becomes clear—from the point of view of the critical philosophy itself—that the absolute dichotomy between regulative and constitutive principles cannot be maintained.[129] It becomes clear, that is, that the critical system requires completion via the *Transition* project, whose task is precisely to establish principles that are at the same time both regulative and constitutive (22, 178.26–33, 240.25–28, 241.19). Hence, it is not unexpected when the aether-deduction of Transition 1–14 attempts to forge an explicit connection between the distributive unity of experience and its collective unity:

> . . . therefore this proof [the aether-deduction] is singular of its kind [der einzige seiner Art], because here the idea of distributive unity of all possible experience in general converges together in a concept with that of collective [unity]. (21, 552.14–17)
>
> This manner of proving the existence of an outer sense-object must strike one as *singular* [*einzig*] in its kind ⟨(which is without example)⟩, but which nevertheless should not be surprising, because its object is also special in that it is *individual* [*einzeln*] and contains in itself not merely *distributive* universality (like other representations from a priori concepts) but rather *collective* universality. (603.4–9)

These considerations—together with the circumstance that the object of the aether-deduction, unlike the transcendental ideal, is a phenomenal or empirical object—make it perfectly intelligible, it seems to me, that the aether-deduction should now aim to establish the actual objective existence of its object.[130]

129. As noted above, there is a suggestion of the need for a necessary connection between the regulative and constitutive domains even in the first *Critique* at A651/B679.

130. For this reason I cannot follow Förster's attempt to view the aether-deduction as concerned merely with a *regulative* idea, and thus with "a transcendental ideal in the critical sense" ([32], p. 226). Nor can I agree with Lehmann's contention that Kant is not really attempting to prove the existence of an empirical object here ([66], pp. 109ff.). Such interpretations seem to me not only to be inconsistent with Kant's repeated attempts to prove a priori the *existence* of the aether, but also to miss the logic of Kant's situation: unless the aether-deduction is both regulative *and* constitutive, it can in no way bridge the relevant "gap" in the critical system.

It is one thing, however, to understand Kant's motivations for attempting to prove the actual objective existence of caloric (which is therefore not merely hypothetical) in the aether-deduction; it is quite another to see how he proposes to accomplish such a proof. For it is by no means obvious that the idea of a complete science must be supplemented in precisely this way. Granting, that is, that the constitutive procedure of the *Metaphysical Foundations* must be extended further into the domain of the properly empirical, so as to specify more exactly the content of the ideal unified system of physics and to secure thereby the desired intersection or continuous connection with the regulative procedure of reflective judgement: Why does it follow that the unification of physics in question should be based on a universally distributed caloric or aether, rather than, say, on a further elaboration of Newtonian action-at-a-distance forces? Why, for example, should not the Boscovichian "atomistic" program, based on point-particles and a single (variously attractive and repulsive) action-at-a-distance force, provide an equally promising basis for an eventual unified physics as does the caloric or aether theory?[131]

The key to Kant's strategy in the aether-deduction, I suggest, is that he is constructing a solution to the problem of a unified and complete science out of what, in his terms, is the statement of that problem. In the terms of the critical philosophy the problem is to find a representation that is somehow both constitutive and regulative—and thus, as we have seen, has both distributive and collective unity or universality. Indeed, the aether-deduction is entirely based on the idea of collective universality or, as Kant sometimes puts it, synthetic-universality:

> ⟨The agreement with the whole (not with all *divisim* but rather with all *coniunctim*) of possible experience.
>
> To execute a direct proof concerning a matter penetrating all bodies in nature (and thus the whole universe) would be a task of physics, for observations from experience would belong thereto. Now it is to occur from a priori principles because the problem belongs to the transition *to* physics. Thus the problem belongs to a priori principles of the possibility of experience as a whole.⟩ (21, 546.22–29)

> The proof of the existence of an all-penetrating and all-moving elementary-material in a system of matter, if it is to proceed from a priori principles, must think all experience as contained in One experience comprising all its objects [Einer aller ihre Objecte umfassenden Erfahrung], and if experiences are spoken of then these are nothing but parts and aggregates of a synthetically-universal experience—and what contradicts the conditions of being an object of possible experience is no existing thing. (549.20–27)

131. For the general opposition between action-at-a-distance and aether programs in the eighteenth century, see, e.g., Heilbron [47], chap. I, §6.

But what exactly distinguishes a synthetically universal or collectively universal representation?

Kant makes the distinction explicit in the following passage:

> Of a universally distributed, all-penetrating, and continuous elementary-material as a non-hypothetical material but rather an object of experience.
>
> There is objectively considered only *One* experience and if subjectively *experiences* are spoken of then these are nothing but parts and lawfully connected aggregate of a synthetically-universal experience.*
>
> The ⟨world-⟩space [⟨Welt-⟩raum] is the totality of the whole of all possible outer experience in so far as it is filled. An absolutely-empty space ⟨of non-being⟩ *in* or *around* it, on the other hand, is no object of possible experience.
>
> *Analytically-universal is a concept through which one in many,—synthetically- through which many in one as together is thought under a concept. (21, 247.2–15)

Thus, in the traditional logical terminology employed by Kant, an analytically universal (and thus distributively universal) representation is a *concept,* which occurs in all the concepts falling *under* it (each member of its extension) as a character or constituent concept (as a constitutent of the intension of each concept falling under it). In analytically universal or conceptual represention, therefore, the representation of the parts (the constituents of the intension of the concept) precedes and makes possible the representation of the whole (the given concept containing these constituent characters *within* itself).[132] Synthetically universal (and thus collectively universal) representation somehow reverses this priority: here the representation of the whole precedes and makes possible that of the parts.[133]

It follows, then, as Kant explains in §77 of the *Critique of Judgement,* that a synthetically universal representation cannot be a product of our (discursive) understanding:

> In fact our understanding has the property, that in its cognitions, e.g., of the cause of a product, it must proceed from the *analytically-universal* (from concepts) to the particular (the given empirical intuition); whereby it there-

132. See the *Jäsche Logik,* Part I, Section One: 9, 91–100.

133. Thus, in the Transcendental Ideal, Kant contrasts the analytically universal (and indeed maximally general) concept of a reality in general with the collectively universal idea of the *ens realissimum:* "The universal concept of a reality in general cannot be subdivided a priori, because, without experience, one is acquainted with no specific types of reality which would be contained under this genus. Thus the transcendental major premise of the thoroughgoing determination of all things is nothing other than the representation of the totality of all reality—not merely a concept which contains all predicates according to its transcendental content *under itself,* but rather which comprehends them *within itself;* and the thoroughgoing determination of any thing rests on the limitation of the *All* of reality . . . " (A577/B605).

> fore determines nothing in relation to the manifold of the latter, but must wait for this determination by the power of judgement for the subsumption of the empirical intuition under the concept (if the object is a natural product). Now, however, we can also think an understanding which, as it is not discursive like ours, but rather intuitive, proceeds from the *synthetically-universal* (the intuition of a whole as such) to the particular, i.e., from the whole to the parts—thus, such that it and its representation of the whole does not contain the *contingency* of the connection of the parts within itself, in order to make possible a determinate form of the whole—as our understanding requires, given that it must advance from the parts as universally-thought grounds to various possible forms to be subsumed under them as consequences. According to the constitution of our understanding, on the other hand, a real whole of nature is only to be considered as the effect of the concurrent moving forces of the parts. (5, 407.13–30)

Indeed, it is for precisely this reason that Kant argues, in the *Critique of Judgement,* that the systematic unity of nature as a whole can in the end only be grounded teleologically on the idea of an intelligent Author of the world.[134]

We see once again, therefore, that in the critical period the systematic unity of nature as a whole can only be represented through an idea of reason: namely, through the transcendental ideal of the *ens realissimum*—"personified" as the archetypal intellect or divine intelligence.[135] Yet this representation cannot fulfill the demands of the *Transi-*

134. See 407.30–408.10: "Thus, if we wish to represent, not the possibility of the whole as depending upon the parts, in accordance with our discursive understanding, but rather the possibility of the parts (according to their constitution and connection) as depending upon the whole, in accordance with the standards of the intuitive (original) understanding: then, according to precisely the above peculiarity of our understanding, this cannot occur in such a way that the whole contains the ground of the possibility of the connection of the parts (which would be a contradiction in the discursive mode of cognition), but rather only in such a way that the *representation* of the whole contains the ground of the possiblity of its form and the connection of the parts belonging thereto. Since, however, the whole would now be therefore an effect, a *product,* whose *representation* is viewed as the *cause* of its possibility, but the product of a cause whose ground of determination is merely the representation of its effect is called a purpose; it follows that it is merely a consequence of the particular constitution of our understanding when we represent products of nature to ourselves as possible according to another type of causality than that of the natural laws of matter—namely, only according to that of purposes and final causes."

135. See the footnote to the Transcendental Ideal at A583/B611: "*This ideal of the most-real being, although it is in fact a mere representation, is therefore first *realized,* i.e., made into an object, thereupon *hypostatized,* and finally, through a natural progress of reason towards completion of the unity, even *personified,* as we shall soon articulate: because the regulative unity of experience does not rest on the appearances themselves (sensibility alone), but rather on the connection of their manifold through the *understanding* (in an apperception); therefore the unity of the highest reality and the complete determinability (possibility) of all things seems to lie in a highest understanding, and thus in an *intelligence.*"

tion project, of course, for it is merely regulative and in no way constitutive: no actual object can ever be given corresponding to the ideal of reason. In the aether-deduction, I suggest, Kant has come to believe that there is one (and only one) *constitutive* representation which, as it were, is the closest possible approximation to the transcendental ideal: namely, the representation of a universally distributed and continuous "All of matter" filling all space. And, I suggest, the privileged status of this particular representation derives from the circumstance that, among all representations available to human cognition, only our representation of space (and time) as a whole possesses the desired synthetic or collective universality.

Kant draws the relevant parallel between the representation of space as a whole and that of the *ens realissimum* already in the Transcendental Ideal:

> For all negations (which are yet the only predicates through which everything else can be distinguised from the most-real being) are merely limitations of a greater, and finally of the highest, reality; and thus presuppose it and are merely derived from it according to their content. All manifoldness of things is only a correspondingly various mode of limiting the concept of the highest reality, which is their common substrate, just as all figures [Figuren] are possible only as different modes of limiting the infinite space. (A578/B606)[136]

Indeed, it is for precisely this reason that Kant argues, in the Transcendental Aesthetic, that the representation of space is not a discursive or conceptual representation:

> Space is no discursive or, as one says, universal concept of relations of things in general, but rather a pure intuition. For, first, one can represent only one single [einigen] space, and when one speaks of many spaces one means thereby only parts of one and the same unique [alleinigen] space.

136. See also A619–620/B647–648: "The ideal of the highest being, according to these considerations, is nothing other than a *regulative principle* of reason, to view all connection in the world *as if* it arose out of an all-sufficient necessary cause, in order to base thereon the rule of a systematic—and in accordance with universal laws necessary—unity in the explanation of such connection. At the same time, however, it is unavoidable that, by means of a transcendental *subreption,* we represent this formal principle as constitutive and think this unity hypostatically. For, just as space, because it originally makes possible all figures [Gestalten], which are simply limitations of space, although it is equally only a principle of sensibility, is nonetheless taken for precisely that reason as an absolutely necessary thing subsisting for itself and an object given a priori in itself—so it also comes about entirely naturally that, since the systematic unity of nature can in no way be erected as a principle of the empirical use of our reason, except in so far as we base this on the idea of a most real being as the highest cause, this idea is represented thereby as an actual object, and this, in turn, because it is the highest condition, as necessary; and thus a *regulative* principle is transformed into a *constitutive* principle . . . . "

> Moreover, these parts cannot precede the single all-inclusive space [dem einigen allbefassenden Raume], as it were as its constituents (though which its composition is possible), but rather they are thought only *in it*. It is essentially singular [einig]: the manifold in it, and thus also the universal concept of spaces in general, rests simply on limitations. (A24–25/B39)

The representation of infinite space as a whole thus itself has synthetic rather than analytic universality: the representation of the whole precedes and makes possible the representation of its parts.[137]

What then fuels the aether-deduction, I suggest, is the conception of the aether or caloric as *realized* or *hypostatized* space:

> This mode of proof of the existence of a peculiar world-material [Weltstoff] penetrating all bodies and permanently agitating them through attraction and repulsion is somewhat strange; for the ground of proof is *subjective,* deriving from the conditions of the possibility of experience, which presupposes moving forces and excludes the void in order to fill space with always agitating matter—which may be called *caloric* ⟨or aether, etc.⟩—and this proposition is to be grounded a priori on concepts *without hypotheses.*—Not merely the right but also the necessity for postulating such universally distributed material is grounded in the concept of the latter as *hypostatically* thought space.—Space (like time also) is a magnitude that cannot exist except as part of a still greater whole. Here, however, it is unsuitable that a thing in itself could exist merely as part, since parts are necessary grounds of the possibility of a whole; for the whole must first be given in order that the manifold in it be thought as part. (21, 221.2–18)

> ⟨There is only One space and only One time and One matter in which all motion is found. The real and objective principle of experience which constitutes a Single [Einiges] whole according to form permits no remaining unfilled space outside itself and inside itself. All moving forces lie in it. This composite [Zusammengesetzte] is not locomotive [ortverändernd] and no body. The beginning of its motion is also the eternity of the latter.
>
> The basis of the whole of the unification of all moving forces of matter is caloric (as it were the hypostatized space itself in which everything moves), the principle of the possibility of the whole [of] possible experience.⟩ (224.3–13)

And, on this basis, the argument of the aether-deduction can finally be understood as follows.

The ground for unifying all moving forces of matter, and thereby making experience possible as a single whole, must, as we have seen, possess

137. Compare B40 and also, from §77 of the *Critique of Judgement:* "the unity of space, which, however, is no real-ground of generation [of natural products] but only their formal condition; although it has one similarity with the real-ground which we are seeking: that in it no part can be determined except in relation to the whole (whose representation therefore grounds the possibility of the parts)" (5, 409.3–8).

both collective and distributive (both synthetic and analytic) universality. As perpetually agitating or vibrating through attraction and repulsion, caloric comprises force—a derivative concept of the understanding (A20–21/B35, A82/B108, A204/B249–250)—and thus possesses distributive or analytic universality. As uniformly distributed in infinite space (and time), however, caloric can be considered as realized or hypostatized space (and time), and thus also possesses collective or synthetic universality.[138] Hence, the representation of a uniformly distributed, perpetually agitating or vibrating caloric or aether is uniquely well suited to play the role of the basis for unifying all moving forces of matter (the basis for an ideal complete science), and thereby making experience possible as a single whole.

By contrast, the alternative "atomistic" program—which would found a unified physics and chemistry on the further elaboration of microscopic action-at-a-distance forces—appears, from Kant's point of view, clearly to lack the crucial element of collective or synthetic universality. Here the representation of the parts necessarily precedes and makes possible the representation of the whole, and, as Kant puts it in the above quoted passage from the *Critique of Judgement,* such an "atomistic" representation is therefore a merely analytically universal product of the discursive understanding—where "a real whole of nature is only to be considered as the effect of the concurrent moving forces of the parts." Indeed, we saw that in the critical period Kant despairs of the possibility of a truly scientific chemistry precisely because he could then envision only a foundation of this latter (merely analytically universal) type:

> So long, therefore, as there is still no concept to be found for the chemical actions of matters on one another that can be constructed—i.e., no law of approach or withdrawal of the parts can be specified . . . then chemistry can be no more than systematic art or experimental doctrine but never a proper science . . . . (4, 470.36–471.6)

Now, in the period of the aether-deduction, Kant has come to see that a unification of physics and chemistry—and thus a truly scientific chemistry—can be elaborated in an entirely different fashion: on the basis of the unique representation meeting the demand, which itself arises naturally and inevitably within the critical philosophy, for both collective and distributive, both synthetic and analytic, universality.

But it is not yet clear exactly *how* the representation of caloric or the

138. See 21, 561.28–30: "⟨There must be a synthetically universal (universally distributed) basis of the moving forces of matter which contains merely the ground of the possibility of the experience of an *existence* in space *(spatium sensibile)*⟩"; and 564.1–4: "But now the principle of the possibility of all experience is ⟨the realization of⟩ space itself as an individual [einzeln] object of the senses (i.e., of empirical intuition)."

aether is actually to ground a scientific chemistry; nor, consequently, is it yet clear exactly what it means to conceive this representation as the *basis* of the moving forces of matter. In one of the later drafts of Transition 1–14 (Transition 12b) Kant illustrates his meaning as follows:

> One can also call *caloric* the *basis* (first cause) of all moving forces of matter: for it is ⟨thought as⟩ the *primitive material* [*Urstoff*] *(materia primaria)* that is immediately moving. By contrast, all other materials (e.g., oxygen, hydrogen, etc.), which must first be moved by means of this material, are moving as *secondary materials* [*Nachstoff*] *(materia secudaria)* and (e.g., light) are only modes of this material. And now the formation of bodies by means of specifically different elements produces composite forms—which, however, must not be placed beside the principle of the possibility of One experience but rather subordinated to it. (21, 605.5–13)

Here caloric, as basis of the moving forces and primitive material, is contrasted with the *elements* of Lavoisier's systematic chemistry—conceived as derivative materials.[139] Since according to Lavoisier elements such as oxygen and hydrogen figure in chemical reactions of composition and decomposition by means of their association with caloric (through which they pass into and out of the gaseous state), the latter is conceived as the primarily or immediately moving material in chemical interactions.[140] Moreover, according to Kant's own version of the caloric theory of the states of aggregation, the formation of solid bodies depends on the action of caloric and on the existence of a variety of specifically different (inhomogeneous) elements. Caloric is the basis of the moving forces of matter, then, in the sense that it is the foundation of both Lavoisier's systematic chemistry and the new conception (based on the theory of latent heats) of the three states of aggregation.

That caloric becomes the basis of the moving forces of matter through Lavoisier's theory of the elements is confirmed in the next major set of outline-drafts from the tenth and eleventh fascicles (1799–1800). There Kant is particularly concerned to distinguish the concept of *matter* [Materie], which is necessarily singular, from that of *material* [Stoff],

139. Accordingly, Kant employs the German version of the new chemical nomenclature: "oxygène" (so-named, from Lavoisier's bastardized Greek, as the principle of acidity) becomes "Säuerstoff," "hydrogène" (so-named as the principle of water) becomes "Wasserstoff," and so on.

140. See 22, 508.26–509.1, where, in the context of a discussion of the elements, Kant speaks of "the separation of two matters from one another, such as hydrogen from water, where then the other part unites with the iron as oxygen in that it simultaneously relinquishes the all-penetrating caloric." Kant is here illustrating the new doctrine of the elements with Lavoisier's experiments on the decomposition of water via iron (see, e.g., *Elements,* Part One, chap. VIII)—where, as in all other types of oxidation, the release of caloric from its union with the oxygen base plays a crucial role.

which can be plural. And it is clear, furthermore, that this distinction is central to the *Transition* project:

> Physics is the science (doctrinal system) of the totality *(complexus)* of empirical cognition of perceptions as the moving forces of matter affecting the subject, in so far as they constitute a system called experience bound together in an absolute whole.
>
> The transition from the metaph. found. of nat. sci. to physics is therefore not an aggregation of empirical representations with consciousness, but rather the concept of their synthetic unity for the sake of the possibility of experience, which is always thought as a system of empirical representations (not an empirical system because that would be a contradiction).
>
> Physics is therefore *empirical-doctrine* (through observation and experiment) of the moving forces of matter. However, since experience (outer as well as inner) as a subjective system of perceptions is always One, it follows that the moving forces in space affecting the sense of the subject are, in virtue of their coexistence in space, already moving in all parts of the latter (for an empty space is no object of possible experience); and it further follows that the parts of matter, as movable ⟨and moving⟩ substances must be thought, not under the name of *matters* (for matter is universally distributed unity of the movable), but under that of *materials* [*Stoffe*]—of which there can be many and various. To be sure, the latter are all in agreement in that they are moving in outer relations through attraction or repulsion, but in the manner in which they modify the composition and separation of matter they furnish ⟨specifically⟩ different ⟨body-forming⟩ moving forces. Each of these materials is, as a *foundation* [*Grundlage*] (basis) of the forces in question, the active cause of the former relations and bears the name of the phenomenon of its action (⟨oxygen, carbon, hydrogen,⟩ nitrogen, etc.)—where the name of caloric [is borne] because [caloric] presents the most universally distributed phenomenon ⟨[of heat]⟩ . . . . (22, 359.15–360.15)

Thus *materials* (plural) are Lavoisier's elements, conceived as principles (bases) or radicals of various types of chemical action:

> ⟨Substances as moving forces of matter in so far as they are different according to their specifically different forces—*Materials (substantiae radicales). Adhaerens* to the *primario.* Such as oxygen, carbon, hydrogen, and nitrogen. Where then the basis of a certain material, e.g., the basis of muriatic acid, is still spoken of.⟩ (450.14–18)

> The principle of experience concerning the actuality of a certain species of matter (material), of which one is *universally distributed,* etc., and of which one contains the *basis* of the other species, e.g., of muriatic acid, or which contains the universal basis of all *primitively moving* forces called *caloric.* (478.26–30)[141]

141. Kant emphasizes that the basis of muriatic acid (Salzsäure), in particular, is as yet unknown: see 351.15–18, 508.22–26, 13.9–11, 93.27–30, 106.7–12.

The importance of Lavoisier's system of chemical classification (together with the fundamental role of caloric therein) is therefore evident.

Lavoisier's system in fact yields a method for the complete classification of the different types of matter. This classification is to proceed by genus and specific difference, as it were combinatorily: each species is characterized by the presence of a substance that serves as the basis or principle of that species, each subspecies thereof contains this substance plus its own distinguishing basis or radical, and so on. Thus, for example, all acids contain oxygen, the acidifying principle, plus a basis or radical characteristic of each particular type of acid: sulphuric acid = oxygen + sulphur, phosphoric acid = oxygen + phosphorous, etc. Moreover, just as oxygen is the acidifying principle, caloric is the "gasifying" principle: each type of gas is a combination of a particular base plus caloric—in particular, oxygen gas = oxygen (base) + caloric. Such a conception, as we have seen, is absolutely central to Lavoisier's system, for it is precisely this caloric theory of gases that enables him to unify chemistry with the new discoveries in the theory of heat—thereby explaining the heat released by or absorbed in chemical reactions. In this sense, the concept of caloric constitutes the foundation for the new system of chemical classification as well.[142]

These last considerations now enable us to connect the aether-deduction with the task of the *Transition* project, as described abstractly and generally in §II above, in a more precise and explicit fashion. In §II we argued that the task of the *Transition* project is to extend the "top down," constitutive, procedure of the *Metaphysical Foundations* even further into the domain of the properly empirical in order ultimately to meet or connect with the "bottom up," merely regulative, procedure of reflective judgement. The latter guides the process of empirical classification or concept formation, the procedure of systematizing lower-level empirical concepts under higher-level empirical concepts. The former, on the other hand, articulates (schematizes) the very highest-level concept of empirical classification—the empirical concept of matter. Moreover, the concept of matter articulated in the *Metaphysical Foundations* is the most general possible: it applies to all types of matter whatsoever, regardless of "specific variety" and state of aggregation. Indeed, as we have seen, the *Metaphysical Foundations* holds that *only* the concept of matter in general can be articulated a priori: all problems concerning either "specific variety" or state of aggregation are physical rather than metaphysical questions.

It is natural, therefore, that an extension of the constitutive procedure

142. For the significance of Lavoisier's system of chemical classification, including the foundational role of caloric as "gasifying" principle, see Metzger [80], especially pp. 38–44.

of the *Metaphysical Foundations* should establish a priori a principle for further specifying the universal concept of matter in general into distinct primitive types of matter—that is, into elements—whose combinations then make possible a complete classification of the concept of matter in general via genus and specific difference. Kant describes this conception of the elements as a "qualitative atomism":

> ⟨Matter as movable substance in general—or also as a particular [one] of a certain quality (therefore one can not speak of *matters* but only of matter, which is thus thought of as everywhere uniform).
>
> Matter as material *(Basis virium moventium)* is the qualitative unity of moving force, not composed out of various heterogeneous [forces] but still belonging to matter as a particular element *(atomistica qualitiva)*, and is to be distinguished from the *medium deferens*. *Materials* can be heterogeneous, but *matter* (which is always only one) is homogeneous.⟩ (402.1–10)

Further, the physics that the *Transition* project is to ground is the theory of such elements:

> ⟨Thus forces can also be thought in matter as *materials*—i.e., as substances belonging to the motion of matter—which constitute the basis of these forces, and physics is a doctrinal system thereof.—These materials, considered in the quality of moving forces, can be enumerated a priori according to principles: as grounded on *attraction and repulsion*, both, however, on penetrating- or superficial [force], acting from whole to part, etc., coercible, etc. May be enumerated and classified a priori according to principles. Basis and matter that is guiding.⟩ (409.2–10)

But we have just seen that the elements or materials in question here are the elements of Lavoisier's systematic chemistry.[143] Hence, what Kant appears to be envisioning in the aether-deduction is an a priori foundation for the new system of chemical classification—based, of course, on the concept of caloric.

If this is correct, then Kant, in the aether-deduction, has continued to be unusually sensitive and faithful to the scientific developments with which he is confronted. For Lavoisier made chemistry into a genuine theoretical science, not, as the *Metaphysical Foundations* would have it, by articulating new microscopic force laws analogous to the law of gravitation, but rather precisely by fundamentally reorganizing the system of chemical classification. This reorganization takes full advantage of the new discoveries concerning the gaseous state, "physicalizes" chemistry by connecting it essentially with advances in the theory of heat and of the

143. The "guiding" or "deferent" matter Kant speaks of here is the aether or caloric: see, e.g., 22, 224.4–6, 394.11–27.

states of aggregation, and has, of course, served as the basis of the science of chemistry ever since. Kant's attempt to deploy the machinery of the critical philosophy so as to ground a priori the crucial hypothetical construct of the new system—the imponderable matter of fire or caloric—is not only a response to a natural and inevitable requirement of the critical philosophy itself (as we saw in §II); it is also a most insightful attempt to come to terms with the central scientific revolution of the eighteenth century.

V · The Fate of the Aether-Deduction

The aether-deduction proceeds, I have argued, by searching for the one and only representation available to the critical philosophy that has both distributive and collective, both analytic and synthetic, universality. For only such a representation can effect an intersection, as it were, of the constitutive domain of the understanding and the *Metaphysical Foundations* with the regulative domain of reflective judgement. Only such a representation, that is, can possibly serve as the basis for a complete and unified science. What makes the aether-deduction work, therefore, is the idea that the representation of a universally distributed, space-filling matter functioning as "the basis of the moving forces of matter" (namely, caloric or the aether)—and this representation alone—combines the required distributive and collective, analytic and synthetic, constitutive and regulative, characteristics.

Serious doubts can be raised, however, concerning the actual constitutive force of these considerations. For it remains doubtful that the real existence of an object corresponding to the representation in question can thereby be inferred a priori. Suppose we grant that *only* the representation of caloric or the aether can possibly ground an ideal complete science. Suppose we grant further that this representation, and this representation alone, possesses both distributive and collective, both analytic and synthetic, universality—and therefore represents an intersection of the constitutive and regulative domains. Why should the representation of such an intersection, and hence the representation of the aether, not itself remain a mere idea or ideal of reason? We know, to be sure, that *if* the ideal complete science is possible, it is possible only on the basis of a universally distributed aether. Yet, since this complete science is never actually achieved but is only continually approximated as an ideal, what entitles us now to assert that its basis—the universally distributed aether—actually and objectively exists? It remains entirely obscure, in other words, how the representation of an intersection of the constitutive and regulative domains itself acquires more than merely regulative force: we can conceive this representation as a goal we must strive to realize, but it is in no way clear how we can ever be in a position to guarantee such a realization.[144]

One way to appreciate the extent of the problem here is to observe that we have been operating so far with only the unspecified idea of an extension of the constitutive procedure of the understanding and the *Metaphysical Foundations* into the domain of reflective judgement, the domain of the properly empirical. We have so far appealed only to the abstract thought of an intersection of the constitutive and regulative domains. Yet the *Metaphysical Foundations* itself has constitutive force with respect to a quite specific concept of matter, which includes, in particular, a foundation for the Newtonian laws of motion and the theory of universal gravitation. The universally distributed "All of matter" of the aether-deduction, however, appears to have nothing whatever to do with these central ingredients of specifically Newtonian physics—and nothing whatever to do, therefore, with the specific concept of matter developed in the *Metaphysical Foundations*. Indeed, as we have seen, the aether is especially appropriate as a foundation for the new chemistry because of precisely its independence from specifically Newtonian physics: its independence from microscopic action-at-a-distance force laws, in particular.[145] How, then, can the aether-deduction actually generate an extension of the constitutive procedure of the *Metaphysical Foundations*? It would seem that this is quite impossible if all we have to go on is the mere (otherwise entirely unspecified) idea of such an extension.

Indeed, in Transition 8, Kant himself appears to acknowledge that something more than the mere idea of an extension of the *Metaphysical Foundations* (and thus the mere idea of an intersection of the constitutive and regulative domains) is required for genuine constitutive force:

> The thought of an elementary system of the moving forces of matter *(cogitatio)* necessarily precedes its perception *(perceptio)* and is given in the subject a priori through reason as subjective principle of the combination of its elementary parts in a whole *(Forma dat esse rei)*.—The whole as object of

144. For this reason, attempts to construe the aether-deduction as in the end merely regulative—such as those of Förster and Lehmann mentioned in note 130 above—are natural and inevitable. I believe that they also miss the full force and complexity of Kant's thought on this issue, however.

145. To be sure, Kant conceives the aether as perpetually oscillating or vibrating through forces of attraction and repulsion, but he provides no considerations that would connect these supposed forces with the fundamental forces of attraction (underlying gravitation) and repulsion (underlying impenetrability) of the *Metaphysical Foundations*. Similarly, when Kant imports the language of attraction and repulsion into the description of chemical affinities (as in the above-quoted passage from 22, 359.15–360.15), this usage is non-technical—if not metaphorical—and has no direct connection with specifically Newtonian forces (defined by the laws of motion).

possible experience, which must therefore not proceed atomistically from the composition of the empty with the full (hence not *mechanically*), but rather must proceed *dynamically* as combination of external forces mutually agitating one another (by originally agitating one another through attraction and repulsion of the elementary material completely and uniformly distributed in space as first instigating all motion and thus continuing constantly to infinity).—This proposition still belongs to the metaphysical found. of N.S. in relation to the whole of One possible experience; for *experiences* can only be thought together as parts of a total experience united according to one principle.

This principle is now subjective for the cosmic observer [Weltbeschauer] *(cosmotherous):* a *basis,* in the idea, of all united forces setting the matter of the whole universe in motion: but it does not prove the existence of such a material (such as that one calls the all-penetrating and permanently moving caloric); and it is so far a hypothetical material. Since, however, its idea first represents space itself (although indirectly) as something perceptible and as an unconditioned-whole (inwardly moved and externally universally moving), it follows that this matter is to be assumed as the first mover *(primum mobile et movens),* subjectively, as the basis of the theory of the highest moving forces of matter for the sake of a system of experience. (21, 552.18–553.17)

This passage suggests that the aether-deduction is to proceed in two distinct stages, which can perhaps be articulated as follows.

In the first stage, we consider the *Metaphysical Foundations* "in relation to the whole of One possible experience"—that is, we consider an ideal extension of the constitutive procedure of the *Metaphysical Foundations* so as to intersect or meet with the regulative procedure of reflective judgement and thus to secure the *collective* unity of experience. We thereby generate the idea of a universally distributed aether, but we are not, at this stage, able to affirm the objective existence of this object. On the contrary, such objective existence can only be "assumed" at a second stage, where we introduce the additional consideration that the idea of the aether "first represents space itself . . . as something perceptible and as an unconditioned whole." The first stage, then, is indeed merely regulative; only the second is truly constitutive.[146]

146. That the *Transition* project is to proceed in two distinct stages, the first merely regulative and the second constitutive, is suggested by the passage from Farrago 1–4 at 22, 178.26–33, quoted in §II above. There "the concept of physics as that towards which the metaphysical foundations has the tendency"—which, I would suggest, is simply the unspecified idea of an extension of the constitutive procedure of the *Metaphysical Foundations*—is characterized as merely regulative. (Förster [32], p. 226, cites the phrase "in the idea" at 21.553.8 as evidence for a purely regulative reading of the aether-deduction—or at least for Kant's "ambivalence" in this regard. My suggestion is that Kant is here only referring to a first step in the proof.)

Now, as we have seen in §I, Kant initiates a fundamental reconsideration of the central concerns of the *Metaphysical Foundations*—the Newtonian argument for universal gravitation, in particular—in the outline-drafts of the tenth and eleventh fascicles written immediately after Transition 1–14 (1799–1800). Kant there contrasts the merely mathematical and mechanical work of Huygens with the truly philosophical and dynamical achievement of Newton, and the point, I have urged, is that only Newton is able to move from a purely kinematical analysis of circular motion to the theory of a genuinely dynamical attractive force acting universally and immediately at a distance. Moreover, this Newtonian procedure requires philosophical foundations of natural science (and not merely mathematical foundations), because it rests on the a priori grounding of the laws of motion and the key properties of universality and immediacy effected by the *Metaphysical Foundations:*

> . . . it belongs to the metaphysical foundations of natural science and thus to philosophy to use the mathematical principles as an instrument for the sake of philosophy in regard to the relations of given forces of matter, and [to proceed] from the Keplerian formulas (the three analogies) to the moving forces that act in accordance with them [and] to ground the system of universal gravitation by the original attraction . . . . (22, 518.12–18)

Finally, Kant observes that Newton's achievement, again in contrast to that of Huygens, belongs to the transition from the metaphysical foundation of natural science to physics:

> ⟨Huygens's transition from the metaph. F. of N.S. to the mathematical [foundations], and Newton's to physics—merely through the concept of gravitational attraction which Kepler did not think of.⟩ (353.10–12)

And this makes sense, for Newton's argument for universal gravitation starts from the merely empirical regularities recorded in Kepler's laws of planetary motion (compare 521.14–21, 528.15–529.5) and thereby builds a bridge between the synthetic a priori laws established in the *Metaphysical Foundations* (the laws of motion and the properties of universality and immediacy) and the domain of the properly empirical.[147]

Since the Newtonian argument for universal gravitation thus builds a bridge between the a priori and properly empirical domains, it follows that considerations of systematicity—that is, considerations belonging to the regulative use of reason or reflective judgement—must necessarily play a role:

147. Kant notes that the theory of gravitation builds a bridge between the a priori concepts developed in the *Metaphysical Foundations* and the domain of the properly empirical already in loose-leaf 5 (1798) at 21, 483.24–29: see note 47 above.

> Instead of Kepler's *aggregation* of motions containing empirically assembled rules, Newton created a principle of the system of moving forces from active causes. Unity[.] (521.19–21)

Indeed, Kant discusses the discovery of universal gravitation under the rubric of the regulative use of reason already in the first *Critique:*

> Thus, for example, when the orbits of the planets are given us through a not yet fully rectified experience as circular, and we find deviations, we then suppose such orbits as the circle can be transformed into through all infinite intermediate degrees in accordance with a constant law—i.e., the motions of the planets, which are not circular, perhaps approximate this property more or less exactly—and we hit upon the ellipse. The comets exhibit a still greater deviation in their paths, since they (as far as observation reaches) never return in their orbits; but we then conjecture a parabolic path, which is still akin to the ellipse, and, when the major axis of the latter is extended very far, it cannot be distinguished from the former in any of our observations. We therefore arrive, under the guidance of these principles, at unity in the kinds [Gattungen] of these orbits in their figure, and thereby unity in the cause of all laws of their motion (gravitation) . . . . (A662–663/B690–691)

It follows, then, that the regulative procedure of reflective judgement is presupposed even in the science grounded and taken as paradigmatic in the *Metaphysical Foundations.*[148]

In any case, however, in the tenth and eleventh fascicles Kant also connects the theory of universal gravitation with the idea of space as a sensible or perceptible object:

> Now *Newton* stepped forth and, as philosopher, introduced ⟨into the universe a moving force called gravitational attraction, identically connected with space itself and to be viewed merely as sensible space, as⟩ universal world-attraction of all bodies through empty space and motions through central forces . . . . (522.1–5)

And this idea is elaborated and clarified in the immediately succeeding outline-draft:

> Space is, to be sure, merely the form of outer intuition and the subjective [element] of the mode of being externally affected, but it is still considered

148. Hence, it is not surprising that Kant comes to characterize the kind of interplay between metaphysical and mathematical foundations of natural science exemplified by the argument for universal gravitation as a *schematism of the faculty of judgement:* "⟨The metaphysical foundations precede the mathematical [foundations]. The former united with the latter yield the schematism ⟨of the faculty of judgement according to the relation of empirical intuition⟩ of space and time⟩" (22, 484.11–14); see also 491.3–7, 494.32–495.1, 495.20–24. I am indebted to Robert Butts for emphasizing to me the importance of reflective judgement in the argument for universal gravitation—the importance of A662–663/B690–691, in particular.

> as something externally given, as real relation in so far as it must be thought as a principle of the possibility of perceptions—but experience still precedes.
>
> In this connection we *must* represent to ourselves matter (the movable in space), but also a moving force of masses of matter which is an action of masses through empty space *(actio in distans)* represented as extending to infinity, which is unlimited; yet every whole of matter is limited (body) and, in fact, by two forces—original attraction and repulsion—without whose combined action there would be absolutely no matter and space would be empty, but still cognized as such—which is contradictory.
>
> However, it is a proposition which is not grounded on physics (empirical doctrine of moving forces), but rather originally grounding physics, that there must be an attraction, even without reacting repulsion, among bodies moving around a common center of motion, in virtue of which and their orbital motions bodies (the heavenly bodies) moved in orbits must move around centers of motion and so finally around an unmoved [center of motion] in the whole of space.
>
> All bodies strive to approach one another through motion in empty space and, in fact, according to the law of proportionality to the quantity of the masses and the inverse proportion to the square of the distances, in virtue of an impulsion *(impulsus)* of attraction. (524.3–27)

It appears, then, that it is gravitational attraction, together with the patterns of orbital motion engendered thereby, which first makes it possible for us to represent "space itself . . . as something perceptible and as an unconditioned whole."

This last idea, it seems to me, can only be interpreted as follows. Universal gravitation, together with the patterns of orbital motion engendered thereby, makes space into an actual object of experience (and not simply a mere form of intuition) by generating a procedure for continually approximating a privileged frame of reference: a frame of reference fixed at the center of mass (center of gravity) of all matter. We start (with Ptolemy) with a frame of reference fixed at the center of the earth; move from there (with Copernicus and Kepler) to a frame of reference fixed at the center of the sun (and in which the fixed stars are at rest); move from there (with Newton) to a frame of reference fixed at the center of mass of the solar system; move from there to a frame of reference fixed at the center of mass of the Milky Way galaxy; and so on. In this way we continually approach an "unmoved" space or frame of reference relative to which all *true* motions are to be ultimately defined—thereby continually approaching an ideal counterpart of Newtonian absolute space. Space is thus transformed from a mere form of intuition into an actual object of experience in which all true motions—and hence all objective laws of motion—are to be set up.[149]

149. This procedure for transforming apparent motions into true motions—and hence

But what do these ideas have to do with the *Transition* project and the aether-deduction? The above-quoted passage continues as follows:

> All bodies strive to approach one another through motion in empty space and, in fact, according to the law of proportionality to the quantity of masses and the inverse proportion to the square of the distances, in virtue of an impulsion *(impulsus)* of attraction. (But how are the distances perceived if the moving forces are to be active in empty space?) In order to determine the distances through perception space must be perceptible; therefore it cannot be empty.—There are therefore mathematical foundations of natural science which simultaneously devolve upon philosophy, because they concern the quality of moving force according to its *causality,* and mathematics here acts as instrument.
>
> *Materials complementa virium moventium materiae.*—The quality of matter can not be thought atomistically but must be thought as dynamically grounded[.]—This grounding is the original attraction of bodies through empty space, which is thus no object of perception, but can merely be thought. The intelligible space is the formal representation of the subject in so far as it is affected by outer things.
>
> From the unity of matter it follows that there is a common principle (basis) of its forces, which constitutes unlimited space as object of the senses *(originaria basis et communis)* or which contains moving force in a particular way *(basis specifica).*—The former is represented as the substance occupying all space, represented a priori for itself, without having particular properties except the occupation of space in itself. This sensible space is assumed as limiting itself through moving forces. (524.24–525.19)

Thus the grounding of space through universal gravitation in fact takes us only part of the way towards space as an object of perception or sensible space. For the empty space through which gravitational attraction acts is "no object of perception, but can merely be thought." We arrive at a truly sensible space only by asking how the distances figuring in the law of gravitation are themselves actually perceived, and this finally leads us to "the substance occupying all space" or the aether.

Kant's meaning becomes clear in the following pages, where, after again asking how distance (and thus the law of gravitation itself) becomes an object of perception, he continues:

> To this Newtonian principle of universal attraction through empty space there now corresponds a similar principle of repulsion *(virium repellentium),* which can similarly be no object of experience for itself, but is only necessary

---

appearances into experience—by, as it were, constructing continual approximations to Newtonian absolute space is discussed in the fourth chapter or Phenomenology of the *Metaphysical Foundations.* Accordingly, absolute space is there characterized as an idea of reason (4, 559.7–9). Compare also the second Observation to Definition 1 of the first chapter or Phoronomy (481.2–482.6). See Chapter 3 above.

to present space as sensible object. It is the characteristic of the matter to act on the senses at a distance, in that which is thereby presented to sensation and empirical *intuition,* not so much the intermediate [matter] affecting the subject, as rather the object by means of this matter. Light and sound (with their colors and tones) are such means of transition [Überschritte] which make representable an action-at-a-distance *(actio in distans)* as immediately possible. We see or hear light and sound, not as immediately touching the eye or ear, but rather as an influence of sensible objects on our organ as distanced from us. (529.27–530.12)

And Kant makes essentially the same point several pages later:

> How is the distance of the point of attraction from the moving body perceived, or in general determinable, in empty space? For one cannot weigh the heavenly bodies and can determine their quantity of matter only relatively—when one is moved through acceleration of another whose position is perceptible through no intuition[.]
>
> The motions of light and sound are media through which distances can be determined through moving forces. Both also have peculiar qualities (subjective) of sensation[.]
>
> Moreover, the distance of no other body in empty space can be determined through experience, because the absolute quantity of matter is not determinable for itself—and thus [is so] only through media (as above) which require a time for their motion. (537.13–24)[150]

Kant's point, then, is that it is impossible for us, as perceiving subjects, actually to employ universal gravitation in the determination of space (as object of experience) unless we have a means for perceiving (and thereby determining) the relative distances or relative positions of the heavenly bodies. This, in turn, cannot be achieved in truly empty space, but only by means of a space-filling medium that somehow conveys signals from the heavenly bodies to our senses. We thereby arrive, once again, at the concept of the aether: more precisely, at the concept of a light-aether.

It is noteworthy, therefore, that the above line of thought is prefigured in the argument of Transition 1–14:

> The very same proposition concerning the existence of an elementary material filling the entire universe (as a continuum) is at the same time the acting cause ⟨of the possibility⟩ of a universal community of the heavenly bodies among one another—not the imperceptible [community] of attraction

150. Kant here appears to allude to the circumstance that in the Newtonian determination of the center of mass of the solar system the masses of the primary bodies are determined from the accelerations and distances of their satellites: if m_p is the mass of the primary body, r is the distance of the satellite, and a_s is the acceleration of the satellite, then $m_p = a_s r^2/G$, where G is the gravitational constant. This determination of masses, implemented in Proposition VIII of Book III of *Principia,* is of course absolutely central to the procedure for continually approximating absolute space described above.

> through empty space, but rather the perceptible [community] by means of that material which we may call caloric or light-matter, whose mutual interaction depending on the first motion makes known its existence in all other [matters]. (21, 565.3–10; compare also 520.30)

Indeed, we even find a hint of this line of thought in the Third Analogy. Kant there gives the earth-moon system as an example of dynamical community (B257), which clearly suggests community effected by gravitation. He then remarks that dynamical community can be either immediate or mediate (A213/B259), and gives the following example of the latter: "the light, which plays between our eyes and the heavenly bodies effects a mediate community between us and them and thereby establishes their simultaneity" (A213/B260). It seems clear, then, that gravitation effects the *immediate* community (and thus simultaneity) of the heavenly bodies in space, whereas light establishes this *mediately*—that is, through the perceputal contact if affords us with gravitation.[151]

Putting all these considerations together, we arrive at a version of the aether-deduction which is in fact a continuation of the constitutive procedure of the *Metaphysical Foundations*—which genuinely engages the specific content of the concept of matter articulated there. The *Metaphysical Foundations,* by considering the conditions under which alone we can construct an objective counterpart of Newtonian absolute space (relative to which true as opposed to merely apparent motions are first objectively defined), provides an a priori foundation for the laws of motion and the two crucial properties of universality and immediacy of gravitational attraction. For only if these are presupposed can we determinately carry out the argument of Book III of *Principia.* The point of our present considerations, however, is that we, as perceiving subjects, cannot even begin this Newtonian construction unless we also presuppose a means for actually making perceptual contact with the heavenly bodies: more precisely, with the observable (Keplerian) "phenomena" with which the Newtonian argument begins. But such perceptual contact presupposes a universally

151. This could perhaps explain why Kant, in the passage from Transition 8 (21, 552.18–553.7) with which we began, asserts that the aether "first represents space itself (although *indirectly*) as something perceptible and as an unconditioned whole" (emphasis added). That the *(immediate)* dynamical community of the heavenly bodies in space is effected by gravitation is supported by the argument of the *Metaphysical Foundations:* there the third analogy is instantiated by the law of action and reaction in the Mechanics (Prop. 4), which is then restated, under the category of necessity, in the Phenomenology (Prop. 3)—where it is then illustrated by gravity as an "active dynamical influence given through experience" (4, 562.12–13). Moreover, I think we may also find an earlier suggestion of the above argument from gravitation (and thus attaction) to light (and thus repulsion) in the Observation to Proposition 5 of the Dynamics: for discussion see [35].

distributed medium—such as a light-matter—and thus presupposes the aether. The aether is therefore an additional necessary precondition of the Newtonian argument, which, as we now see, is only partially grounded by the *Metaphysical Foundations* itself. The complete a priori grounding can only be effected by the aether-deduction, and, in precisely this way, the latter acquires genuinely constitutive force.

Unfortunately, this version of the aether-deduction is also fatally flawed. For the aether thereby grounded—a universally distributed medium of perceptual contact—has no intrinsic connection with the heat-aether or caloric serving as the basis for the new chemistry and experimental physics. From the point of view of this version of the aether-deduction, then, the heat-aether or caloric of the new chemistry and experimental physics must itself have a merely hypothetical status. Indeed, Kant explicitly asserts this immediately following our passage from 22, 524–525:

> Matter is the external sensible object in general, in so far as it can be only One and unlimited, as opposed to empty space. Its moving forces as specifically different types of matter are called *materials (Materies, materiei)*—parts of matter to which specifically different forces therefore also pertain—and are movable substances (e.g., nitrogen, carbon). One of these so-called materials, which, as assumed to be everywhere present and all-penetrating, is the *guiding* material [and] is merely hypothetical (caloric): namely, is suited for the motion and division of all materials—and, in addition, may also well be mere quality of motion. (525.20–526.4)

From our present point of view, in other words, an aether playing the specific role of caloric, heat-matter, or *chemical*-aether now has the status of a mere hypothetical construct; consequently, it is no longer certain that the material as opposed to the mechanical theory of heat is in any way privileged.

Kant is quite consistent in thus contrasting the light-aether and the chemical-aether or caloric throughout the remainder of the *Opus* For example, in the next major series of outline-drafts assembled in the seventh fascicle (1799–1800), we find:

> Of the Newtonian concept: ***Philos.*** *nat. princ. mathematica.*—Transcendental philosophy makes such a thing possible without μεταβασις εις αλλο γενος; because one, for the forces, determines space, in which and according to which laws they are to act. The forces here lie already in the representation of space.
>
> One can also (but only in a conditional fashion) postulate a priori the existence of a light-matter dispersed throughout the entire universe, because we would not otherwise perceive the objects in space at all distances. According to the rule of identity. With heat the situation is not such—namely, that there must be such a material—because it is something merely subjective,

and the extension of bodies through heat is only for the eyes and thus for light; and it is therefore only inferred as the effect of a cause. (22, 84.5–17)[152]

And, in the final series of outline-drafts constituting the first fascicle (1800–1803), Kant writes:

⟨Is there caloric as a particular material or is heat the mere capacity for perceiving the force of repulsion of matter (of materials in general)?

Whether the characteristic of things *to be visible* does not have analogy with the property in space *to be sensible for heat,* and can a material for this be assumed?

There are only *sound* and *light-rays* as means of cognition—both *in straight lines.*⟩ (21, 110.4–11)

Thus heat, unlike light, does not serve as a "means of cognition"—as a general medium for putting us into perceptual contact with the objects of our knowledge. Therefore heat, unlike light, can generate no a priori aether-deduction and is at best a hypothetical material.[153]

But it now follows, it seems to me, that the aether-deduction must fail for the light-aether or universally distributed medium of perception as well. For recall that we were earlier able to arrive at a universally distributed aether (as opposed to a foundation based on action-at-a-distance forces, say) only by appealing to the representation of the ideal complete and unified science. Such a representation, in the critical system, must function as a kind of intersection between the constitutive and regulative domains, and must therefore possess both distributive and collective, both

152. Compare, e.g., 56.16–57.2: "⟨How are laws of unified space- and time-determinations of moving forces possible a priori? Newton's work. Immediate *actio in distans* (through empty space).—Of the mutually acting motion of light in full space, yet without dispersion, since the divergence of rays and, simultaneously, Roemer's time-condition act mutually opposed to their motion.—Of the magnet—Heat, as inwardly moving force of the body, is, in that it expands and disperses matter, a hypothetical material—and can well be the mere action of the repulsion of a matter set in oscillation⟩." Kant's talk of the intimate connection between forces and the representation of space is, I think, to be understood in terms of the relation between the force of gravitation and the determination of Newtonian absolute space (via approximations thereto) discussed above. Compare also 22, 299.15–18: "For, with the unity of space in the relation of the unification of motion, unity of combining forces is also thought in the same synthetic concept [of the relation of the unification of motion]"; and 107.16–21: "⟨Space is not merely, as in pure intuition, a mathematical object (4 conic sections), but is also filled with forces for physics as a *sensible* space, and matter is stratified through attraction and repulsion—as the ponderable materials of the planets manifest their distances in so far as they are attractive⟩." This last point again appears to be an allusion to the determination of the different masses (and thus different densities) of the planets as in note 150 above.

153. Kant explicitly asserts that caloric is a hypothetical material in a letter to C. G. Hagen of April 2, 1800 (12, 301.20–21). See also, e.g., 22, 534.11–15, 51.9–11, 62.10–11, 106.7–12, 128.4, 129.22.

analytic and synthetic, universality. The one and only representation possessing these features is that of a universally distributed "All of matter" serving as the basis for a unified system of the moving forces of matter, and the caloric or chemical-aether underlying the new chemistry and experimental physics can then plausibly be viewed as an actual instantiation of this Kantian "All of matter." What, however, does light—or, in general, a means for our perceptual contact with objects of cognition—have to do with the "All of matter" serving as the basis for a system of the moving forces of matter? Why should our means for establishing perceptual contact with objects, whatever this may turn out to be, *also* constitute a basis for the ideal complete science—and therefore possess both distributive and collective, both analytic and synthetic, universality?

Thus, for example, Kant's conception of a light-aether clearly presupposes the wave theory of light as opposed to the particle theory.[154] But how is the latter to be ruled out a priori? To be sure, if we could somehow associate the supposed light-aether with a possible basis for the unified system of the moving forces—such as we have seen caloric or the chemical-aether to be—an argument for the privileged status of the wave theory would indeed be forthcoming. Yet precisely this association has now been dissolved: our means for establishing perceptual contact with objects has been explicitly split off from the one possible basis for a unified system of chemistry and physics that we can envision, and the latter, in fact, has now been characterized as merely hypothetical. How, then, can we possibly conclude that our means for establishing perceptual contact with objects must be universally distributed, and thus must be an aether?

Kant's attempt to infuse genuine constitutive force into the aether-deduction therefore appears to founder on a dilemma. On the one hand, we can model the object of our deduction on the heat-matter or chemical-aether of the new chemistry and experimental physics. We then have a plausible and promising basis for the ideal complete science, and we can thus engage the Kantian conceptions of constitutive and regulative domains, and of distributive and collective or analytic and synthetic universality. We thereby arrive at the representation of a universally distributed "All of matter"—and, moreover, we arrive at this representation a priori. Yet, since the heat-matter or chemical-aether of the new chemistry and experimental physics has no direct connection with the Newtonian laws of motion and theory of universal gravitation, we do not really make contact with the specific content of the concept of matter articulated in the *Metaphysical Foundations*,[155] and it is therefore hard to see how the

154. As Kant explicitly states at 22, 32.28–29: "That light is no shooting motion *(eiaculatio)* of a matter but a wave motion *(undulatio)*."

155. Lavoisier's experimental method is firmly based on the principle of the conservation

aether-deduction can acquire genuine constitutive force. On the other hand, however, we can model the object of our deduction on the light-aether of Euler's wave theory of light. This aether, as universal medium for affording us perceptual contact with the heavenly bodies, can then plausibly be taken as an additional presupposition of the Newtonian argument for universal gravitation. We thereby engage the specific content of the concept of matter articulated in the *Metaphysical Foundations,* and we can thus infuse our argument with genuine constitutive force. Yet, since this light-aether has now been explicitly split off from the heat-aether or caloric of the new chemistry and experimental physics, we are no longer concerned with a possible basis for the ideal complete science; and it is therefore hard to see how we can deploy the Kantian conceptions of distributive and collective or analytic and synthetic universality. That our means for establishing perceptual contact with objects must be precisely an aether (a universally distributed matter) can thus no longer be established a priori.

It follows, then, that neither a light-aether nor a heat-matter or chemical-aether can be established a priori as actually existing objects. Our grounds for postulating either or both can only be empirical or hypothetical—in virtue of their explanatory power in the context of the best available theories of the phenomena in question.[156] To be sure, we can establish a priori that only the representation of a universally distributed "All of matter"—as the one and only representation possessing both distributive and collective universality—can possibly serve to unite the constitutive and regulative domains and thereby ground the ideal complete science. But we can in no way establish a priori the objective existence of an object corresponding to this representation. From the point of view of the critical philosophy, therefore, such a representation remains a mere idea or ideal of reason possessing only regulative force; and, if the success of the chemical revolution now allows us to assert the existence of a corresponding object, this can now only be viewed as a merely empirical and hypothetical assertion. The aether-deduction—and hence the *Transition* project—must in the end be considered a failure.

---

of the quantity of matter, as measured by the balance, and thus this experimental method certainly engages specifically Newtonian concepts (see note 83 above and also Part I of Metzger [79]). Moreover, the crucial importance of conservation of quantity of matter here is certainly not lost on Kant (see note 124 above). This circumstance can hardly favor the chemically based version of the aether-deduction, however, for heat-matter or caloric is necessarily *imponderable* and thus constitutes a conspicuous exception to the conservation principle (that is, to the idea that all materials involved in chemical reactions can be precisely "tracked" by means of the balance).

156. It is no wonder, therefore, that Kant, in the period we are now considering, is at times inclined to the view that *both* caloric and the light-aether are merely hypothetical: see, e.g., 22, 92.29–93.26 (note) together with 81.16–18.

Our understanding of the underlying problem here can be deepened and put into perspective if we now ask how the status of the aether-deduction really differs from that of the supposedly constitutive argument of the *Metaphysical Foundations*. Is not the concept of matter articulated in the *Metaphysical Foundations* itself an empirical concept? Is it not the case, then, that like all empirical concepts its "possibility must either be known a posteriori and empirically, or it cannot be known at all" (A222/B269–270)? Does not Kant, for precisely this reason, explicitly assert that "all means escape us for *constructing* this concept of matter and for presenting as possible in intuition what we think universally" (4, 527.10–12)? Moreover, in the very first Definition of the *Metaphysical Foundations,* Kant characterizes matter as the movable in space and then, in the second Observation to this Definition, remarks:

> Finally I observe: that, since the *movability* of an object in space cannot be cognized a priori and without instruction through experience, it could not be enumerated by me in the *Critique of Pure Reason* under the pure concepts of the understanding for precisely this reason; and this concept, as empirical, can only find a place in a natural science as applied metaphysics, which is concerned with a concept given through experience, although according to a priori principles. (482.7–13)

Since the objective existence of an object corresponding to the empirical concept of matter can only be established a posteriori, does not this concept actually have the same (merely empirical and hypothetical) status as the idea of the aether or universally distributed "All of matter"? And, if this is the case, how does the argument of the *Metaphysical Foundations* itself acquire genuine constitutive force?[157]

I think that a fundamental reconsideration of our last question can be discerned in the so-called *Selbstsetzungslehre* or doctrine of self-positing developed in the last fascicles of the *Opus*. What I have in mind here is that aspect of the *Selbstsetzungslehre* according to which self-positing occurs in pure, as opposed to empirical, intuition.[158] This aspect of the doctrine

157. From this point of view, therefore, it begins to look as if only mathematical concepts (for which alone objects can be constructed a priori in pure intuition) and pure concepts of the understanding (whose application to objects of experience is established by transcendental deduction) actually possess constitutive force; for only in the case of such concepts can objective reality be established a priori. Indeed, with respect to the latter concepts Kant says: "Thus only thereby, in that these concepts a priori express the relations of perceptions in every experience, does one cognize their objective reality—i.e., their transcendental truth—and, in fact, independent of experience, but still not independent of all relation to the form of an experience in general and the synthetic unity in which alone objects can be empirically known" (A221–222/B269).

158. Interpreters such as Lehmann and Förster concentrate on an aspect of the *Selbstsetzungslehre* essentially involving *empirical* intuition and, in particular, my consciousness of

emerges already in the tenth and eleventh fascicles:

> The subject affects itself and itself becomes object in the appearance in the composition of the moving forces for the grounding of experience as determination of an object as completely determined (existing) thing. Therefore it is not empirical intuition, and still less empirical concept generated from perception, but rather an act of synthetic cognition a priori (transcendental) that makes experience possible subjectively—through which the subject engaged in composition [das zusammensetzende Subject] becomes itself an object, but only in the appearance according to the formal principle [of making] the aggregate of perceptions (as empirical representations with consciousness) into a subjective system of their connection in experience, which experience of sense objects can only be One according to the principle of identity. (22, 364.24–365.6)
>
> ⟨The formal [element] of pure (not empirical) intuition is in a priori representation in the appearance: i.e., represents self-determination as the subject affects itself.⟩ (480.21–23)
>
> Consciousness of my self does not commence from the material [element],—i.e., not from sensible representation as perception—but from the formal [element] of the a priori synthesis of the manifold of pure intuition—not from the object of knowledge but from the coordination of possible sensible representations in the subject affected by objects: i.e., from the cognition of objects as appearance. (448.16–22)
>
> ⟨The subject posits itself in pure intuition and constitutes itself as object.⟩ (452.23–24)

Of course it is so far entirely obscure what it could possibly mean for the subject thus to determine itself, or to affect itself, or to posit itself a priori in pure intuition.

The nature of a priori self-positing in pure intuition becomes clearer, I think, in the seventh fascicle. Kant there introduces the idea of an a priori intuition of oneself as follows:

> The first thought proceeding from the power of representation is the intuition of oneself and the category of synthetic unity of the manifold in the appearance, i.e., the pure (not empirical) representation which precedes all perception under the a priori principle: how are synthetic propositions possi-

---

my own body as an organism: see Lehmann [66], especially pp. 115–122; Förster [32], especially pp. 230–235. Although I myself am unable to see how any truly a priori principles can be generated thereby, it is to be especially emphasized that I focus on only one aspect of Kant's thought here. Moreover, in the last fascicles of the *Opus* Kant broadens his discussion considerably (for example, there is also a morally-practical *Selbstsetzungslehre*, reflections on God and the nature of transcendental philosophy, a new theory of the ideas of reason, and so on); indeed, it is arguable that the *Transition* project itself plays a relatively minor role here. (Compare note 1 above.)

ble a priori? The answer is: they are identically contained in the unconditioned unity of space and time as pure intuitions, whose quality consists in the circumstance that the subject posits itself as given *(dabile)*, but whose quantity consists in the act of composition as unlimited progression *(cogitabile)* as thinkable whole containing the intuition of an infinite whole. (22, 11.2–12)

A few pages later, however, this very same a priori self-positing is closely connected with the representation of motion (as the describing of a space):

> Space, time, and motion in the describing of the former according to the three dimensions of volume, surface, and the point, which are the first, and indeed mathematical principles (axioms) of intuition: i.e., they are no objects of perception (empirical representation with consciousness) as given existing things, but rather are formal [principles of] composition of the manifold in pure intuition made by the subject itself, and ground the task of transcendental philosophy: "How are synthetic cognitions a priori possible?"—whereby the subject constitutes itself as an object not derived from something given. (16.3–12)[159]

The connection between a priori self-positing and the representation of motion is also characterized as follows:

> Space, time, the describing of a space in a given time, and motion according to the categories of quantity, quality, relation, and modality of the object of intuition (of a given object), according to which the subject constitutes itself as object of intuition—which is not empirical but pure intuition a priori, as appearance. (68.20–25)[160]

But how are we to interpret this connection?

What we have here, I suggest, is a reconsideration of the doctrine of a priori self-affection already articulated in the first *Critique* in §24 of the second edition transcendental deduction under the rubric of "figurative synthesis" or "transcendental synthesis of the imagination." These latter notions are introduced as follows:

> This *synthesis* of the manifold of sensible intuition, which is possible and

159. Compare the marginal comment recorded below at 16.19–25: "⟨Space, time, and determination of the manifold in the intuition of objects *(descriptio)* given in space and time are a priori principles of synthetic cognition a priori (of transcendental philosophy) as axioms of intuition in which the construction of concepts is mathematical, but the principle of synthesis therein as appearances on which sensible cognition is to be based is philosophical⟩."

160. On the immediately preceding page we find: "⟨*Spatium, Tempus, Positus* are not objects of intuition but themselves forms of intuitions that proceed synthetically from the faculty of cognition a priori⟩" (68.7–9), and "⟨Space, and time, and the categories of determination of the object of intuition in space and time—all a priori whereby the object [*sic*] posits itself⟩" (68.13–15). It is clear, then, that there is at least an intimate connection between *Positus* and the describing of a space or motion.

> necessary a priori, can be called *figurative (synthesis speciosa),* in distinction from that which is thought in the mere category in relation to the manifold of an intuition in general and is called combination of the understanding *(synthesis intellectualis);* both are *transcendental,* not merely because they themselves precede a priori, but also because they ground the possibility of other cognition.
>
> But the figurative synthesis, if it unfolds merely from the original synthetic unity of apperception, i.e., the former transcendental unity, must, in distinction from the mere intellectual combination, be called the *transcendental synthesis of the imagination.* (B151)

Kant then asserts that the transcendental synthesis of the imagination is "an action [Wirkung] of the understanding upon sensibility and its first application (at the same time the ground of all the rest) to objects of our possible intuition" (B152).

The following paragraph considers the "paradox" thereby engendered: "namely, because we only intuit ourselves as we are inwardly *affected,* which appears to be contradictory, in that we must relate towards ourselves as passive" (B152–153). The paradox is then resolved as follows:

> That which determines inner sense is the understanding and its original capacity to combine the manifold of intuition, i.e., to bring it under an apperception (as that on which its possibility itself rests). Now, since the understanding in us men is itself no faculty of intuition and could, even if these were given in sensibility, still not take these up *in itself*—in order, as it were, to combine the manifold of *its* own intuition; it follows that its synthesis, when it is considered for itself alone, is nothing other than the unity of the action [Handlung] of which it is also conscious, as such, without sensibility—but through which it is itself capable of determining sensibility inwardly with respect to the manifold, whatever may be the form according to which it is given. [The understanding] therefore exerts, under the name of a *transcendental synthesis of the imagination,* this action [Handlung] on the *passive* subject, whose *faculty* it is—whereupon we correctly assert that inner sense is thereby affected. (B153–154)

Transcendental synthesis or figurative synthesis is therefore an act of a priori self-affection whereby the (active) faculty of understanding affects the (passive) faculty of sensibility.

From the present point of view, however, what is most striking about §24 is the way in which Kant then illustrates the doctrine of figurative synthesis:

> This we also always observe in ourselves. We can think no line without *drawing* it in thought, no circle without *describing* it. We can absolutely not represent the three dimensions of space without *setting* three lines at right-angles to one another from the same point. And even time itself we cannot represent without attending in the *drawing* of a straight line (which

> is to be the outer figurative representation of time) merely to the action [Handlung] of synthesis of the manifold by which we successively determine inner sense—and thereby attend to the succession of this determination in it. Motion, as action [Handlung] of the subject (not as determination of an object*), and thus the synthesis of the manifold in space—if we abstract from the latter and attend merely to the action [Handlung] by which we determine *inner* sense according to its form—even produces the concept of succession in the first place. The understanding, therefore, does not find some such combination of the manifold in [inner sense], but rather *produces it,* in that it *affects* [inner sense]. (B154–155)

Thus Kant, in §24 of the second edition transcendental deduction, already illustrates the doctrine of a priori self-affection (self-determination) with the representation of motion.

Moreover, as Kant makes clear in the appended footnote to B155, the representation of motion in question here is an a priori representation in pure intuition (the motion of a mere mathematical point), which, as such, belongs to both pure mathematics and transcendental philosophy:

> *Motion of an *object* in space does not belong to a pure science, hence not also to geometry; because, that something is movable cannot be cognized a priori but only through experience. But motion, as the *describing* of a space, is a pure act [Aktus] of successive synthesis of the manifold in outer intuition in general through the productive imagination, and belongs not only to geometry but even to transcendental philosophy.[161]

The motion realizing the idea of figurative synthesis indeed belongs to transcendental philosophy—because it is "an action of the understanding upon sensibility and its first application (at the same time the ground of all the rest) to objects of our possible intuition," because it "even produces the concept of succession in the first place," and because "in order to present *alteration,* as the intuition corresponding to the concept of *causality,* we must take as our example motion, as alteration in space—indeed, it is even the case that we can thereby alone make intuitible to ourselves alterations, whose possibility no pure understanding can conceive" (B291).

It follows, therefore, that the representation of motion in pure intuition—conceived as the realization of figurative synthesis—first

161. That the motion in question here is that of a mere mathematical point follows from the Observation to Definition 5 of the Phoronomy of the *Metaphysical Foundations:* "In phoronomy, since I am acquainted with matter through no other property than its movability, and therefore it may be considered only as a point, motion is considered as *the describing of a space,* but still in such a way that I do not merely attend, as in geometry, to the space described, but also to the time in which, and thus the speed with which, a point describes a space" (4, 489.6–11).

grounds the possibility of synthetic a priori knowledge. It is thus not surprising when Kant explicitly asserts this in the eleventh fascicle of the *Opus:*

> *Space, time,* and what combines both ⟨intuitions (outer and inner)⟩ into One: *motion,* ⟨i.e., the act [Act]⟩ of describing ⟨a space⟩ in a certain time, are no ⟨given⟩ things as objects of perception (empirical representation with consciousness) given for themselves outside the subject; they are mere forms of sensible intuition which belong to the subject a priori and contain the general problem of transcendental philosophy: "How are synthetic propositions a priori possible?" The objects are here given only in the appearance as subjective forms of intuition on which the possibility of a synthetic cognition a priori is also grounded. (22, 440.16–25)

Moreover, in the final (first) fascicle Kant makes it clear that this grounding of synthetic a priori knowledge via space, time, and motion takes place precisely through the mathematical theory of motion—and thus, above all, through the mathematical physics of Newton:

> Newton with his mathematical foundations of *natural science: space, time,* and *motion* in space of a body (matter) in *space* and *time.* (21, 129.10–12)

> ⟨*Pure mathematics* as presentation *(intuition in space and time)* and *motion* (pure intuitions) constitutes the basis or posits the entrance on the scene . . . Newton's⟩ [.] (130.27–131.2)

> *Natural philosophy,* thought as united in its *mathematical* and *physical mechanical* relations (systematically), is *transcendental philosophy*[.] (132.4–6)

> *Natural philosophy* (according to Newton) united in its mathematical and physical-mechanical relations for the exercise of forces (unitable by rational beings) is *transcendental philosophy*[.] (132.11–13)

> Mathematics belongs under the title of philosophy. For the former rests (in so far as it is pure) also on *space, time,* and (both in relation) *motion in space* and time. (156.22–24)

Pure mathematics becomes essentially dependent on transcendental philosophy through the circumstance that the latter grounds Newtonian mathematical physics—and *thereby* grounds experience.[162]

162. Compare also 68.4–13: "Transcendental philosophy is objectively neither philosophy nor mathematics but what *subjectively* represents both united—philosophical as well as mathematical *cognition*—the doctrine of the grounding of synthetic cognition a priori from principles—the subjective principle—not merely from concepts. ⟨Therefore, it also contains mathematics.⟩ The possibility of such principles is an idea whose validity is susceptible to no proof: just as little as that of axioms of mathematics. Transcendental philosophy is a philosophy in so far as it makes use of mathematics as an instrument to guide its concepts

We can now see how the procedure of the *Metaphysical Foundations* acquires genuine constitutive force. It is true that the concept of matter articulated therein, the concept of the movable in space, is an empirical concept; for, as we have just seen, motion "of an *object* in space" can only be given empirically.[163] Yet motion, considered as "the *describing* of a space," is *also* a pure a priori representation and thus "belongs not only to geometry but even to transcendental philosophy." Indeed, this pure representation of motion is an act of self-positing or self-determination of the subject, whereby the understanding first determines sensibility and prepares the ground for all synthetic a priori cognition. Hence, the empirical concept of matter articulated in the *Metaphysical Foundations* is inextricably connected with (is, as it were, an empirical realization of) a pure a priori representation; and the latter not only has direct constitutive force (for, as a mathematical concept, it can actually be constructed), it is in fact the ground of all constitutive principles as such. For the pure representation of motion is the ground of that mathematical physics which concretely realizes the abstract categories of the understanding and thus makes experience possible.

From the point of view of transcendental philosophy, then, we can now explain the starting point of the *Metaphysical Foundations* (which is simply taken for granted in that work). In particular, we can explain why "the fundamental determination of something that is to be an object of the outer senses must be motion," so that "natural science is throughout an either pure or applied *doctrine of motion*" (4, 476.9–477.2), and we can understand why Kant hopes to see his metaphysical doctine of body brought "into unity with the mathematical doctrine of motion" (478.30–31). We can explain why matter is defined as *the movable* in space. Thus, in the *Opus,* Kant conceives the *Selbstsetzungslehre* as providing a transcendental foundation for the *Metaphysical Foundations:*

> ⟨Metaphysics and transcendental philosophy are distinguished from one another in that the *first* contains already given a priori principles of natural science, but the latter contains such as contain the ground of the possibility even of metaphysics and its synthetic a priori principles[.]

---

and principles, in order to present them in One system"—which passage is immediately followed by the example of Newton's derivation of the inverse-square law from Kepler's harmonic law (68.14–18).

163. Compare A41/B58: "Finally, that transcendental aesthetic can contain no more than these two elements, namely space and time, is clear; because all other concepts belonging to sensibility, even that of motion, which unites both elements, presuppose something empirical. For the latter presupposes the perception of something movable. But in space, considered in itself, there is nothing movable: therefore the movable must be something, which is found *in space only through experience,* and is thus an empirical datum."

There one begins not from objects but from the system of the possibility of constituting one's own thinking subject—and one is oneself author of one's own power of thought⟩ [.] (22, 79.1–8)

Transcendental philosophy is the docrine of the necessity of a system of synthetic a priori principles from concepts erected on behalf of metaphysics. It is not a science expounding (objectively) principles concerning objects, but rather concerning the subject of cognition [and] expounding the latter's extent and limits. It precedes the metaphysical foundations of all other philosophical sciences[.] (21, 63.30–64.4)

Accordingly, Kant, in the last fascicle, speaks of a "transition from the metaphysical foundations of natural science to transcendental philosophy" or a "progression from the metaphysical foundations of natural science to transcendental philosophy" (see, e.g., 45.1–2, 73.23–24).

But what does all this have to do with the transition from the metaphysical foundations of natural science to *physics?* First of all, from the point of view of transcendental philosophy, we can see that the application of the concept of the movable in space to objects of experience requires a distinction between true and apparent motion:

⟨True alteration of place can only be grounded on dynamical principles, e.g., of attraction, but even here it is not in relation to space in general[.]
Alteration of the place of A is not always motion of the body A. For if B is moved then *so is* the place of A also *altered,* but A *does not move* (does not alter its place).⟩ (22, 441.27–442.5)

⟨Space, time (as intuitions), motion; synthetic unity in *relation* of intuitions as appearances and *cause* of motion, moving force: these are all conditions of sensible objects. Principles of the possibility of experience.⟩ (442.12–15)

The application of the concept of the movable in space to objects of experience therefore also implies the application of the concept of force: specifically, the application of the Newtonian theory of universal gravitational attraction.[164]

Second, however, we can now see, from the point of view of transcendental philosophy once again, that Newtonian universal gravitation—as

164. Compare the fourth chapter or Phenomenology of the *Metaphysical Foundations.* There matter is defined as "the movable, in so far as it, as such a thing, can be an object of experience" (4, 554.6–7); and Kant then explains that "the movable, as such a thing, becomes an object of experience, when a certain *object* (thus here a material thing) is thought as *determined* in relation to the *predicate* of motion" (554.13–15). Thus "if the movable, *as such a thing,* namely according to its motion, is to be thought as determined—i.e., for the sake of a possible experience—it is necessary to indicate the conditions under which the object (matter) must be determined one way or another through the predicate of motion" (555.2–5). There follows a procedure for "reducing all motion and rest to absolute space" (560.5–6), which, I have argued, is modeled on the argument of Book III of *Principia.*

an immediate action-at-a-distance in empty space—cannot by itself alone constitute an object of perception. Indeed, the theory of universal gravitation, considered in abstraction from the rest of physics, describes the motions of a system of mere mass-points, which latter is more a mathematical object in pure intuition than a true physical system:

> Not the object of intuition of space but the subject is first determined for universal gravitation through the concept of the composition of the manifold as the schema of moving forces according to their form. The attractions of bodies which are not yet there in space are, like the signs of figures for geometry, not existing moving forces in space previously given, but rather conditions of intuition under which they may be given: namely, that of attraction in inverse ratio of the square of the distance of objects of possible perception if objects were to be given thereto.
>
> How is perception of the attraction of bodies through empty space possible? This question presupposes: how is perception of the relations of place of bodies ⟨and their alterations⟩ possible in empty space? (536.22–537.4)

Thus, attraction at a distance through empty space, which has just been seen to be *necessary* for the empirical application of the concept of motion from the standpoint of transcendental philosophy, is now seen to be *insufficient* from this very same standpoint.

It follows that some additional, properly physical, content is needed:

> Space and time are not objects but subjective forms of sensible intuition as unity of the composition of the possibility of perceptions according to three dimensions[.]—Therefore Newton established moving forces of attraction a priori, before the bodies that exert them are yet given. Attraction *(actio in distans)* immediately in empty space.
>
> But the formation of bodies is preceded by the concept of the matter through which any matter forms bodies, and repulsion of the parts thereof in the contact of their surfaces must be thought—without which space would still be empty, i.e., no sensible object, and thus absolutely no object of possible perception.—Consequently, the idea of a universally-filling expansum also belongs to the cognition of space, by means of which perception of bodies at a distance must be possible; hence it must be represented also as *actio in distans* extended in all spaces [Weiten] and as [acting] on the senses through moving forces according to the inverse ratio of the square of the distances (if it is also only represented through groping [herumtappen]) and requires time (according to Römer's discovery of light-aberration). (21, 59.28–60.15)[165]

In order to perceive the action of universal gravitation, and hence actually to apply the concept of motion, we therefore require that space be occu-

165. Kant here alludes to the inverse-square law for the propagation of light-intensity (diminishing on spherical surfaces)—discussed, for example, in the *Metaphysical Foundations* at 4, 518.35–519.28.

pied by real physical bodies (and not mere mathematical mass-points) and also that there be some means of real physical influence (for example, light) putting us into perceptual contact with these bodies.

This means, however, that the *Metaphysical Foundations* is now seen as radically incomplete from the standpoint of transcendental philosophy. For the *Metaphysical Foundations*, as we have observed, provides an a priori foundation only for the theory of universal gravitation and deliberately refrains from a philosophical consideration of both the formation of bodies (that is, the problem of cohesion and solidity) and of the phenomena treated by the emerging new experimental sciences of heat, light, electricity, and chemistry. From the point of view of transcendental philosophy it now appears that the *Metaphysical Foundations* calls for an extension of its own argument into precisely this previously excluded empirical domain. Hence, transcendental philosophy now institutes a transition from the metaphysical foundations of natural science to physics:

> Transcendental philosophy is autonomy, i.e., a prescribing reason determining its synthetic principles, extent, and limits in a complete system.
> It commences from the *metaphysical foundations of natural science* [and] contains the a priori principles of the *transition* of the latter *to physics* and the formal [element] thereof; and without becoming heteronomous it steps over to physics as a principle of the possibility of experience, through which an aggregate of perceptions becomes a whole of experience, and finally to experience as asymptotic approximation to proof *from experience* itself.. (59.17–26)

In other words, Kant now sees the *Transition* project as proceeding from metaphysics, through transcendental philosophy, and finally to physics.[166]

Yet it also appears that Kant's last reconsideration of the relationship between transcendental philosophy and the *Metaphysical Foundations* has indeed led to the problem of the *Transition* project, but not, sadly, to its solution. To be sure, we now see that the questions of body formation (cohesion and solidity) and of a physical influence (such as light) proceeding from bodies to our sense organs are inextricably connected with the method and purpose of the *Metaphysical Foundations* itself. But how are these questions possibly to be settled by, or even investigated under the guidance of, a priori principles? It is certainly natural, as we have seen, that Kant should seek such a priori guidance in the form of a universally distributed heat-matter or caloric—conceived as a medium of perception or light-aether as well. Unfortunately, however, we have also seen that

166. Compare 22, 86.10–11: "The transition from the metaphysical foundations of N.S. to transcendental philosophy and from the latter to physics"; and 119.7–9: "(1) Transition from metaphysics to transcendental philosophy. (2) From transcendental philosophy to physics through mathematics in pure intuition of space and time."

Kant's attempt at an a priori aether-deduction founders on a dilemma: a dilemma in which the role of medium of perception or light-aether is necessarily (and explicitly) disassociated from the role of heat-matter or chemical-aether. It follows that Kant has finally failed to provide a priori considerations that would, for example, favor a wave theory of light over a particle theory—or, to take another example, favor an explanation of cohesion and solidity via the actions of an aether over one framed in terms of action-at-a-distance forces. In the end, the transition from the metaphysical foundations of natural science to physics is and can be effected solely by the actual historical progress of natural science itself, and thus by considerations which, from a Kantian point of view, are merely empirical and hypothetical.

With the benefit of hindsight—that is, after having actually gone through the historical progress of natural science with which Kant was struggling—it is of course easy to understand why the *Transition* project must inevitably have failed. As an examination of the *Selbstsetzungslehre* has made clear, Kant's most fundamental analysis of the a priori grounding of experience rests on the mathematical theory of motion as elaborated in Newton's *Principia.* Yet the emerging experimental sciences Kant is attempting to accommodate in the *Transition* project have no direct connection whatever with this mathematical theory of motion. Lavoisier's chemical revolution, in particular, is effected not by a new mathematization of chemical theory (such as an elaboration of new microscopic force laws), but rather by a thoroughgoing reorganization of chemical classification. Kant, as one would expect, exercises great insight, sensitivity, and resourcefulness in attempting to accommodate this circumstance, but, in the end, it simply cannot be reconciled with his critical theory of a priori knowledge.

It is ironic, then, that the mathematization of chemistry was to begin in earnest in the early years of the nineteenth century with the introduction of the concept of atomic weight by John Dalton. This concept arose from careful experimental work on combining proportions and established the theoretical framework on which all modern chemistry rests.[167] In particu-

167. The experimental work resulting in the concept of atomic weight is of course a refinement of the *experimental method* established by Lavoiser: see notes 83, 124, and 155 above. In this connection, it is particularly interesting to note that one of the most important early contributors to the theory of combining proportions, J. B. Richter, received his doctorate at Königsberg in 1789 with a dissertation on *De usu matheseos in chymia;* his most important work was then *Anfangsgründe der Stöchyometrie oder Messkunst chymischer Elemente* (1792–1794)—see Partington [95], vol. 3, pp. 674–688. However, although Kant certainly knew of Richter in his capacity as rector (see 12, 417.28–36; 13, 584–585), I know of no evidence that suggests an appreciation of Richter's work on Kant's part: on the contrary, the absence of any discussion of combining proportions in the *Opus* strongly suggests that Kant did not know it—or at least did not grasp its importance.

lar, when integrated with the concept of atomic number (introduced in the 1860s), it led to the modern periodic table of the elements. Moreover, Dalton's atomic theory also constituted the basis for the electro-chemical researches of Jöns Berzelius in the early 1800s, which, refined and extended by such thinkers as H. Kolbe and A. Kekulé, eventuated in the modern electronic theory of valency.

During this same period, of course, the modern mathematical theory of electro-magnestism was constructed: beginning with Charles-Augustin Coulomb's formulation of the inverse-square law of electro-static force in 1785,[168] and culminating in James Clerk Maxwell's theory of the electromagnetic field formulated in the 1870s. Moreover, as we know, Maxwell was also able to explain the behavior of light as an electro-magnetic wave, and the idea of a light-aether was thereby given a secure mathematical foundation incorporating the phenomena of electricity and magnetism as well. Finally, the theory of the electron and the Bohr model of the atom made it possible to explain the electronic properties of atoms in terms of orbital motion in a Coulomb field, and hence to explain valency and chemical bonding. Chemistry was thus fully integrated with the theory of electro-magnetism as well, and therefore finally engaged with the mathematical theory of motion.

What is perhaps most ironic of all, however, is that this same theory of electro-magnetism led inevitably to Albert Einstein's special theory of relativity, wherein the very foundations of Newtonian kinematics—including, in particular, the concept of absolute simultaneity—are definitively overthrown. Moreover, in attempting to integrate this new theory with gravity, Einstein was forced to deviate from the classical conceptions of space and time in an even more radical fashion. For, precisely because it essentially involves immediate attraction at a distance (and thus absolute simultaneity), the Newtonian theory of gravitation cannot possibly be valid in a relativistic context. In Einstein's new theory of gravitation—the general theory of relativity—not only may space (more precisely, space-time) have a non-Euclidean structure, but space (space-time) in fact has no fixed structure at all independent of the matter and energy distributed therein; and it follows that there is no longer a sense in which the structure of space (space-time) can be viewed as fixed a priori. But this means that we have come full circle: through the development of the experimental sciences since Kant's time—and, in particular, through the attempt finally to integrate these sciences with the theory of gravitation—we have now

168. As far as I have been able to determine, Kant himself was never acquainted with Coulomb's work, which remained largely unknown in Germany until after 1800: see Heilbron [47], p. 227.

reached a point at which the most fundamental commitment of the critical philosophy, the idea of a fixed a priori spatio-temporal structure serving as the foundation of the exact sciences and indeed of all human knowledge, can no longer consistently be maintained. I continue to find myself most impressed, however, not so much by the circumstance that Kant was in error here as by the absolutely astonishing penetration, dedication, and scope of his effort philosophically to comprehend the development of the exact sciences—which effort, as we have seen, continued undiminished almost to his death.

References and Translations for Kant's Writings

All references to Kant's writings and correspondence, except references to the *Critique of Pure Reason*, are given by volume, page, and line number of the Akademie edition of *Kant's gesammelte Schriften* (Berlin, 1902–); the *Critique of Pure Reason* is cited by the standard A and B pagination of the first (1781) and second (1787) editions respectively. I have also made use of the *Theorie-Werkausgabe Immanuel Kant*, Hrsg. W. Weischedel (Frankfurt, 1968); and the *Kritik der reinen Vernunft*, Hrsg. R. Schmidt (Hamburg, 1956).

All translations from Kant's German are my own, although I have closely consulted the standard English translations given below. In rendering Kant's Latin I have relied heavily on the available English translations and also on Weischedel's translations into German. The translations I have consulted are as follows:

Handyside, J. *Kant's Inaugural Dissertation and Early Writings on Space* (Chicago, 1929).

Hastie, W. *Kant's Cosmogony* (Glasgow, 1900).

England, F. *Kant's Conception of God* (New York, 1968).

Beck, L. *Kant's Latin Writings* (New York, 1986).

Abbott, T. K. *Kant's Introduction to Logic and His Essay on the Mistaken Subtilty of the Four Figures* (New York, 1963).

Treash, G. *The One Possible Basis for a Demonstration of the Existence of God* (New York, 1979).

Kerferd, G., and Walford, D. *Kant: Selected Pre-Critical Writings and Correspondence with Beck* (Manchester, 1968).

Carus, P. *Prolegomena to Any Future Metaphysics* (La Salle, 1902).

Beck, L. *Prolegomena to Any Future Metaphysics* (Indianapolis, 1951). Follows Carus.

Ellington, J. *Prolegomena to Any Future Metaphysics* (Indianapolis, 1977). Follows Beck and Carus.

Lucas, P. *Prolegomena to Any Future Metaphysics* (Manchester, 1953).

Kemp Smith, N. *Immanuel Kant's Critique of Pure Reason* (London, 1929).

Ellington, J. *Metaphysical Foundations of Natural Science* (Indianapolis, 1970).

Bernard, J. *Critique of Judgement* (New York, 1951).
Haden, J. *First Introduction to the Critique of Judgement* (Indianapolis, 1965).
Greene, T., and Hudson, H. *Religion within the Limits of Reason Alone* (La Salle, 1934).
Gregor, M. *The Doctrine of Virtue* (Philadelphia, 1964).
Gregor, M. *Anthropology from a Pragmatic Point of View* (The Hague, 1974).
Allison, H. *The Kant-Eberhard Controversy* (Baltimore, 1973).
Hartman, R., and Schwarz, W. *Logic* (Indianapolis, 1974).
Zweig, A. *Kant: Philosophical Correspondence 1759–99* (Chicago, 1967).

General Bibliography

1. Adickes, E. *Kant als Naturforscher* (Berlin, 1924).
2. Allison, H. *Kant's Transcendental Idealism* (New Haven, Conn., 1983).
3. Arnoldt, A. *Kritische Excurse im Gebiete der Kant-Forschung* (Königsberg, 1894).
4. Beck, L. *Studies in the Philosophy of Kant* (Indianapolis, 1965).
5. Beck, L. *Early German Philosophy: Kant and His Predecessors* (Cambridge, Mass., 1969).
6. Beth, E. "Über Lockes 'Allgemeines Dreieck'," *Kant-Studien* 49 (1956–57): 361–380.
7. Black, J. *Lectures on the Elements of Chemistry, delivered in the University of Edinburgh,* ed. J. Robison (Edinburgh, 1803).
8. Bonola, R. *Non-Euclidean Geometry* (New York, 1955).
9. Boyer, C. *The History of the Calculus and Its Conceptual Development* (New York, 1949).
10. Brittan, G. *Kant's Philosophy of Science* (Princeton, 1978).
11. Brittan, G. "Constructibility and the World-Picture," in G. Funke and T. Seebohm, eds., *Proceedings of the Sixth International Kant Congress,* vol. II/2 (Washington, D.C., 1989).
12. Buchdahl, G. *Metaphysics and the Philosophy of Science* (Oxford, 1969).
13. Buchdahl, G. "The Conception of Lawlikeness in Kant's Philosophy of Science," in L. Beck, ed., *Kant's Theory of Knowledge* (Dordrecht, 1974).
14. Calinger, R. "The Newtonian-Wolffian Confrontation in the St. Petersburg Academy of Sciences," *Cahiers d'Histoire Mondiale* 11 (1968): 417–435.
15. Calinger, R. "The Newtonian-Wolffian Controversy (1740–1759)," *Journal of the History of Ideas* 30 (1969): 319–330.
16. Cantor, G., and Hodge, M., eds. *Conceptions of Ether* (Cambridge, 1981).
17. Carnap, R. "Intellectual Autobiography," in P. Schilpp, ed., *The Philosophy of Rudolf Carnap* (La Salle, 1963).
18. Cassirer, E. *Kant's Life and Thought,* trans. J. Haden (New Haven, 1981). Originally published in 1918.
19. Cohen, I. B. *The Newtonian Revolution* (Cambridge, 1980).

20. DiSalle, R. "The 'Essential Properties' of Matter, Space, and Time," in P. Bricker and R. I. G. Hughes, eds., *Philosophical Perspectives on Newtonian Science* (Cambridge, Mass., 1990).
21. Einstein, A. *Sidelights on Relativity* (London, 1922).
22. Enderton, H. *A Mathematical Introduction to Logic* (New York, 1972).
23. Erdmann, B. *Martin Knutzen und seine Zeit* (Leipzig, 1876).
24. Erxleben, J. C. P. *Anfangsgründe der Naturlehre,* Hrsg. G. C. Lichtenberg (Göttingen, 1784).
25. Euler, L. *Lettres de L. Euler à une Princesse d'Allemagne,* ed. A. Cournot (Paris, 1842).
26. Euler, L. *Letters of Euler on Different Subjects in Natural Philosophy: Addressed to a German Princess,* ed. D. Brewster (New York, 1833).
27. Euler, L. "Réflexions sur l'espace et le temps," *Mémoires de l'Académie des Sciences de Berlin* (1748), in *Leonhardi Euleri Opera omnia,* Series Tertia II: 376–383.
28. Euler, L. "Reflections on Space and Time," in A. Koslow, ed., *The Changeless Order* (New York, 1967).
29. Eves, H. *A Survey of Geometry* (Boston, 1963).
30. Fox, R. *The Caloric Theory of Gases from Lavoisier to Regnault* (Oxford, 1971).
31. Förster, E. "Is There 'a Gap' in Kant's Critical System?" *Journal for the History of Philosophy* 25 (1987): 533–555.
32. Förster, E. "Kant's *Selbstsetzungslehre,*" in E. Förster, ed., *Kant's Transcendental Deductions* (Stanford, 1989).
33. Frege, G. *Die Grundlagen der Arithmetik* (Breslau, 1884).
34. Frege, G. *The Foundations of Arithmetic,* trans. J. Austin (Oxford, 1950). Includes [33] on facing pages with the original pagination.
35. Friedman, M. "Kant and Newton: Why Gravity Is Essential to Matter," in P. Bricker and R. I. G. Hughes, eds., *Philosophical Perspectives on Newtonian Science* (Cambridge, Mass., 1990).
36. Friedman, M. "Causal Laws and the Foundations of Natural Science," in P. Guyer, ed., *The Cambridge Companion to Kant* (Cambridge, 1992).
37. Gehler, J. S. T. *Physikalisches Wörterbuch* (Leipzig, 1798–1801).
38. Gloy, K. *Die Kantische Theorie der Naturwissenschaft* (Berlin, 1976).
39. Grabiner, J. *The Origins of Cauchy's Rigorous Calculus* (Cambridge, Mass., 1981).
40. Guerlac, H. *Lavoisier—The Crucial Year* (Ithaca, 1961).
41. Guerlac, H. "Chemistry as a Branch of Physics: Laplace's Collaboration with Lavoisier," *Historical Studies in the Physical Sciences* 7 (1976): 193–276.
42. Hankins, T. *Science and the Enlightenment* (Cambridge, 1985).
43. Harman, P. *Metaphysics and Natural Philosophy* (Brighton, 1982).
44. Harper, W. "Kant on the A Priori and Material Necessity," in R. Butts, ed., *Kant's Philosophy of Physical Science* (Dordrecht, 1986).
45. Heath, T., ed. *Appolonius of Perga: Treatise on Conic Sections* (Cambridge, 1896).
46. Heath, T., ed. *The Thirteen Books of Euclid's Elements* (Cambridge, 1926).
47. Heilbron, J. *Elements of Early Modern Physics* (Berkeley, 1982).

48. Hilbert, D. *Grundlagen der Geometrie* (Leipzig, 1899).
49. Hilbert, D. *Foundations of Geometry*, ed. L. Unger (La Salle, 1971).
50. Hintikka, J. "On Kant's Notion of Intuition (Anschauung)," in T. Penelhum and J. MacIntosh, eds., *The First Critique* (Belmont, Calif., 1969).
51. Hintikka, J., ed. *The Philosophy of Mathematics* (Oxford, 1969).
52. Hintikka, J. *Logic, Language-Games and Information* (Oxford, 1973).
53. Hintikka, J. *Knowledge and the Known* (Dordrecht, 1974).
54. Hopkins, J. "Visual Geometry," *Philosophical Review* 82 (1973): 3–34.
55. Hoppe, H. *Kants Theorie der Physik* (Frankfurt am Main, 1969).
56. Keill, J. *Introduction to the True Physics* (London, 1726).
57. Kemp Smith, N. *A Commentary to Kant's 'Critique of Pure Reason'* (London, 1923).
58. Kirwan, R. *An Essay on Phlogiston and the Constitution of Acids—To which are added NOTES, Exhibiting and defending the Antiphlogistic Theory; and annexed to the French Edition of this Work; by Messrs. de Morveau, Lavoisier, de la Place, Monge, Bertholett, and de Fourcroy* (London, 1789).
59. Kitcher, P. "Kant and the Foundations of Mathematics," *Philosophical Review* 84 (1975): 23–50.
60. Kitcher, P. "Kant's Philosophy of Science," in A. Wood, ed., *Self and Nature in Kant's Philosophy* (Ithaca, 1984).
61. Koyré, A. *Newtonian Studies* (Chicago, 1965).
62. Kuhn, T. *The Essential Tension* (Chicago, 1977).
63. Lavoisier, A. *Traité Élémentaire de Chimie,* in *Oeuvres de Lavoisier,* Tome Premier (Paris, 1864).
64. Lavoisier, A. *Elements of Chemistry,* trans. R. Kerr (Edinburgh, 1790), ed. D. McKie (New York, 1965).
65. Laywine, A. "Physical Influx and the Origins of the Critical Philosophy" (Ph.D. dissertation, University of Chicago, 1991).
66. Lehmann, G. *Kants Tugenden* (Berlin, 1980).
67. Leibniz, G. W. *Die philosophischen Schriften von Gottfried Wilhelm Leibniz,* Hrsg. C. Gerhardt (Berlin, 1875–1890).
68. Leibniz, G. W. *Opuscules et Fragments Inédits de Leibniz,* ed. L. Couterat (Paris, 1903).
69. Leibniz, G. W. *Theodicy,* ed. A. Farrer (London, 1951).
70. Leibniz, G. W. *The Leibniz-Clarke Correspondence,* ed. H. Alexander (Manchester, 1956).
71. Leibniz, G. W. *New Essays on Human Understanding,* ed. P. Remnant and J. Bennett (Cambridge, 1981).
72. Leibniz, G. W. *Philosophical Essays,* ed. R. Ariew and D. Garber (Indianapolis, 1989).
73. Martin, G. *Kant's Metaphysics and Theory of Science,* trans. P. Lucas (Manchester, 1955). Originally published in 1951.
74. Martin, G. *Arithmetic and Combinatorics: Kant and His Contemporaries,* trans. J. Wubnig (Carbondale, Ill., 1985). Originally published in 1972.
75. McKie, D., and Heathcote, N. *The Discovery of Specific and Latent Heats* (London, 1935).
76. McMullin, E. *Newton on Matter and Activity* (Notre Dame, 1978).

77. Melnick, A. *Kant's Analogies of Experience* (Chicago, 1973).
78. Menzel, A. "Die Stellung der Mathematik in Kants vorkritischer Philosophie," *Kant-Studien* 16 (1911): 139–213.
79. Metzger, H. *Newton, Stahl, Boerhaave, et la Doctrine Chimique* (Paris, 1930).
80. Metzger, H. *La Philosophie de la Matière chez Lavoisier* (Paris, 1935).
81. Mueller, I. *Philosophy of Mathematics and Deductive Structure in Euclid's 'Elements'* (Cambridge, Mass., 1981).
82. Newton, I. *Isaac Newton's 'Philosophiae Naturalis Principia Mathematica'*, ed. A. Koyré and I. B. Cohen (Cambridge, Mass., 1972).
83. Newton, I. *Mathematical Principles of Natural Philosophy,* trans. A. Motte (1729), rev. F. Cajori (Berkeley, 1934).
84. Newton, I. *Opticks* (London, 1931). Based on the fourth edition (1730).
85. Newton, I. "Quadrature of Curves" (1692), in D. Whiteside, ed., *The Mathematical Works of Isaac Newton,* vol. I (New York, 1964).
86. Newton, I. *The Mathematical Papers of Isaac Newton,* ed. D. Whiteside (Cambridge, 1974).
87. Newton, I. *The Correspondence of Isaac Newton,* vol. V: 1709–1713, ed. A. Hall and L. Tilling (Cambridge, 1975).
88. Okruhlik, K. "Kant on the Foundations of Science," in W. Shea, ed., *Nature Mathematized* (Dordrecht, 1983).
89. Palter, R. "Absolute Space and Absolute Motion in Kant's Critical Philosophy," in L. Beck, ed., *Kant's Theory of Knowledge* (Dordrecht, 1974).
90. Parsons, C. "Mathematics, Foundations of," in P. Edwards, ed., *The Encyclopedia of Philosophy* (New York, 1977).
91. Parsons, C. "Objects and Logic," *Monist* 65 (1982): 491–516.
92. Parsons, C. *Mathematics in Philosophy* (Ithaca, 1983).
93. Parsons, C. "Remarks on Pure Natural Science," in A. Wood, ed., *Self and Nature in Kant's Philosophy* (Ithaca, 1984).
94. Parsons, C. "Arithmetic and the Categories," *Topoi* 3 (1984): 109–121.
95. Partington, J. R. *A History of Chemistry* (London, 1962).
96. Pasch, M. *Vorlesungen über neure Geometrie* (Leipzig, 1882).
97. Plaass, P. *Kants Theorie der Naturwissenshaft* (Göttingen, 1965).
98. Polonoff, I. *Force, Cosmos, Monads and Other Themes of Kant's Early Thought* (Kant-Studien Ergänzungsheft Nr. 107, 1973).
99. Ricketts, T. "Frege, the *Tractatus,* and the Logocentric Predicament," *Noûs* 19 (1985): 3–15.
100. Ricketts, T. "Objectivity and Objecthood: Frege's Metaphysics of Judgment," in L. Haaparanta and J. Hintikka, eds., *Frege Synthesized* (Dordrecht, 1986).
101. Ricketts, T. "Generality, Meaning, and Sense in Frege," *Pacific Philosophical Quarterly* 67 (1986): 172–195.
102. Russell, B. *The Principles of Mathematics* (Cambridge, 1903).
103. Schäfer, L. *Kants Metaphysik der Natur* (Berlin, 1966).
104. Stein, H. "Newtonian Space-Time," *Texas Quarterly* 10 (1967): 174–200.
105. Stein, H. "Some Philosophical Prehistory of General Relativity," in J. Ear-

man, C. Glymour, and J. Stachel, eds., *Minnesota Studies in the Philosophy of Science,* vol. VIII (Minneapolis, 1977).

106. Stein, H. "Eudoxus and Dedekind: On the Ancient Greek Theory of Ratios, and Its Relation to Modern Mathematics," *Synthese* 84 (1990): 163–211.
107. Tait, W. W. "Reflections on the Concept of A Priori Truth and Its Corruption by Kant," in M. Detlefsen, ed., *Proof and Knowledge in Mathematics* (London, 1991).
108. Thompson, M. "Singular Terms and Intuitions in Kant's Epistemology," *Review of Metaphysics* 26 (1972–73): 314–343.
109. Timerding, H. "Kant und Euler," *Kant-Studien* 23 (1919): 18–64.
110. Tonelli, G. *Elementi Metodologici e Metafisici in Kant dal 1745 al 1768* (Torino, 1959).
111. Tonelli, G. "Der Streit über die Mathematische Methode in der Philosophie in der ersten Hälfte des 18 Jahrhunderts und die Entstehung von Kants Schrift über die 'Deutlichkeit'," *Archive für Philosophie* 9 (1959): 37–66.
112. Torretti, R. *Philosophy of Geometry from Riemann to Poincaré* (Dordrecht, 1978).
113. Tuschling, B. *Metaphysische und transzendentale Dynamik in Kants Opus postumum* (Berlin, 1971).
114. Tuschling, B. "Kants 'Metaphysische Anfangsgründe der Naturwissenschaft' und das Opus postumum," in G. Prauss, ed., *Kant: Zur Deutung seiner Theorie von Erkennen und Handeln* (Köln, 1973).
115. Tuschling, B. "Apperception and Ether: On the Idea of a Transcendental Deduction of Matter in Kant's *Opus postumum,*" in E. Förster, ed. *Kant's Transcendental Deductions* (Stanford, 1989).
116. Vleeschauwer, H. de. *The Development of Kantian Thought,* trans. A. Duncan (London, 1962). Originally published in 1939.
117. Vuillemin, J. *Physique et Métaphysique Kantiennes* (Paris, 1955).
118. Warda, A. *Immanuel Kants Bücher* (Berlin, 1922).
119. Westfall, R. *Force in Newton's Physics* (New York, 1971).
120. Whiteside, D. "Patterns of Mathematical Thought in the Later Seventeenth Century," *Archive for History of Exact Sciences* 1 (1961): 179–388.
121. Wilson, C. "From Kepler's Laws, So-called, to Universal Gravitation: Empirical Factors," *Archive for History of Exact Sciences* 6 (1970): 89–170.
122. Wilson, C. "The Great Inequality of Jupiter and Saturn: from Kepler to Laplace," *Archive for History of Exact Sciences* 33 (1985): 15–290.
123. Wittgenstein, L. *Tractatus Logico-Philosophicus* (London, 1922).
124. Young, J. "Kant on the Construction of Arithmetical Concepts," *Kant-Studien* 73 (1982): 17–46.

Index